SYSTEM DYNAMICS:
Modeling and Response

SYSTEM DYNAMICS:
Modeling and Response

Ernest O. Doebelin

The Ohio State University

CHARLES E. MERRILL PUBLISHING COMPANY
A Bell & Howell Company
Columbus, Ohio

Published by
Charles E. Merrill Publishing Co.
A Bell & Howell Company
Columbus, Ohio 43216

International Standard Book Number: 0-675-09120-9

Library of Congress Catalog Card Number: 77-187802

3 4 5 6 – 77

Printed in the United States of America

PREFACE

Since System Dynamics courses are not "traditional" in any branch of engineering (in the sense that a Thermodynamics course is traditional in Mechanical Engineering curricula), it is necessary not only to justify the approach of this text, but also to make a case for the subject matter itself. At the author's school, System Dynamics courses were introduced quite recently as required courses in Mechanical Engineering. While the author had been "lobbying" for such courses for some time, their actual introduction had to await a general curricular reorganization required by a change in the College of Engineering from a five-year to a four-year program. Such a 20% reduction in available time acted as a strong incentive for increased efficiency in the entire curriculum, and this was one factor, of several, which led to the introduction of System Dynamics courses. Most engineering schools operate on four-year programs; therefore, System Dynamics courses should have a broad appeal because of their intrinsic merit and also because they can significantly improve efficiency.

As early as 1955, The Ohio State Mechanical Engineering Department had instituted required Instrumentation and Automatic Control courses, which naturally tend to include much of the material now considered as System Dynamics. Also, required courses in Machine Dynamics (including vibration) and Electrical Circuits have been traditional in many Mechanical Engineering departments. Certain other branches of engineering exhibit similar patterns. Out of these diverse roots grew the notion that it would be pedagogically sound to gather together the common foundation of these various subject areas and present it to students, at an elementary level, as a unified whole and very early in the curriculum. They would thus be made aware that apparently unrelated devices and processes can have many common aspects of behavior and yield to similar analytical and design schemes. The introduction of System Dynamics was not intended to eliminate all the above-mentioned courses (in fact only one course, as detailed below, was eliminated). Rather, it frees the later courses from the burden of needless repetition (some repetition is of course desirable) of common background material at the expense of penetration into practical engineering applications.

v

The widespread introduction of System Dynamics into engineering curricula has proceeded along two different paths. First, some schools have introduced so-called core curricula, which run for as much as the first two or three years, and which all students take as a common core, specialization by department occurring quite late, if at all. An interdisciplinary System Dynamics course at the sophomore level fits quite naturally into such a scheme. At many other schools, where only the freshman year is common, the introduction of System Dynamics may rest with the individual departments, even though the *course* may still have a strong interdisciplinary flavor. Those departments interested in developing such courses may well prefer to operate independently. As long as there are sufficient students to fill the sections, little efficiency is gained by several departments sharing a single course, while the compromises necessary for a joint venture may reduce the course effectiveness for *all* the students.

At the author's school, System Dynamics is now the first Mechanical Engineering course the student encounters. The material of this text, through Chapter 8, is presented in a course of 3 hours lecture, 2 hours laboratory per week for one quarter to students who are second- or third-quarter sophomores and have had conventional physics, statics and dynamics, and digital computer courses. They have also had, or are taking concurrently, a differential equations course. Chapters 9 and 10 are presented in the first third of a course (Mechanical Engineering Analysis) of similar lecture/lab format, which immediately follows System Dynamics, and which includes in addition a variety of applied mathematics topics of use to mechanical engineers. For semester-length courses, the entire text could be covered with 3 to 5 hours lecture per week. While laboratory development and operation always consume large amounts of faculty time when properly done, the author is convinced that the effort is most worthwhile and particularly effective in System Dynamics. The text, of course, does not presume a laboratory and can thus be used with a strictly lecture format.

Institution of System Dynamics coincident with changing from a 5- to a 4-year program was associated with the following curricular changes in Mechanical Engineering. The traditional 3-course Electrical Engineering sequence in Circuits, Machinery, and Electronics was reduced to Machinery and Electronics, since System Dynamics gives adequate circuit analysis competence for non-specialists. The treatment of elementary circuit analysis within a Mechanical Engineering course is unconventional, but actually helps to motivate students to learn this "alien" material, since they can see that their own department considers it of sufficient importance to teach it themselves. That is, students may be more impressed with the importance of a multi-disciplinary point of view if they actually see it in their own teachers. Many schools may wish to retain their present Electrical Engineering treatment; if so, the students will just be better prepared in this area because of the repetition and extra time spent. Separate courses in Kinematics and Machine Dynamics were combined into a single course with reduced emphasis on vibrations, since System Dynamics treats this topic adequately for non-specialists. A new elective in Vibration was introduced for machine design specialists. This course can go into practical applications quite thoroughly, since System Dynamics has provided a sound base. The required courses in Instrumentation and Control were significantly augmented with more practical applications, since less time was needed for basic background.

The following comments should be made regarding the text's point of view. First, while System Dynamics especially lends itself to abstract mathematical formulations of great generality, the author prefers to leave this for graduate-level treatments; thus this text is intended specifically for sophomores or early juniors. Experience indicates that many engineering students, by their sophomore year, are getting somewhat disenchanted with a diet of pure science and are wondering when they will see some "real engineering." Freshman "design" courses are one possible answer to this problem; a practically-oriented System Dynamics course may be another. Thus, this text emphasizes applications and practical realities with the laboratory allowing hands-on contact with real hardware. An early System Dynamics course with this viewpoint can thus serve as a motivating influence for subsequent engineering courses in addition to fulfilling its specific technical objectives.

In line with this practical emphasis, the real forms of the system elements (dampers, springs, capacitors, etc.) are discussed at more length than usual. This somewhat lengthens the text, but this material is mainly descriptive and can be left for outside reading. While the theoretical foundations of System Dynamics will always emphasize linear systems, the ready availability of powerful and easily-learned digital simulation programs now allows significant exposure of undergraduates to the more realistic nonlinear behavior. Familiarity with this type of program greatly enhances the students' capability for creating useful models of real systems. While analog computers offer similar advantages, they do not provide the means to handle large numbers of students and problems at reasonable cost and with the time available. Analog techniques and lab experience with analog equipment are, however, still an important part of the course.

A rather lengthy, but mainly descriptive and quickly read first chapter attempts to place the overall subject of mathematical modeling in some perspective. This is important since the subsequent heavy emphasis on lumped, linear models could create too narrow a view. While a sophomore may not at this time appreciate all the mathematical implications of the more complex models, he will at least be made aware that other approaches exist and will see where the approach of this particular course fits into the broader picture. After the introductory chapter, the subject is developed systematically, starting with definitions and discussions of the system elements, mechanical, electrical, fluid, and thermal. As mentioned earlier, considerable detail on practical matters is included to inject realism; however these descriptive sections can largely be left for outside reading. A brief chapter on energy conversion allows later treatment of mixed systems and points up the interaction between various engineering subject areas and disciplines.

Chapter 6 presents methods of equation solution utilizing analytical, analog and digital simulation techniques. Analytical methods are limited to linear and linearized nonlinear models while the versatility of computer methods in handling realistic nonlinear models is heavily emphasized. Chapters 7 and 8 thoroughly treat the modeling and response of first-order and second-order systems of all types. Frequent use of digital simulation to compare linearized and nonlinear models establishes confidence in linear methods while also indicating their limitations and the power of the computer in developing models which closely represent real system behavior.

Chapter 9 relates the behavior of the basic first- and second-order systems to that of lumped, linear models of arbitrarily high order in terms of transient response, stability, and frequency response. At this point the *writing* of equations for complex systems has become quite easy and access to a simply used digital simulation program (such as CSMP) puts their *solutions* well within the reach of sophomores. The significance of this development in computer application is great and cannot be appreciated until one has actually used such a program. Capability for analytically handling driving inputs is extended to periodic functions using Fourier Series and frequency response. Again, digital simulation easily gets both the final steady state and also the starting transient. In chapter 10 the transition from lumped to distributed models is demonstrated in terms of two specific examples and then generalized in a qualitative sense. Upon completing this last chapter it may be useful to return briefly to chapter 1 where the model-classification scheme was first presented, since this material should now be more completely appreciated.

Throughout the text there is a parallel development of linear system analytical techniques and general-purpose computer simulation via *easily used* digital programs. The first provides the conceptual base for developing the "gut feeling" for the *general nature* of system response so necessary to one who will be effective in designing new and useful devices and processes. The second provides the powerful computational methods necessary to determine quickly the behavior of *specific systems*, first in linearized form and then with added nonlinearities or higher-order linear models of almost unlimited complexity. The development of confidence in the application of this dual capability should stand the student in good stead for any later courses and ultimately in an engineering career of social usefulness.

In acknowledging the contributions of others, let me first express appreciation to department chairmen S. M. Marco and D. D. Glower for providing an environment wherein one who preferred to spend his time almost exclusively in teaching could not only survive but prosper. The manuscript was ably typed by Carol Settles and Judy Althoff. The usual apologies to wife and family for the unavoidable periods of neglect associated with writing seem somehow unnecessary, since they were so cheerfully accepted.

May, 1972   *Ernest O. Doebelin*

CONTENTS

ix

APPENDIXES

INDEX   489

1

INTRODUCTION

1–1. WHAT IS SYSTEM DYNAMICS?

The words *system* and *dynamics* are both subject to such varied inter-
pretations that their combination as a title of a textbook or an engineering
course may leave the reader with some doubts as to just what it is that he
is about to be exposed to. A system is thought of as an arrangement of
parts or components; dynamics refers to a situation which is changing
with time. Within these broad definitions it might appear that the subject
of system dynamics includes much of the engineering science basic to the
established disciplines such as mechanical, electrical, and chemical engineering. Such a viewpoint could certainly be defended; however, it is of
little help in clarifying the particular definition of system dynamics which
is today widely accepted and which forms the subject matter of this book.

   To begin to formulate the restricted definition of system dynamics
which we wish to consider here, let us investigate the meaning of the words
modeling and *response* used in the text sub-title. In engineering, the word
modeling has two principal meanings; one associated with *physical* models,
the other with *mathematical* models. By a physical model we here mean
an assemblage of actual hardware, constructed according to appropriate
scaling laws, such that it will behave in a manner predictably related to
the behavior of the full-scale device or system. The model need not neces-

1

sarily reproduce *all* aspects of the full-scale system's behavior; different models may be constructed for evaluating aerodynamic effects and structural vibrations, for example, in booster rockets used for launching space vehicles. A one-fifth scale model of the Saturn rocket[1] constructed to study vibration problems, for instance, needed only to simulate the proper distribution of inertia and flexibility in order to predict the vibrational frequencies; other aspects of the model could deviate from the full-scale system. In this example the scaling laws showed that if the model were made of the same materials as the full-scale but with all linear dimensions reduced to one-fifth, then the frequencies measured on the model would be five times those of the full-scale.

The engineers responsible for the design of the Saturn rocket also calculated *theoretically* the vibrational frequencies. To do this they had to develop a mathematical model of the rocket structure. This does not involve the building of any hardware in a model shop, rather it requires the judicious application of the appropriate physical laws and engineering judgment to the development of a set of mathematical equations which (hopefully) will adequately describe the vibratory motions. Solution of these equations allows prediction of the response of the structure to any vibration-exciting forces which might be present. Here we have our first example of system dynamics which serves to set the pattern for all our later work; *modeling* is the process of describing the physical system in mathematical terms, and *response* refers to the solution of the equations which show the behavior of the system. The first activity requires creative engineering of the highest order and does not necessarily lead to a unique result. That is, a given physical system may be modeled in an infinite variety of ways since such analysis *always* requires the application of simplifying assumptions and each engineer will not, in general, make the same assumptions. Once the equations are developed, however, then there is only one possible solution, thus the second activity (finding the response corresponding to a *given* equation) is usually a matter of mathematical routine, a routine which must, however, be learned.

In the case of the Saturn rocket vibrations referred to above, the chronology of events might have been as follows. The basic design concepts of the rocket are developed from consideration of the mission to be performed, which is to get three men to the moon and back with acceptable risks, with limited resources, and by a specified date. A tremendous amount of preliminary planning, analysis, and decision-making with regard to the *overall* system for achieving this goal (not just the rocket vehicle itself, but ground-support systems, tracking stations, manufacturing logistics,

---

[1] J. S. Mixson, J. J. Catherine, and A. Arman, *Investigation of the Lateral Vibration Characteristics of a 1/5 Scale Model of Saturn SA-1*, NASA TN D-1593, Jan. 1963.

etc.) finally result in the definition of the vehicle's configuration. Structures engineers then design a structure to accommodate the functions of the various subsystems. Such design leans heavily on extrapolation of past experience to arrive at preliminary designs which can then be analyzed (mathematically modeled) to determine their adequacy with regard to the pertinent design criteria. Modifications and re-analysis occur until a design which is optimum in some practical sense is achieved. One of the important analyses to be made involves the vibration frequencies mentioned earlier. Due to the importance and great cost of the full-scale vehicle and the doubt which always accompanies even a careful theoretical analysis, a decision was made to construct a one-fifth scale physical model for vibration testing. Since the full-scale vehicle was eventually built and vibration tested,[2] the data shown in Fig. 1–1 are available showing

	Predicted or Measured Vibration Frequencies, Hertz	
	First Mode	Second Mode
Mathematical Model	2.80	5.72
1/5 Scale Model Test	2.80	5.20
Full-Scale Test	2.83	5.68

FIGURE 1–1

Comparison of Theoretical, Scale Model, and Full-Scale
Vibration Studies

a comparison of theoretical predictions, model testing results, and full-scale test results. The data shown exhibit quite good agreement; however, the model tests also revealed several other natural frequencies not predicted at all by the theory then being used, and indicated the changes needed in the mathematical model to get better agreement with the observed behavior.

While the use of scale models is an important engineering tool in many areas, and has a well-developed theoretical background,[3] we do not in this book choose to treat this material and restrict ourselves to questions

---

[2] J. S. Mixson, J. J. Catherine, *Comparison of Experimental Vibration Characteristics Obtained from a 1/5 Scale Model and a Full-Scale Saturn SA-1*, NASA TN D-2215, Nov. 1964.

[3] H. Langhaar, *Dimensional Analysis and the Theory of Models* (New York: John Wiley & Sons, Inc., 1951).

of *mathematical* modeling and solution and interpretation of the resulting equations.

1-2. PROBLEMS OF MODELING.

We have seen that System Dynamics involves the mathematical modeling of an assemblage of components so as to arrive at a set of equations which represent the dynamic behavior of the system and which can be solved to determine the response to various sorts of stimuli. This definition is still, however, too broad to suit our present purpose, since it places no restriction whatever on the nature of the mathematical description of the system. While the study of the unsteady behavior of an assemblage of components by *any* mathematical means would certainly be a study of the dynamics of a system, we would like to suggest that only for a narrowly restricted type of mathematical model does a unified, highly developed, and widely applicable body of knowledge exist. This mathematical model is a set of ordinary, linear, differential equations with constant coefficients, and is the basis for our study of System Dynamics.

The Input/System/Output Concept   To appreciate the significance of this limitation to a particular class of model, let us explore briefly what other *types* of models might reasonably be assumed. The framework for this discussion will be the input/system/output concept. What we mean by this is displayed graphically in Fig. 1–2. Inputs are physical quantities which are applied to a system and thereby cause certain variables (outputs) associated with the system to respond. The simple examples of Fig. 1–2 show systems with a single input and a single output, whereas in general, a system may simultaneously have several of each. Note also that a given system may be shown with different sets of input/output quantities depending on the interest of the analyst in one or the other aspect of the system's behavior. A more complex and practically interesting example of the concept is given in Fig. 1–3.

From our intuitive ideas about the cause-and-effect nature of the physical world, it is clear that if we precisely define a model of a physical system and subject it to specific known inputs, the outputs are completely determined. We might point out that, just as it is necessary to formulate simplified models of real systems, so is it also necessary to work with simplified versions (*models*) of the real inputs. The main reason for this is that the actual inputs which the system will see in service are always to a certain extent (and sometimes very much) unpredictable. Referring to Fig. 1–3 for example, the designer of an automobile suspension system cannot really know the profile of every road which a given car will be

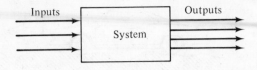

General Input/System/Output Configuration

(a)

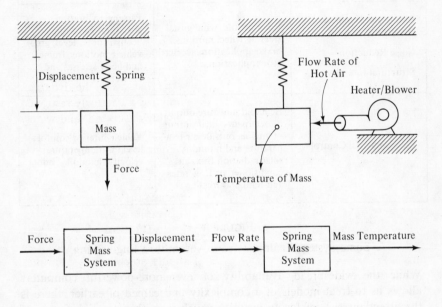

Example of Same System with Different Input/Output Variables

(b)   (c)

FIGURE 1-2

The Input/System/Output Concept

driven over in its lifetime, just as an aircraft designer is uncertain as to the exact nature of the turbulent wind gusts which might cause structural failures of a wing. These uncertainties in the inputs, coupled with the need for a simplified model of the system, will of course lead to predicted responses (outputs) which must also have some uncertainty. It is a most difficult and important task of the engineer to formulate analytical models which will be sufficiently sophisticated to provide results of practical significance while at the same time not being so complex as to obscure basic relationships or be uneconomic in terms of time and/or money.

Engineering Group With Design Responsibility	Inputs of Interest to this Group	Corresponding Outputs
Riding and Handling Qualities Noise Reduction Structural Integrity	Road profile, wind gusts, steering wheel motions, brake application, accelerator application.	Passenger vibration, structural vibration, acoustic noise level, gross vehicle motions, frame and body stresses and deflections.
Environmental Control	Heat and moisture output of occupants, heat output of engine, outside air temperature and humidity, solar radiation flux, air conditioner air flow rates and temperatures.	Vehicle interior comfort level (air temperature, velocity, humidity, odor)

FIGURE 1–3

Some System Examples from Automotive Engineering

While the widespread availability of ever-more-powerful computers allows us to treat models of a complexity undreamed of earlier, there is the attendant risk of losing intuitive touch with the essential nature of the system. The development of this intuition or "feel" is a vital ingredient in the growth of the individual engineer from one who can analyze the creations of others to one who himself engages in the creative design of new and useful systems.

Models of Inputs   While our main concern is the different types of *system* models which one might use, it is also appropriate and useful to attempt some generalization and classification of input types. By an *input*, we mean some agency which can cause a system to respond. One useful categorization consists of breaking inputs into two broad types, initial energy storage and external driving, as in Fig. 1–4. In Fig. 1–2b, for example, initial energy storage could be realized by displacing the mass from its static equilibrium position, thereby storing additional potential (strain) energy in the spring. If the mass is now released (without any external driving forces) it will respond by oscillating in a characteristic

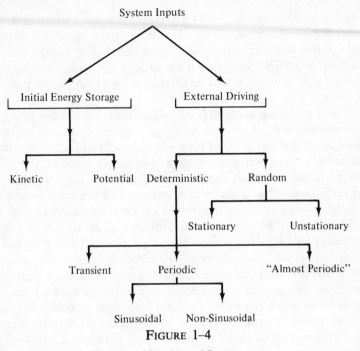

FIGURE 1–4

Classification of Inputs

fashion. Alternatively, the mass could be given an initial velocity (kinetic energy) and again be released, free of external driving forces, to perform its motion. Of course, initial potential and kinetic energies could be applied simultaneously also. In general, initial energy storage refers to a situation in which a system is put into a state different from some reference equilibrium state and then "released," free of external driving agencies, to respond in its characteristic way.

The term *external driving* implies that we have conceptually set up an envelope, or boundary, around some assemblage of components, and defined the interior of the envelope as our *system* and the exterior as the system's *environment*. External driving agencies are physical quantities which pass from the environment through the envelope (or *interface*) into the system, and cause it to respond. In practical situations there may sometimes be interactions between the environment and the system; however, since we are here dealing with the problem on a relatively introductory level, we will often use the concept of an ideal source in describing inputs. For example, in Fig. 1–2b, we indicate an external driving force acting on the mass without being explicit as to the exact physical means for providing this force. We are simply taking the viewpoint that we *choose* to study the response of the system to a force input. Hopefully, there will be some (perhaps many) practical situations which correspond to this idealized configuration. The (unspecified) means for providing such inputs is called an *ideal source*. In the case of Fig. 1–2b, it is an ideal force source, which means that it can supply whatever force we choose to assign to it, and it is totally unaffected by being coupled to the system it is driving.

Following the diagram of Fig. 1–4, we see that external driving agencies are first classified into deterministic or random types. As mentioned earlier, all real inputs have at least some element of randomness or unpredictability; thus deterministic models of inputs are always simplifications of reality, although they are quite adequate for many purposes. A deterministic input is one whose complete time history is given, for example a force $F = 10 \sin 100t$, where t is time in seconds, and the number 10 has units of pounds force. Deterministic inputs are further classified as *transient* (occur once and then die out), *periodic* (repeat over and over in a definite cycle, ideally forever), and *almost periodic* (continuing functions which are completely predictable but do not exhibit a strict periodicity; example, amplitude-modulated sine waves). Random inputs have time histories which cannot be predicted before the input actually occurs, although *statistical* properties of the input can be specified. For example, if an airplane is flown over a certain uniform terrain at a certain altitude and speed, it will encounter air turbulence (vertical gust velocity) which could be measured and recorded as a function of time, as in Fig. 1–5. This time history shows no periodicity, and is a good example of a random input. While there is no analytical mathematical function available to describe this phenomenon, one *can* compute certain *statistical* properties such as the average value, mean-square value, etc., from the recorded time history. If the airplane now turns around and retraces its first flight path through the turbulence, the time history of $V(t)$ will *not* be a duplicate of the first flight; however, the *statistical* properties should be essentially the same (if the weather has not changed, etc.). Thus, when working with random inputs, there is no hope of ever calculating a specific time history before it actually occurs; however, *statistical* predictions can be made and can have practical usefulness. When mathematically studying the response of a system to random inputs, it is generally necessary to work with a simplified mathematical model of the real random input. The science of probability and statistics provides many such models, and it is necessary to find one which approximates the physical situation with adequate fidelity in each specific application. The so-called "normal" distribution law (Gaussian probability function) has found wide use in such studies. Returning to the aircraft turbulence example, suppose we consider a complete coast-to-coast flight (Fig. 1–6), which will of course

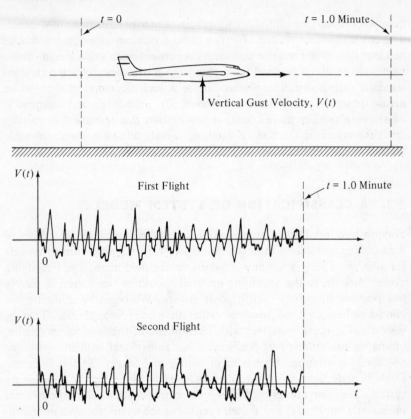

FIGURE 1–5

Stationary Random Input

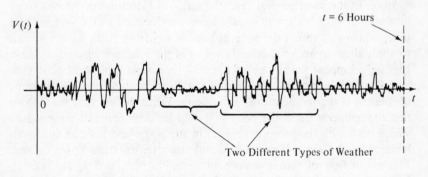

FIGURE 1–6

Unstationary Random Input

take place at varying altitudes and will encounter various types of weather. The entire six-hour record of $V(t)$ will be a random input, but it should be clear that now even the *statistical* properties (mean value, mean-square value, etc.) will not be fixed throughout the entire period. Such a random signal is called *unstationary* and can be a realistic representation of an actual physical phenomenon; however, its mathematical treatment is most complex. The more common assumption that statistical properties are time-invariant (*stationary* random signal) allows a more tractable mathematical treatment.

1-3. A CLASSIFICATION OF SYSTEM MODELS.

Having now established some initial understanding of the problems of formulating suitable models for inputs, we turn our attention to models for systems. There is not any standard or accepted method of classifying system models, so the upcoming material should be considered as merely one possible (hopefully useful) point of view. We begin by confining our considerations to a macroscopic rather than microscopic scale. That is, we do not concern ourselves with phenomena at the level of molecules, atoms, or sub-atomic particles/waves, but rather deal with the gross (or so-called "continuum") behavior of matter and energy. At this level, the fundamental "laws of nature" applicable to the study of engineering systems consider matter and energy as being continuously (though not necessarily uniformly) distributed over the space within the system boundaries. Description of the spatial extent of a system in mathematical terms requires setting up a coordinate system, such as, for example, the familiar rectangular XYZ coordinates of Fig. 1-7.

Since, in the most general case, the physical quantities of interest to us in a system vary both with regard to space (location in the system) and time, when we express the natural laws in equation form we are forced to apply them to an *infinitesimal element* of the system, such as $dx\,dy\,dz$ in Fig. 1-7. Suppose the system of Fig. 1-7 represents a physical body subjected to various inputs in the form of heat flows, and the response (output) of interest is the temperature of the body. When we say that we wish to find *the* temperature of the body it should be clear that this temperature varies both with the location (x, y, z) in the body, and also, at any given location, with time t. If we call the unknown temperature T, we see that T is a function of four independent variables, that is $T = T(x, y, z, t)$. Let us consider the body of Fig. 1-7 to be solid and with no internal heat generating mechanisms such as electric heating or chemical reactions. The basic physical laws pertinent to this problem are then the Fourier

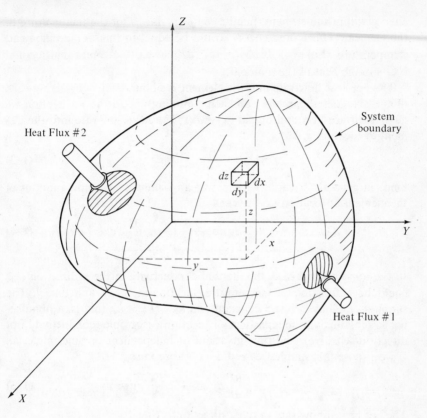

FIGURE 1–7

Heat Transfer System

Law of Heat Conduction and the Conservation of Energy. The heat conduction law says that the instantaneous rate of heat flow per unit area in a given direction is directly proportional to the rate of change of temperature, with respect to distance, in that same direction. Mathematically,

$$q_s = -k_s \frac{\partial T}{\partial s}, \frac{\text{Btu/ft}^2}{\text{sec}} \tag{1–1}$$

where

$$k_s \triangleq \text{thermal conductivity of the material in the } s \text{ direction} \tag{1–2}$$

The rate of change of temperature with respect to distance is written as a *partial* derivative since temperature is generally a function of *several* independent variables. The law of conservation of energy, applied to the element $dx\ dy\ dz$, says that, over a time interval dt, the difference between the inflow and outflow of heat energy must appear as additional stored energy within the element, manifested as a rise dT of the temperature of the element. This stored energy would be equal to (mass) (specific heat) (temperature rise) $= (\rho \, dx \, dy \, dz) \, (c) \, (dT)$ where $\rho \triangleq$ mass density and $c \triangleq$ specific heat of the material.

The net heat flow rate into the element is obtained by considering each of the six sides separately and then summing up. In the x direction we have a surface at x and one at $(x + dx)$. The heat flow rate into the element at location x is

$$Q_x = -k_x \frac{\partial T}{\partial x} \, dy \, dz \qquad (1\text{–}3)$$

Since in general Q_x varies with x, we can write the heat flow rate out of the element at location $(x + dx)$ as

$$Q_{x+dx} = \underbrace{-k_x \frac{\partial T}{\partial x} \, dy \, dz}_{\text{value at } x} + \underbrace{\frac{\partial}{\partial x}\left[-k_x \frac{\partial T}{\partial x} \, dy \, dz \right] dx}_{\text{change across } dx} \qquad (1\text{–}4)$$

In the most general case, the thermal conductivity in the x direction (k_x) could be a function of temperature, location (x, y, z), and time t. The simplest case has k_x just a constant and $k_x = k_y = k_z$, that is, a homogeneous (conductivity independent of location), isotropic (conductivity not direction sensitive), and time-invariant (k independent of time) material. If we choose this simplest case, Eq. (1–4) becomes

$$Q_{x+dx} = -k \frac{\partial T}{\partial x} \, dy \, dz - k \frac{\partial^2 T}{\partial x^2} \, dy \, dz \, dx \qquad (1\text{–}5)$$

The net heat inflow rate in the x direction is then

$$Q_x - Q_{x+dx} = k \frac{\partial^2 T}{\partial x^2} \, dx \, dy \, dz, \frac{\text{Btu}}{\text{sec}} \qquad (1\text{–}6)$$

Similarly for the y and z directions,

$$Q_y - Q_{y+dy} = k \frac{\partial^2 T}{\partial y^2} \, dx \, dy \, dz \qquad (1\text{–}7)$$

$$Q_z - Q_{z+dz} = k \frac{\partial^2 T}{\partial z^2} \, dx \, dy \, dz \qquad (1\text{–}8)$$

Conservation of energy then gives

$$k \, dx \, dy \, dz \left[\frac{\partial^2 T}{\partial x^2} + \frac{\partial^2 T}{\partial y^2} + \frac{\partial^2 T}{\partial z^2} \right] dt = \rho c \, dx \, dy \, dz \, dT \qquad (1\text{–}9)$$

$$\frac{dT}{dt} = \frac{k}{\rho c} \left[\frac{\partial^2 T}{\partial x^2} + \frac{\partial^2 T}{\partial y^2} + \frac{\partial^2 T}{\partial z^2} \right]$$

and since T is a function of more than one variable, the time derivative should be written as

$$\frac{\partial T}{\partial t} = \frac{k}{\rho c} \left[\frac{\partial^2 T}{\partial x^2} + \frac{\partial^2 T}{\partial y^2} + \frac{\partial^2 T}{\partial x^2} \right] \qquad (1\text{--}10)$$

Equation (1–10), a partial differential equation, is a relationship among the unknown temperature T, the location coordinates x, y, z, the time t, and system parameters k, ρ and c. It is not, however, a *complete* description of a heat transfer problem, since we have nowhere said anything about the location of the *boundaries* of the system or conditions existing at those boundaries. Also, the state of the system at the initial instant of time ($t = 0$) must be given. For example, in Fig. 1–7, it might be that all surfaces of the body (except for the shaded areas over which the input heat flows number one and two exist) are perfectly insulated. It would also be necessary to state mathematically the nature of the two heat flow inputs and the areas over which they act. The initial temperature distribution would be given as a mathematical function $T(x, y, z, 0) = $ a known function; for example, the simplest case would assume the temperature throughout the body initially uniform at, say, zero degrees ($T(x, y, z, 0) = 0$). If all the information of the above types were given, then the problem would be completely defined and a solution would theoretically exist, though it might be impossible to find analytically unless the conditions were quite simple. If a solution *could* be found, this would mean that we would have $T(x, y, z, t)$ as a definite known function such that if someone specifies *any* point x, y, z in the body and *any* time t, we can tell them what the temperature is. That is, having given the inputs (initial temperature distribution and heat flows number one and two) and the system (the differential equation and boundary conditions) we can then hope to find the outputs (temperature/time histories at any point in the body).

Application of the basic physical laws pertinent to various other types of dynamic physical phenomena (solid mechanics, fluid mechanics, electromagnetics, etc.) at the macroscopic level will in general lead to problem formulations (models) similar to the one just developed, that is, partial differential equations. This is basically true because we specify *dynamics* problems (thus, time is automatically an independent variable) and consider the system variables as quantities which change continuously from point to point in the system, thus giving three spatial independent variables. When the dependent variables (outputs) depend on more than one independent variable, we are bound to get partial differential equations when we apply the physical laws, because these laws generally involve rates of change. These types of models are also called field models or *distributed-parameter* models or *continuous-system* models. In their most general form (least number of simplifying assumptions), most people would agree that such models would behave almost exactly like the real systems at the macroscopic level. Unfortunately, these models can only be solved in a small number of special and simple cases, thus engineers find it necessary and desirable to work with less exact models in most instances. We now wish to give a systematic enumeration of the major classes of models in terms of the nature of the equations employed.

The table of Fig. 1–8 is an attempt to present a listing of model types in a form to facilitate comparison and comprehension of the overall situation. Note that, in general, models toward the top of the table tend to be closer to reality but are mathematically complex, while those toward the bottom involve more simplifying assumptions. The most realistic model treats the physical medium as being continuous and nonlinear with system parameters varying with space in an arbitrary deterministic fashion and with time in a random fashion. (It should be pointed out that the table is not claimed to be either unique or exhaustive. One can come up with a different *method* of classification or, if he sticks with the method shown, can think up model types not given in the table. The table is simply a result of the author's judgment with regard to striking the best compromise between completeness, compactness, and relevance to practical usage.) Using the heat transfer problem discussed earlier, a type 1 model might, for example, have $k = k_0(1 + aT)$ (making it nonlinear), $\rho = \rho_0(1 - \sqrt{x^2 + y^2 + z^2})$ (giving it a space-varying parameter), and have c vary randomly with time about same average value and according to some given statistical model. Equation (1–10) would then become

$$\frac{\partial T}{\partial t} = \frac{k_0(1 + aT)}{\rho_0(1 - \sqrt{x^2 + y^2 + z^2})c} \left[\frac{\partial^2 T}{\partial x^2} + \frac{\partial^2 T}{\partial y^2} + \frac{\partial^2 T}{\partial z^2} \right] \qquad \textbf{(1–11)}$$

$$c \triangleq \text{random variable with specified statistical properties} \quad \textbf{(1–12)}$$

This specific example (with suitable boundary and initial conditions) turns out to be mathematically unsolvable, as would most, if not all, models of type 1.

To convert this example to type 2, let, for instance, $c = c_0(1 + b \sin t)$. For type 3, let $c = c_0 =$ a constant. Types 4, 5, and 6 may be obtained by letting $\rho = \rho_0$ and c be respectively random, $c_0(1 + b \sin t)$ and c_0. All these models would also in general be unsolvable, due mainly to the nonlinearity (boundary conditions could also cause difficulties).

Models 7 through 12 differ from 1 through 6 by reason of their linearity. Linearity requires that dependent variables and their derivatives appear only as additive combinations of first-power terms. Thus the terms T^2, $\sin T$, $\sqrt{\partial T/\partial x}$, $ln(\partial^2 T/\partial x^2)$, $T(\partial T/\partial y)$, e^T are all nonlinear. The difference in "solvability" between nonlinear and linear differential equations is a great one (in favor of linearity); however, one must still go to the type 12 model in the 7-to-12 group before encountering a large and useful class

Model Type Number	Nature of the Medium						Time-Variation of System Parameters		
	Continuous (Field Problems)	Discrete (Network Problems)	Space-Variation of Parameters		Nonlinear	Linear	Random	Deterministic, Variable	Deterministic, Constant
			Variable	Constant					
1	X		X		X		X		
2	X		X		X			X	
3	X		X		X				X
4	X			X	X		X		
5	X			X	X			X	
6	X			X	X				X
7	X		X			X	X		
8	X		X			X		X	
9	X		X			X			X
10	X			X		X	X		
11	X			X		X		X	
12	X			X		X			X
13		X			X		X		
14		X			X			X	
15		X			X				X
16		X				X	X		
17		X				X		X	
18		X				X			X

Least realistic, easiest to solve ← Most realistic, most difficult to solve

FIGURE 1–8

Classification of System Models

of solvable equations; that is, the class of linear partial differential equations with constant coefficients. Even here, the boundary conditions must correspond to relatively simple geometrical shapes (spheres, cylinders, slabs, etc.) if a solution is to be obtained. The classical theories of elasticity, vibrations, acoustics, heat conduction, fluid flow, etc., are to a large extent based on such equations. Having said that many such equations may be solved, we should add that the solutions are often in the form of an infinite series of terms, which, while usable, is not as convenient as a closed-form solution.

In models 13 through 18, the medium is considered as discrete, rather than continuous. Such models are often called *lumped-parameter* (as opposed to distributed-parameter) or *network* (as opposed to field) types. The main simplifying assumption that changes a distributed-parameter model into a lumped-parameter model is that the system dependent variables are now assumed uniform over *finite* regions of space rather than over infinitesimal elements. To clarify this distinction, let us consider the one-dimensional heat conduction problem of Fig. 1–9. The slim metal rod buried in perfect insulation is initially all at 0°Fahrenheit, when at

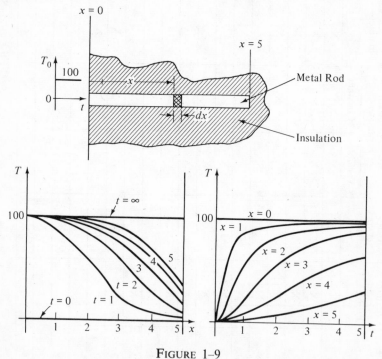

FIGURE 1–9

Distributed-Parameter Heat Conduction Model

time $= 0$ the left end ($x = 0$) is suddenly raised to 100°F and left there forever after. We wish to know how temperature varies with time at every point in the rod. Choosing an infinitesimal element of length dx and applying Fourier's heat conduction law and conservation of energy leads to the type 12 model

$$\frac{\partial T}{\partial t} = \frac{k}{\rho c} \frac{\partial^2 T}{\partial x^2} \qquad (1\text{--}13)$$

where we have assumed $k/\rho c$ as constant and neglected temperature variations in the y and z directions. This model can be solved for $T(x, t)$ in the form of an infinite series; however, the series converges rapidly and the first few terms are sufficient to get a very accurate approximation. These results can be displayed in two ways, graphs of T versus t at various x's (we may choose *any* x at all) or graphs of T versus x at any times t we choose. Fig. 1–9 shows a typical set of graphs.

In a lumped-parameter (discrete) model of this system, we break the rod into a finite number of finite-size "lumps"; in Fig. 1–10 six lumps

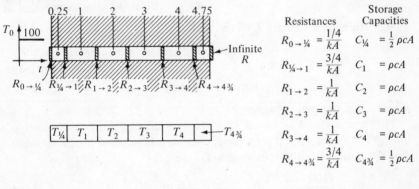

Resistances		Storage Capacities	
$R_{0 \to \frac{1}{4}}$	$= \dfrac{1/4}{kA}$	$C_{\frac{1}{4}}$	$= \frac{1}{2}\rho c A$
$R_{\frac{1}{4} \to 1}$	$= \dfrac{3/4}{kA}$	C_1	$= \rho c A$
$R_{1 \to 2}$	$= \dfrac{1}{kA}$	C_2	$= \rho c A$
$R_{2 \to 3}$	$= \dfrac{1}{kA}$	C_3	$= \rho c A$
$R_{3 \to 4}$	$= \dfrac{1}{kA}$	C_4	$= \rho c A$
$R_{4 \to 4\frac{3}{4}}$	$= \dfrac{3/4}{kA}$	$C_{4\frac{3}{4}}$	$= \frac{1}{2}\rho c A$

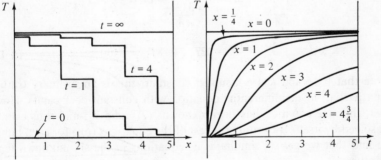

FIGURE 1–10

Lumped-Parameter Heat Conduction Model

have been used. Within a given lump the temperature varies continuously
with time, but at a given instant is uniform throughout that lump. The
material of the rod exhibits two basic actions, a resistance to heat flow
and the capability of storing thermal energy. The heat transfer resistance
is related to k, while the heat storage is related to ρc. In a distributed
parameter model these actions are both distributed throughout the rod;
in a lumped parameter model the resistance is concentrated at the ends
of the lumps and the lump itself serves as storage. We cannot at this stage
pursue the details of how these resistance and storage parameters are
actually calculated; we merely state the following facts.

$$R_t = \text{Resistance of a length } L \text{ of rod} \triangleq \frac{L}{k(\text{cross section area})}$$

$$\triangleq \frac{°F}{\text{Btu/sec}} \tag{1-14}$$

Heat flow through a resistance

$$= \frac{\text{Temperature drop across the resistance}}{R_t} = \frac{\text{Btu}}{\text{sec}} \tag{1-15}$$

energy storage of a length L of rod

$$= [\rho(L(\text{cross section area})c)] \text{ (temperature rise)}$$
$$= \text{Btu} \tag{1-16}$$
$$= C_t \text{ (temperature rise)} \tag{1-17}$$

where $C_t \triangleq$ thermal capacitance.

The analysis of the model consists of applying the conservation of
energy law to each lump separately, over an infinitesimal time interval dt.
For the lump whose temperature is $T_{1/4}$ we get

$$\frac{(100 - T_{1/4})}{\dfrac{1/4}{kA}} dt - \frac{(T_{1/4} - T_1)}{\dfrac{3/4}{kA}} dt = \frac{\rho cA}{2} dT_{1/4} \tag{1-18}$$

$$\frac{3\rho c}{k} \frac{dT_{1/4}}{dt} + 32T_{1/4} - 8T_1 = 2400 \tag{1-19}$$

Note that if ρc and k are assumed constant, this is an ordinary (rather
than partial) linear differential equation with constant coefficients. However, this one equation has *two* unknowns ($T_{1/4}$ and T_1) and thus cannot
be solved by itself. By writing the rest of the equations (one for each lump)
we find that we get six simultaneous equations in the six unknown temperatures.

$$\frac{3\rho c}{k} \frac{dT_1}{dt} + 7T_1 - 4T_{1/4} - 3T_2 = 0 \tag{1-20}$$

$$\frac{\rho c}{k} \frac{dT_2}{dt} + 2T_2 - T_1 - T_3 = 0 \tag{1-21}$$

$$\frac{\rho c}{k} \frac{dT_3}{dt} + 2T_3 - T_2 - T_4 = 0 \tag{1-22}$$

$$\frac{3\rho c}{k} \frac{dT_4}{dt} + 7T_4 - 3T_3 - 4T_{4\,3/4} = 0 \tag{1-23}$$

$$\frac{3\rho c}{k} \frac{dT_{4\,3/4}}{dt} + 8T_{4\,3/4} - 8T_4 = 0 \tag{1-24}$$

This set of equations can be solved (with the six initial conditions $T_{1/4} = T_1 = T_2 = T_3 = T_4 = T_{4\,3/4} = 0$ at $t = 0$) to give an explicit and closed-form solution for each of the temperatures. Figure 1–10 shows plots of typical solutions. Note that the T versus t curves for a given x are smooth, since t varies continuously in the ordinary differential equations. The T versus x curves at given t's are "stepped" since the basic model assumes a uniform temperature for each lump. (If one associates the temperatures $T_{1/4}$, T_1, etc., with the center *points* of each lump, he can then of course "eyeball" in a smooth curve or set of straight line segments connecting these points. These eyeballed curves would probably be better approximations than the stepped curves; however, they do not correspond to the lumped model assumed in the derivation.)

It is probably intuitively clear that as one uses a model with more (and thus smaller) lumps, the lumped model results will approach ever more closely the "exact" distributed parameter results. The price paid for this increasing accuracy is the greater computational effort of solving a larger set of simultaneous equations. While no universal rule can be given to decide how many lumps are "enough," in the above example if one takes the T versus t curves at $x = 1, 2, 3,$ and 4 as the criterion of comparison, the six-lump model shown gives almost perfect results.

Having now clarified the distinction between continuous and discrete models, it should be apparent that, just as models 1 through 12 corresponded to various types of *partial* differential equations, 13 through 18 represent ordinary differential equations. Again, nonlinear (13, 14, 15) and variable-coefficient (16, 17) equations present, in many cases, insurmountable mathematical difficulties. We thus finally arrive at the type 18 model, the ordinary, linear differential equation with constant coefficients. Closed form analytical solutions may be found by routine methods for this class of equation no matter how high the order of the equation or how many simultaneous equations might appear in a set, such as Eq. (1–19) to (1–24). A wealth of *generalized* analytical tools, which give great insight and intuitive feel for the behavior of even complex systems, has been developed for this type of model. One might initially be concerned about the practical validity of such models, since, according to Fig. 1-8, they are the *least* realistic of the model types there listed. Fortunately, many years of experience with such models, in diverse fields, has shown them to be sufficiently accurate in a large number of instances. Even when they are somewhat inaccurate, they are still often used in preliminary stages of design and analysis since they give the *essential* features of behavior in a form simple enough for human comprehension and manipulation. When system design is farther along, nonlinearities and other troublesome features can be brought in through computer studies.

In view of the above discussions, it should not be surprising that many areas of engineering science (electric circuit theory, vibration theory, automatic control theory, etc.) depend heavily on the use of the type 18 model. This text, in common with most others in the system dynamics field, will deal almost exclusively with this form of model. The main reasons for having a separate course in System Dynamics early in a curriculum (usually *before* the courses in vibration, control, circuits, etc.) are efficiency and breadth of understanding. Efficiency comes about since the material of a System Dynamics course is applicable to all the later courses of the types just mentioned. It is thus not necessary that *each* of these later courses present and duplicate this basic background, but instead, they can start at a rather high level and emphasize the special aspects of the particular subject. Perhaps more important, a System Dynamics course, as presented in this text, shows the essential similarities of behavior of a wide variety of physical systems and thus develops a breadth of understanding difficult to achieve in courses devoted to specific types of physical systems. Since one of the outstanding features of contemporary engineering practice is the need to deal with larger and more complex systems, not restricted to a single engineering discipline and often including important nontechnical aspects, the importance of developing breadth of understanding cannot be overemphasized.

1–4. THE VIEWPOINT OF THE SYSTEM ENGINEER.

The term *system engineer* is fraught with the same semantic difficulties encountered with "System Dynamics." Clearly, an engineer who works with systems is a system engineer, but what engineer does *not* work with systems? Perhaps we can help to clear up this situation by talking, not about system engineers, but rather about what it means for *any* engineer to adopt a system engineering viewpoint. We would like to suggest that while some engineers actually have the title "System Engineer" and spend most of their time on questions of overall system behavior, many others will adopt a system engineering viewpoint when it is appropriate to do so, and concentrate on component details the rest of the time. Whether they are called system engineers or not, when they adopt a systems viewpoint they are doing system engineering. This has been going on for as long as there have been engineers, even though the title "System Engineer" is of relatively recent origin. Of course, it is true that, just as other aspects of life get continually more complex, so do the systems that engineers work with; thus present-day system engineering can be exceedingly complex and require a specialized set of tools unnecessary in earlier days.

Just what is this "system engineering viewpoint?" It has several aspects, perhaps the most outstanding being a willingness to forego details in an attempt to gain overall understanding. In the area of dynamic behavior this leads rather naturally to the use of linear, constant-coefficient ordinary differential equations as system models, since this is the only class which can be expanded to comprehend systems with many components while still preserving a large measure of qualitative and quantitative understanding. These major advantages are of course gained by sacrificing detail in system description. Another aspect involves a considerable use of experimentally derived component characteristics. This is often necessary to ensure adequate accuracy, since in a system of many components, individual errors of theoretical prediction will be compounded to result in intolerable overall inaccuracy. Fortunately, this experimental approach at the component level is often possible since many large-scale systems are fabricated from components which already exist, and are thus amenable to experimental study. Finally, there is a considerable interest in *optimization* of system behavior by judicious choice of configuration and numerical values of parameters. That is, ideally, components are chosen and defined based on *total system* performance rather than isolated consideration of individual behavior.

When we state that the system engineering viewpoint requires a consideration of *overall* system behavior we should be clear that it is still necessary to establish a boundary which forms the interface between the system (which we wish to study) and the environment (which we ignore). A characteristic of the development of modern system engineering is that the boundaries of the system which it is necessary and desirable to study tend to expand as time goes by. In the early days, a mechanical engineer concerned with the operation of a power generating station might have been able to perform adequately by considering individual turbines, boilers, and generators as small isolated systems. As power stations became larger and were put under automatic control, it became necessary to expand the system boundary to include most of the equipment in the plant and carefully consider all the interactions. When power stations in several cities or states were interconnected to form vast power grids, system studies were needed which included interactions of many complete power stations. More recently, increasing public concern with environmental pollution requires that power station design studies include consideration of the effect of plant operation on the ecology of the surrounding environment. Furthermore, the choice between nuclear and fossil fuels for such power plants involves safety, psychological, and political factors. We see then that the evolution of technology and society leads naturally to the requirement for consideration of system problems of increasing scope. We do not pretend that this text will prepare one for operation on such a grand scale but perhaps at least a beginning can be made.

1-5. SOME TYPICAL SYSTEM DYNAMICS STUDIES.

The *electro-pneumatic transducer*[4] shown in Fig. 1–11 is a type of device widely used in industrial control systems when part of a system is electrical while another part is pneumatic and these two parts must "talk" to each other. It converts an electrical signal (say in the range 3 to 15 v) to a closely proportional air pressure signal (say in the range 3 to 15 psig.). While the electro-pneumatic transducer is thought of by the designer of the large control system as a *component* which can be purchased "off-the-shelf" and which exhibits desirable operating characteristics, *someone* had to design the transducer so it would behave properly, and by this engineer the transducer is legitimately considered a system, albeit a small-scale one. (Techniques of system dynamics are applicable at both the small-scale and large-scale levels; the large-scale system designer, however, usually *assumes* that components with certain properties are available and concentrates on finding an optimum arrangement and numerical parameter values to meet system requirements.) The operation of the transducer is briefly as follows. The input voltage e_{in} is applied to the coil of wire, causing a current i_c to flow. Since this current lies in the magnetic field of the permanent magnet, the coil feels a magnetic force f_m. This force causes the coil to move, bringing the flapper closer to the air nozzle and raising the output pressure p_o. This pressure, acting over the area A_p, causes a force f_p which opposes f_m. For a steady input voltage, the system will produce an output pressure such that the forces are balanced and equilibrium exists. Desirable performance characteristics of such a transducer include linearity (the voltage/pressure calibration curve should be a straight line within, say, 0.5%) and adequate speed of response (after a sudden change in voltage, the pressure should come to its correct value

---

[4] Moore E/P 77, Moore Products, Springhouse, Pa.

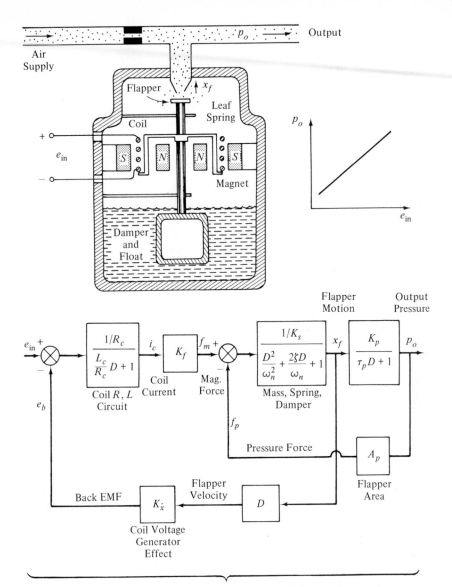

All the above can be collapsed into a simplified approximate overall model when numerical values are properly chosen

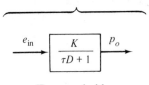

FIGURE 1–11

Electro-Pneumatic Transducer

23

within 0.1 seconds). Application of modeling techniques such as those found later in this text leads to the block diagram of Fig. 1–11. Analysis of this model for static and dynamic response allows comparison of the actual behavior with that required by the specifications. The designer, guided by his analysis, can then "juggle" the numerical values of parameters to get the desired performance.

Note that this practical example (and it is typical of many) requires a closely integrated application of principles from the fields of solid mechanics, fluid mechanics, electricity, and magnetism, etc. If one is to be effective in the analysis (and even more in the creative design) of new systems, he must cultivate this interdisciplinary view. Of course, it is impossible for a mechanical engineer, for example, to be an *expert* in another field such as electrical engineering. However, he should be familiar with certain fundamental physical ideas, try to stay aware of useful hardware developments, and be willing and able to converse with a specialist in another field when it is necessary to augment his understanding of a particular phenomenon or device.

In Fig. 1–12 we see a somewhat larger scale system which, in fact, uses the electro-pneumatic transducer just described as one of its components. The purpose of this system is to regulate the temperature T_C of fluid leaving a process vessel. This temperature is measured with a temperature-sensitive resistor (thermistor) and compared with a desired-temperature setting effected by manual adjustment of the resistance R_V. If the two temperatures (and thus, resistances) are not equal, the bridge circuit produces an error voltage e_E, which is amplified and applied to an electronic controller. The controller output voltage e_M is applied to the electro-pneumatic transducer, whose output pressure p_M positions the steam-flow control valve. Manipulation of steam flow-rate through the heat exchanger allows control of process-fluid temperature T_C. Note that in the block diagram of this system, the detailed operation of the electro-pneumatic transducer is not made apparent; only its overall input/output relation is included. A very careful dynamic analysis of systems such as this is necessary since the presence of "feedback" means that improper design can lead to violent and destructive oscillations.[5]

Some early papers by Paynter[6,7] show interesting examples of system dynamics studies applied to problems of hydroelectric power plants and modeling of river networks for use in flood control. Paynter says, "The

---

[5] E. O. Doebelin, *Dynamic Analysis and Feedback Control* (New York: McGraw-Hill Book Company, 1962).

[6] H. M. Paynter, *Surge and Water Hammer Problems*, ASCE Trans., Paper 2569, Vol. 118, 1953.

[7] H. M. Paynter, "A Palimpsest on the Electronic Analog Art," G. A. Philbrick Researches, Boston, 1955, pp. 217–245.

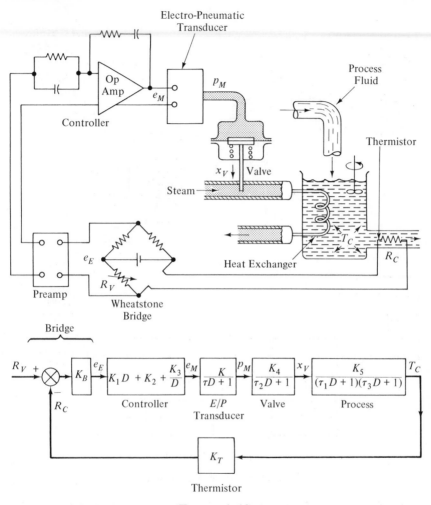

Electro-Pneumatic
Transducer

Process
Fluid

Thermistor

Heat Exchanger

Wheatstone
Bridge

FIGURE 1–12

Temperature Control System

interconnection of hydraulic and steam power plants within modern electric power networks gives rise to numerous complex problems concerning both the influence of load fluctuations and frequency control equipment on the stable operation of the generating units, and, conversely, the effects of the transient behavior of the hydraulic and mechanical components of the units on the performance of the electrical network." Figure 1–13 gives a schematic and block diagram sketch of the type of system considered by Paynter. To provide AC power at a given frequency, such as 60 Hz, the electrical generator must have accurate speed control

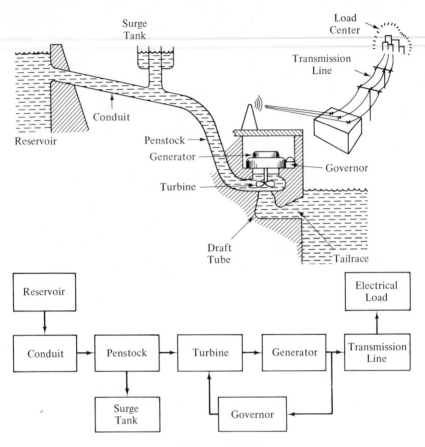

FIGURE 1–13

Hydroelectric Power System

in the face of various disturbances such as changes in electrical demand. This speed control is provided by a governor system which measures the speed and opens or closes the gate of the water turbine to increase or decrease the torque applied to the generator and thus manipulate generator speed. The hydraulic, mechanical, and electrical components of such a system all interact dynamically in a complex fashion and it is necessary to model and analyze the response during the design stage to be sure that the system will function in a stable fashion for all anticipated disturbances and not allow disastrous "power blackouts" to occur. Figure 1–14 shows one of Paynter's sketches representing the fluid flow and storage phenomena occurring when one traces the path of the input rainfall P through various surface and underground flow paths to finally arrive at the output runoff R in a river. Valid models for such phenomena

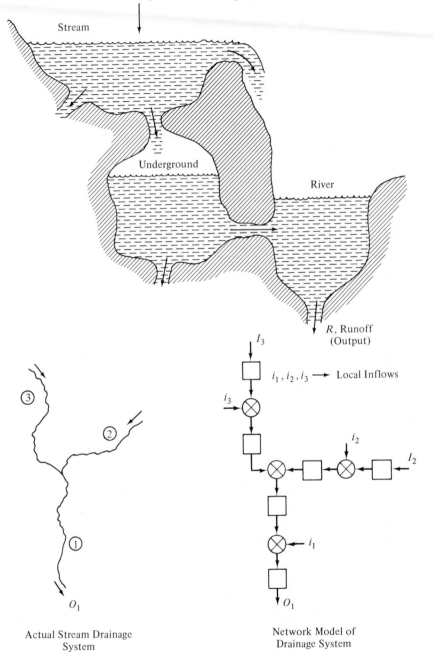

P, Point Rainfall (Input)

Stream

Underground

River

R, Runoff
(Output)

I_3

i_1, i_2, i_3 ⟶ Local Inflows

i_3

i_2

I_2

i_1

O_1

③

②

①

O_1

Actual Stream Drainage
System

Network Model of
Drainage System

FIGURE 1–14

Modeling of Hydrological Systems

27

would be helpful in predicting the timing and magnitude of flood crests from measured rainfall inputs. If the system includes some man-controllable features such as flood-control dams, the model could be used to determine the best way of manipulating these controls.

The application of system dynamic techniques of modeling and response prediction to biological systems has become of increasing interest to both life scientists and engineers. Life scientists find such models of great usefulness in trying to understand complex biological phenomena. Engineers and medical doctors are working together to design and construct artificial hearts, kidneys, and limbs. Engineers concerned with man/machine systems such as piloted aircraft or space vehicles have developed mathematical models of human operator behavior for restricted situations such as tracking and steering. The proper design of water-cooled or heated space suits for astronauts requires a thermal model of human body heat transfer, storage, and metabolism. Since the long period of evolution has enabled nature to develop extremely efficient mechanisms for performing bodily functions, engineers are studying these in an attempt to adapt some of nature's techniques to man-made systems.

Figure 1–15 shows some results of recent attempts to model the neuromuscular actuation system.[8] McRuer et al. state:

> We are interested in engineering descriptions of such systems in three respects:
>
> *Manual Control Engineering*—The basic dynamics of the human operator and the precision of manual control are critically limited by the properties of the neuromuscular system. An understanding of this system has important practical ramifications in determining the effects of control system nonlinearities and sensitivities on manual control.
>
> *Control Theoretic*—The neuromuscular system is an archetypical adaptive actuation system which, if understood operationally, might serve as the inspiration for analogous inanimate systems with similarly useful properties.
>
> *Physiological System Description*—The study of the neuromuscular system as a biological servomechanism provides a framework for the interpretation and elaboration of neurophysiological data.

The above reference attempts to analyze some of the *detailed* workings of the neuromuscular system. *Overall* descriptions of human operators in piloting or tracking tasks have been available for several years now, and are regularly and successfully used [9] in man/machine system design studies.

---

[8] D. T. McRuer et al., *A Neuromuscular Actuation System Model*, NASA CR 73128, 1967.

[9] J. J. Adams and M. W. Goode, *Application of Human Transfer Functions to System Analysis*, NASA TN D-5478, 1969.

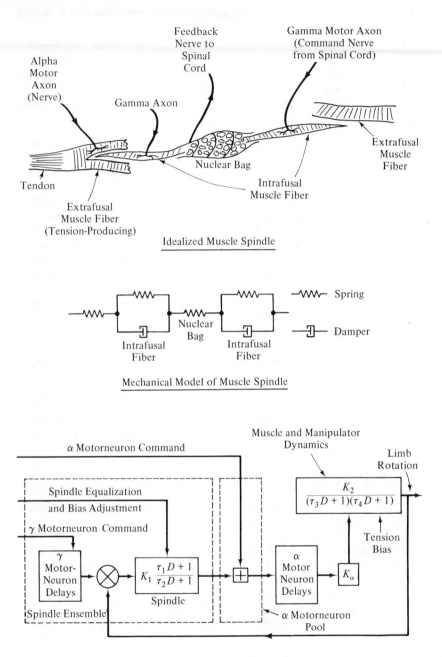

Feedback
Nerve to
Spinal
Cord

Gamma Motor Axon
(Command Nerve
from Spinal Cord)

Alpha
Motor
Axon
(Nerve)

Gamma Axon

Limb
Rotation

Extrafusal
Muscle
Fiber

Nuclear Bag

Tendon

Intrafusal
Muscle Fiber

Extrafusal
Muscle Fiber
(Tension-Producing)

Idealized Muscle Spindle

—WW— Spring

Nuclear
Bag

—☐— Damper

Intrafusal
Fiber

Intrafusal
Fiber

Mechanical Model of Muscle Spindle

Muscle and Manipulator
Dynamics

Limb
Rotation

α Motorneuron Command

Spindle Equalization
and Bias Adjustment

γ Motorneuron Command

$$\frac{K_2}{(\tau_3 D + 1)(\tau_4 D + 1)}$$

γ
Motor-
Neuron
Delays

$K_1 \dfrac{\tau_1 D + 1}{\tau_2 D + 1}$

Spindle

α
Motor
Neuron
Delays

K_α

Tension
Bias

Spindle Ensemble

α Motorneuron
Pool

Neuromuscular System Model

FIGURE 1–15

Neuromuscular Model

29

Physiologists[10] have long been interested in the body temperature regulation system of man. More recently, engineers designing water-cooled space suits[11] for astronauts found they needed a thermal model of man to carry out their design studies effectively. To give some idea as to the nature and usefulness of such models, let us quote directly from Stolwijk and Hardy.

In common with many branches of physiology, the study of thermo-regulation is reaching a stage in which investigative work of a descriptive nature will yield progressively less in the way of improved insight and in which quantitative and analytical approaches will become more and more necessary. Especially in thermoregulation in man, where it is not possible to manipulate separate components of the total system, such quantitative studies have to be performed on the complete intact system with all the control loops in operation. Such a system, even at the present limited state of our knowledge, becomes so complex that the design and interpretation of experimental data on a qualitative and intuitive basis becomes increasingly difficult.

An analog model as described in the present work can be of considerable assistance in both the planning and interpretation of such experimental work. Even though the model represents an over simplification of the actual system, it describes with good accuracy the data on heat production, heat flow and heat loss in the body. This makes it possible to repeat in the model almost any properly documented experiment and to test the quantitative consequences of various proposed conceptual schemes of thermoregula-tion. . . . An additional benefit which derives from the model may not be as immediately apparent, although anyone who has constructed a mathe-matical model will be very familiar with it; the discipline imposed by the quantitative character of the model is very helpful to the insight of the constructor.[12]

Stolwijk and Hardy describe their model as follows:

The human body is represented by three cylinders: the head, the trunk and the extremities. Each cylinder is divided into two or more concentric layers to represent anatomical and functional differences insofar as they are of primary importance in thermoregulation. Heat flow between adjacent parts is by conduction, and all layers exchange heat by convection with a central blood compartment. All three skin layers exchange heat with the environment by conduction, convection, radiation and evaporation. Signals which are proportional to temperature deviations in the brain and to devia-tions in the average skin temperature are supplied to the regulator portion of the model. The regulator then causes evaporative heat loss, heat produc-

---

[10] J. A. J. Stolwijk and J. D. Hardy, *Temperature Regulation in Man—A Theoretical Study*, Pflugers Archiv 291, 1966, pp. 129–162.

[11] P. Webb et al., *Automatic Control of Water Cooling in Space Suits*, NASA CR-1085, 1968.

[12] Stolwijk and Hardy, *Temperature Regulation in Man.*

tion by shivering or changes in the peripheral blood flow to occur in the appropriate locations in the body.[13]

Figure 1-16 is a sketch of this thermal model. The investigators wrote differential equations for this model and then simulated these equations on an electronic analog computer.

It is hoped that the above few examples will be helpful to the reader in illustrating the broad range of problems which has been treated by a system dynamics approach. Many others from diverse fields could have been given and may be found in the literature.

1-6. SYSTEM DESIGN.

While our discussions so far have emphasized the *analysis* of existing systems, it should be pointed out that the major overall function of engineering is the *design* of new products and services which will be useful to society. Of course a large part of design consists of detailed analysis and evaluation of competing concepts; however, we should not lose sight of the fact that analysis is only a part of the overall process. While we will in this text, wherever appropriate, attempt to introduce various concepts and viewpoints pertinent to design, the major emphasis in such an introductory treatment must necessarily be on analytical techniques. Hopefully, as the student matures in progressing through his curriculum, he will encounter increasingly more realistic treatments of design situations as preparation for the complexities of the industrial scene.

It may be helpful, even at this early stage, to try to develop some of this overall perspective and in particular to place the subject matter of this book in proper context relative to the larger view. The "flow charts" of Fig. 1-17 are offered as an aid in organizing our thinking along these lines.

BIBLIOGRAPHY

1. American Society of Mechanical Engineers, *Journal of Dynamic Systems, Measurement and Control*, published quarterly.

2. Cannon, R. H., *Dynamics of Physical Systems*. New York: McGraw-Hill Book Company, 1967.

---

[13] Stolwijk and Hardy, *op. cit.*

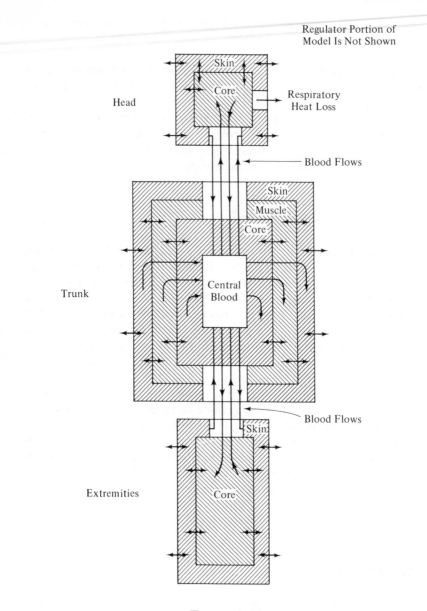

FIGURE 1–16

Thermal Model of Man

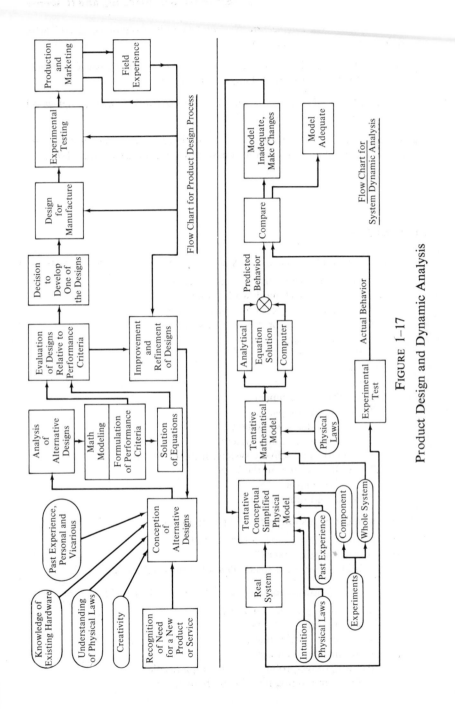

Flow Chart for Product Design Process

Flow Chart for
System Dynamic Analysis

FIGURE 1-17

Product Design and Dynamic Analysis

33

3. Haberman, C. M., *Engineering Systems Analysis*. Columbus, O.: Charles E. Merrill Publishing Company, 1965.

4. Martens, H. R. and D. R. Allen, *Introduction to Systems Theory*. Columbus, O.: Charles E. Merrill Publishing Company, 1969.

5. Reswick, J. B. and C. K. Taft, *Introduction to Dynamic Systems*. Englewood Cliffs, N. J.: Prentice-Hall, Inc., 1967.

6. Shearer, J. L., A. T. Murphy, and H. H. Richardson, *Introduction to System Dynamics*. Reading, Mass.: Addison-Wesley Publishing Co., Inc., 1967.

PROBLEMS

1–1. Identify possible input and output quantities for the following systems:

    a. An airplane flying through rough weather.
    b. An airplane touching down for a landing.
    c. An airplane taking off on a rough runway.
    d. An electric generator driven by a steam turbine.
    e. A pressure gauge.
    f. A thermometer.
    g. A voltmeter.
    h. A university.
    i. The federal government.
    j. A manufacturing plant.
    k. The human digestive system.
    l. The human respiratory system.
    m. The health-care delivery system in the U. S.
    n. The welfare system in the U. S.
    o. An electric motor.
    p. A household refrigerator.

1–2. What might be a source of periodic inputs for

    a. A car traveling over a concrete road?
    b. A train on steel tracks?
    c. A bridge carrying a steady flow of high-speed traffic?

1–3. Classify the following inputs according to the scheme of Fig. 1–4.

    a. Sonic boom pressure on a house.
    b. Wind on a tall smokestack.
    c. Waves on an ocean liner.
    d. Unbalanced tire on car.
    e. Record-groove-profile on phonograph needle.
    f. Noise from riveting machine on human ear.

    g. Frictional heating of truck's brake drums.

    h. Solar heating on communications satellite.

1–4. In the system of Fig. 1–11 the intended input is e_{in}, however p_o may be affected unintentionally by various environmental influences. Discuss the effect on p_o of:

    a. Temperature.

    b. Vertical acceleration; as if aboard a vehicle.

2

SYSTEM ELEMENTS, MECHANICAL

2-1. INTRODUCTION.

Chapters 2 through 4 will introduce the basic building blocks of lumped-parameter modeling, the so-called *system elements*. These might be considered as analogous to the chemical elements of the periodic table; that is, any system encountered in nature can be "built up" from a suitable combination of the system elements just as the chemical elements are combined to form any of the natural or synthetic materials found in the universe. To develop facility in lumped-parameter modeling of real systems, one must become thoroughly familiar with the basic system elements and their behavior.

The system elements are grouped chapter-wise as mechanical, electrical, fluid, and thermal, mainly as a matter of convenience. In practical applications we sometimes encounter systems which are essentially or entirely, say, mechanical or electrical, but we also find cases where several different forms occur in a single system. The classical area of mechanical engineering called "shock and vibration," for example, deals almost entirely with mechanical elements, while electrical circuit analysis deals with electrical elements. The practical design of automatic control systems for machine tools, on the other hand, invariably involves simultaneous consideration of mechanical, electrical, and possibly fluid elements. In such "mixed" systems we find, in addition, energy conversion devices which couple, say, a mechanical to an electrical element or subsystem. For example, an electric motor (an electromechanical energy converter) can be driven by current coming from an electrical circuit and itself provide torque to drive a mechanical system. Chapter 5 will give a brief survey of such devices.

Turning now to the subject of this chapter, the basic mechanical elements are used in modeling those parts of systems involving the motion of solid bodies. Clearly, this encompasses a vast array of practically important situations ranging from the timing accuracy of a wristwatch movement to the stability and riding qualities of a rapid-transit train. Only three elements are required to model the essential features of such systems: the spring element, the dashpot (damper) element, and the inertia element. All of these are found in both translational and rotational versions corresponding to the type of motion occurring. In addition to the three *passive* (non-energy-producing) elements, we also consider the *driving inputs* of mechanical systems, i.e., the forces and motions which cause the elements to respond.

2–2. THE SPRING ELEMENT.

In a modern technological society even the layman has a fair understanding of the concept of "springiness," thus a long discourse on the topic may initially appear superfluous. However, we will see that the subject (just as with any physical concept when critically examined) is replete with significant subtleties.

In our discussion of all the system elements, we will be using the terms *pure* and *ideal*. A real-world spring, for example, is neither pure nor ideal. That is, if the engineer designs or selects from available stock a part which he intends to perform the function of a spring alone, he will find that his "spring" also has some inertia and damping (energy dissipation) which are not at all necessary to the function of the system, but which nevertheless exist. The term *pure* thus refers to an "unadulterated" system element (spring, damper, inertia), that is, one which has *only* the named attribute. The term *ideal* could be replaced by *linear* since it requires that the input/output relationship which defines the element characteristics be strictly linear. Linear elements are considered ideal mainly from a mathematical viewpoint; they lead to equations which are readily solved. From a functional engineering viewpoint, *nonlinear* elements may sometimes actually be more "ideal"; however, they lead to mathematical difficulties which one must be prepared to accept. An element can be pure without being ideal.

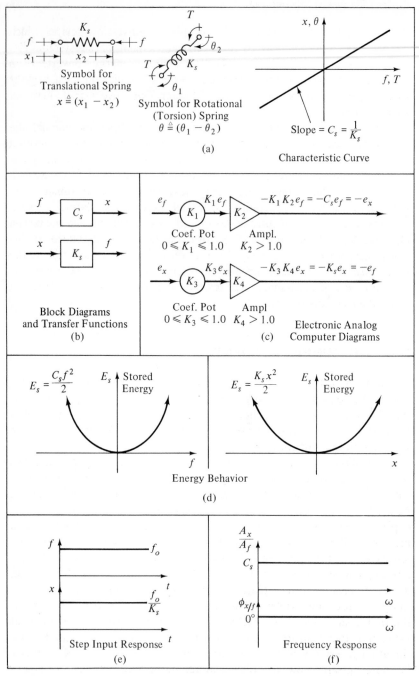

FIGURE 2–1

The Spring Element

38

The definition of the (pure and ideal) translational spring element is contained in the input/output relation (see Fig. 2-1a)

$$f = K_s(x_1 - x_2) = K_s x \tag{2-1}$$

where

$f \triangleq$ force applied to ends of spring, lb_f

$x_1 \triangleq$ displacement of one end, inch

$x_2 \triangleq$ displacement of other end, inch

$K_s \triangleq$ spring constant, $lb_f/inch$

$x \triangleq x_1 - x_2 \triangleq$ relative displacement of ends, inch

$\triangleq$ means "equal by definition"

The origins for coordinates x_1 and x_2 must be such that the spring is at its "free length" (zero force) condition when $x_1 = x_2$. The symbols $+\longrightarrow$ indicate the assumed positive direction (sign convention) for the particular force or displacement. For rotary motions we have

$$T = K_s(\theta_1 - \theta_2) = K_s\theta \tag{2-2}$$

where

$T \triangleq$ torque (moment) applied to ends of spring, inch-lb_f

$\theta_1 \triangleq$ angular displacement of one end, radian

$\theta_2 \triangleq$ angular displacement of other end, radian

$K_s \triangleq$ spring constant, inch-lb_f/radian

$\theta \triangleq \theta_1 - \theta_2 \triangleq$ relative angular displacement of ends, radians

The spring constant K_s might also be called the stiffness, since a large value of K_s corresponds to a stiff spring. The spring could also be described in terms of its *compliance* C_s which is defined as the reciprocal of K_s ($C_s \triangleq 1/K_s$) and is clearly a "softness" parameter. Using compliance one could write the spring description as

$$x = C_s f \tag{2-3}$$

or

$$\theta = C_s T \tag{2-4}$$

Engineers have found the use of *block diagrams* most useful in the design and analysis of all kinds of systems. In a block diagram, rather than showing a "picture" of the system hardware, we show blocks containing mathematical descriptions (*transfer functions*) of the hardware and connect these blocks with lines which denote the input and output signals for each block. At this stage we consider a transfer function as an output/input ratio. Since in some practical problems the known input is force or torque, while in others it is displacement, we can define two kinds of transfer functions for a spring element.

$$\text{force-input transfer function} \triangleq \frac{\text{output}}{\text{input}} = \frac{x}{f} = C_s, \quad \frac{\text{inch}}{\text{lb}_f} \quad (2\text{-}5)$$

$$\text{motion-input transfer function} \triangleq \frac{\text{output}}{\text{input}} = \frac{f}{x} = K_s, \quad \frac{\text{lb}_f}{\text{inch}} \quad (2\text{-}6)$$

and similarly for the rotational case. Using the transfer function concept, one can obtain the output of an element or system by multiplying the input times the transfer function (see Fig. 2–1b).

Computers of all types (analog, digital, and hybrid) are widely used in system studies and we wish to prepare for this as soon as possible. In electronic analog computers (which we will discuss in detail later) a particular component is used to simulate transfer functions of the form (2–3)–(2–6). In such a computer *all* the input and output variables are represented by voltages (full-scale range is usually ± 10 or ± 100 volts) and we thus need a component which will accept an input voltage e_f representing, say, an input force and produce an output voltage e_x representing displacement and numerically equal to $C_s e_f$. This can be done with a coefficient potentiometer and amplifier, and Fig. 2–1c shows the standard symbols employed. Note that the overall arrangement also changes the algebraic sign. This is an unavoidable peculiarity of the amplifier electronics which appears when we want coefficients (such as C_s) with numerical values greater than 1.0. For values of 1.0 or less the amplifier is not needed and the sign change is avoided. When a *digital* computer is used for system dynamics problems, we ordinarily make use of a digital simulation language (in this text we employ the IBM CSMP) which allows one to program problems in a simple fashion similar to analog computer techniques. Simulation of the spring element would be accomplished with the simple Fortran statement X = 4.57*F if C_s were 4.57 inch/lb$_f$ or X = CS*F if the numerical value of CS (C_s) were given elsewhere in the program by a data card.

If a force is gradually applied to a spring element and then maintained constant, we find that the force has done work in deflecting the spring and this energy is now *stored* in the spring and could be recovered when the spring is allowed to relax. We will be interested in such energy considerations for all system elements, so let us pursue this with the spring. Figure 2–2 shows the force application process graphically. Since the instantaneous power taken from the force source and put into the spring is given by the definition of power as (force) (velocity), we can write instantaneous power = (instantaneous force) (instantaneous velocity) =

$$\left(\frac{f_0 t}{t_1}\right)\left(\frac{C_s f_0}{t_1}\right) \quad (2\text{-}7)$$

$$\text{power} = \frac{C_s f_0^2}{t_1^2} t \quad (2\text{-}8)$$

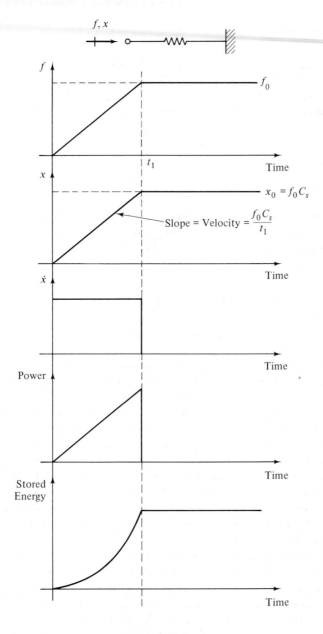

FIGURE 2–2

Energy Storage in Spring Elements

Now the total energy put into the spring is given by

$$\text{stored energy} = \text{work done} = \int (\text{power}) \, dt = \int_0^{t_1} \frac{C_s f_0^2}{t_1^2} t \, dt \quad \textbf{(2-9)}$$

where we integrate only to t_1 since the velocity, and thus the power, is zero thereafter. Carrying out the integration gives the stored energy E_s as

$$E_s = \frac{C_s f_0^2}{2} \quad \textbf{(2-10)}$$

or, alternatively,

$$E_s = \frac{K_s x_0^2}{2} \quad \textbf{(2-11)}$$

Actually, this result is independent of the particular time variation of force or motion used in getting to the final force f_0 or position x_0. From the characteristic curve of Fig. 2–1a we can write for an infinitesimal motion dx

$$\text{work done} = f \, dx = K_s x \, dx \quad \textbf{(2-12)}$$

and thus for a total displacement x_0

$$\text{work done} = \int_0^{x_0} K_s x \, dx = \frac{K_s x_0^2}{2} \quad \textbf{(2-13)}$$

Figure 2–1d displays these energy relations.

The dynamic behavior of spring elements is especially simple since the output/input relation is an algebraic (rather than differential) equation. From Eq. (2–3), for example, we can conclude that the response motion x caused by *any* force input f is *instantaneous*. For example, if $f = 0$ for $t < 0$ and $f = 6 + 2t$ for $t \geq 0$, we can calculate the resulting motion x as $x = C_s f = 6C_s + 2C_s t$ with the time histories as shown in Fig. 2–3a. Clearly, for any input waveform, even the random pattern of Fig. 2–3b, the output motion is an exact "replica" of the driving force.

Two specific dynamic inputs, the step input and the sinusoidal input, will be of considerable general interest for all elements and also complete systems. In a step input, the input quantity instantly jumps from some constant value (can usually be taken as zero) to a different constant value, which is then maintained forever. Figure 2–1e shows the step response of a spring element. For a sinusoidal force input $f = f_0 \sin \omega t$ the resulting displacement is $x = C_s f_0 \sin \omega t$. The response to a sinusoidal input is particularly useful, since we can let the frequency ω take on any value we like, small values corresponding to slow oscillations, and large values to fast oscillations. Recall that ω has units of radians per second and the number of *cycles* of oscillation per second is $\omega/2\pi$. The frequency in cycles per second is given the symbol f (rather than ω) and the standard unit is the Hertz (Hz), that is, one Hertz is one cycle per second. The

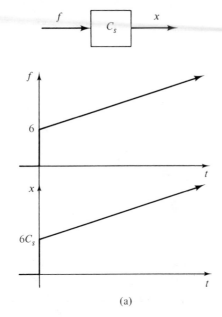

(a)

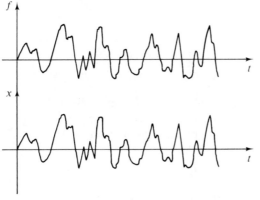

(b)

FIGURE 2–3

Instantaneous Time Response of Spring Elements

response to sinusoidal driving is called the *frequency response* and is completely described by giving the amplitude ratio between output and input quantities and the phase angle between these two sine waves for all frequencies from zero to infinity. Figure 2–4 shows the input and output of a spring element for a typical driving frequency ω. It is clear that the two sine waves are exactly in phase (phase angle = 0) for *any* frequency

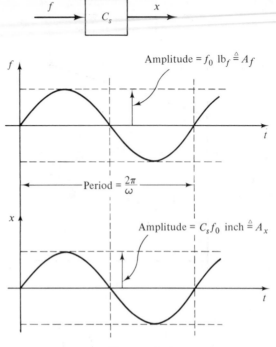

FIGURE 2–4

Frequency Response of Spring Elements

and the amplitude ratio of output displacement divided by input force will be C_s inch/lb$_f$, no matter what the frequency. It is conventional to display the variation of phase angle ϕ and amplitude ratio with frequency graphically, as in Fig. 2–1f.

2–3. REAL SPRINGS: LINEARIZATION.

While carefully made and used springs can closely approach the ideal linear *static* characteristic curve of Fig. 2–1a, they must always deviate, at least slightly, in at least two respects. (For *dynamic* operation, real springs also exhibit *inertia*; that is, they are not *pure* spring elements.) These deviations consist of a nonlinearity of the force/deflection curve and a non-coincidence of loading and unloading curves. The precise sources and degree of nonlinearity vary with the form of spring, the material, etc., but fundamentally are explained by recognizing that *any* rigorous mathematical relation (whether linear or nonlinear) between two *physical* quantities is incapable of precise realization. When, as is sometimes the case, the nonlinearity is severe, we often employ *linearization* in one form or another. Figure 2–5 shows a typical nonlinear force/deflection curve. If a mechanical system using this spring were known to operate with small motions in the neighborhood of the operating point f_0, x_0 we might be tempted to replace the actual nonlinear characteristic with a linearized approximation consisting of the tangent line to the curve at the operating point. This technique is the basis of the widely used and successful method of *small-signal or perturbation analysis*, which is of course not limited to force/deflection relations, but is applicable to *any* nonlinear phenomena represented by smooth curves.

When the characteristic curve is obtained from experimental data, no mathematical formula is available and we must fit the tangent graphically. For known nonlinear *functions* the linearization can be carried out analytically using a truncated Taylor Series expansion. Recall from calculus the Taylor Series expansion for a function $y = f(x)$ about the point x_0:

$$y = f(x_0) + \frac{dy}{dx}\bigg|_{x_0} (x - x_0) + \frac{d^2y}{dx^2}\bigg|_{x_0} \frac{(x - x_0)^2}{2} + \ldots \qquad (2\text{--}14)$$

To get an *exact* representation of y, the complete series (infinite number of terms) must be used. To get an approximation, the series may be truncated after a finite number of terms. To get a *linear* approximation it must be truncated after the first two terms, giving

$$y \approx f(x_0) + \frac{dy}{dx}\bigg|_{x_0} (x - x_0) \qquad (2\text{--}15)$$

which will be seen to be precisely the equation of the tangent line at x_0.

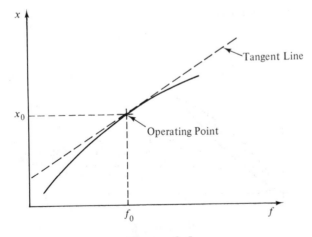

FIGURE 2–5

Linearization of Nonlinear Spring

As an example consider the nonlinear force/deflection relation

$$f = K_1 x + K_2 x^3 = 2x + 5x^3 \tag{2-16}$$

in the neighborhood of the point $x = 1.0$. We have

$$f \approx (2 + 5) + (2 + 15x^2)|_{1.0}(x - 1.0) = -10 + 17x \tag{2-17}$$

The linearized value of the spring constant K_s would be the slope of the tangent line, $df/dx = 17$ lb$_f$/inch. The accuracy of such approximations clearly improves as the motions away from the operating point become smaller; however it is not clear how small is "small enough." When we later study complete systems we will do some computer studies which will give some feeling for such questions.

The other deviation from ideal behavior, the non-coincidence of loading and unloading curves, is displayed in Fig. 2-6. The second law of thermodynamics guarantees that the area under the loading f-x curve (work put

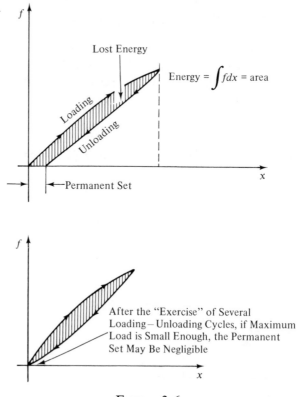

FIGURE 2-6

Energy Losses in Real Springs

into the spring during loading) *must* be greater than that under the un-loading *f-x* curve (work recovered from the spring during unloading). That is, it is impossible to recover 100% of the energy put into *any* system. This behavior of real springs indicates the presence of energy-dissipating mechanisms within the spring, whereas pure spring elements have only energy storage and no dissipation. In many springs these energy losses are quite small and it actually requires careful experimental technique to find a difference between loading and unloading curves; however, such a difference must always exist. For precision springs such as are needed in measuring instruments, special alloys such as Iso-Elastic[1] have been developed. This material exhibits a hysteresis (maximum difference be-tween loading and unloading curves) of less than 0.05% of maximum deflection. On the other hand, a rubber spring (widely used in shock mounts) as in Fig. 2–7d might have a hysteresis of 3 or 4%. This energy loss might actually be beneficial in such an application, since it would tend to damp the vibrations.

While the coil or helical spring of Fig. 2–7a is perhaps most familiar, a wide variety of different geometrical forms can be and are used for spring functions. In fact, almost any physical object will exhibit springlike behavior in that when you press on it, a deflection nearly proportional to the applied force will occur. Figure 2–7 shows some of the more common forms actually used for springs. The hydraulic spring shown depends on the compressibility of oil for its principle of operation and provides high energy storage in a small space.[2] Typical applications are as return springs ("strippers") in punch-and-die assemblies of punch presses. Air springs (Fig. 2–7g) have many desirable properties for automotive suspension systems. The most desirable force/deflection characteristic for this appli-cation is nonlinear as shown in Fig. 2–8.[3] The air spring was designed so as to actually achieve this desirable characteristic. While a piston-cylinder could theoretically be used, the rubber rolling-diaphragm of Fig. 2–7g gives a simpler and better air seal without critical manufacturing tol-erances. Air springs can also be "pumped up" to automatically re-level the car when large loads are carried.[4]

To demonstrate that most any object, even of peculiar shape, exhibits spring-like behavior, the "spring" of Fig. 2–9 was constructed of alumi-num and experimentally calibrated with dead weights and a micrometer,

---

[1] John Chatillon and Sons, New York, U. S. Patent No. 2,174,171.

[2] L. L. Johnson, "The Hydraulic Spring," *Machine Design* (May 26 1960).

[3] V. D. Polhemus and L. J. Kehoe, "The Development of the General Motors Air Spring," *Gen'l. Motors Eng. Jour.* (July–Aug.–Sept. 1957).

[4] *Goodyear Air Suspensions*, Brochure 862-947-16. Akron, Ohio: Goodyear Tire and Rubber Co.

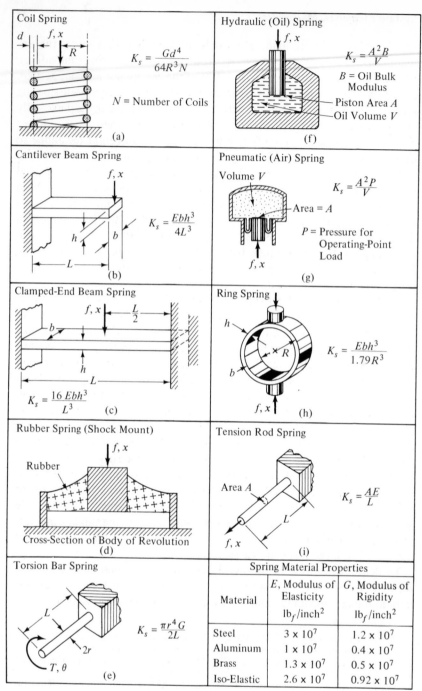

FIGURE 2–7

Several Types of Practical Springs

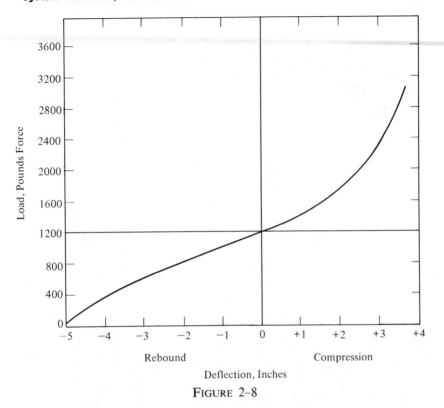

FIGURE 2–8

Automotive Air Spring Characteristic

giving the results shown. Plot this data to see if this "spring" is linear and, if so, find its spring constant. This concept is quite important since many machine parts, such as the crankshaft of an engine, are not *intended* to be springs but their unavoidable springiness can cause vibration problems. We conclude this section with a few more unconventional spring effects. In Fig. 2–10a the horizontal tail ("elevator") of a flying aircraft is deflected by an angle θ to the relative wind, to cause the aircraft to pitch upward. If one were to measure the torque due to wind pressure on the shaft for various values of θ, the graph shown would be obtained. We see that this aerodynamic torque exhibits spring-like behavior and would be modeled as a spring if we were studying the dynamic motion of the elevator. In Figs. 2–10b and c, the action of gravity provides a spring effect in that any motion of the pendulum or liquid column away from their static equilibrium position is accompanied by a restoring force or torque. The buoyancy spring effect of Fig. 2–10d is related to similar phenomena which influence the dynamic behavior of ships. Magnetic and electrostatic

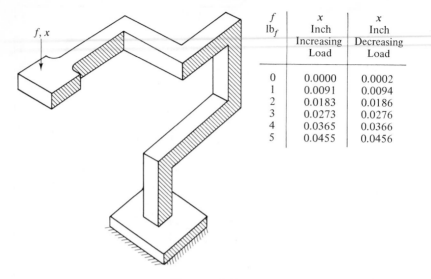

f lb_f	x Inch Increasing Load	x Inch Decreasing Load
0	0.0000	0.0002
1	0.0091	0.0094
2	0.0183	0.0186
3	0.0273	0.0276
4	0.0365	0.0366
5	0.0455	0.0456

FIGURE 2–9

Complex-Shape Spring

springs have been used to "levitate" objects, that is, to support them with no other bodies actually touching them. The magnetic version has been used as a high-speed "frictionless" bearing[5] and to support aircraft models in wind tunnels, while the electrostatic is the basis of sophisticated gyroscope instruments[6] which exhibit extremely low friction effects, since the moving parts are supported electrically rather than in bearings. Centrifugal spring effects, Fig. 2–10g, are somewhat similar to gravity springs in that, in each case, a force "field" creates a preferred position for the object and any displacements away from this position give rise to restoring forces or torques. The pneumatic tires used on vehicles influence the riding and handling qualities of the vehicle and exhibit some interesting spring properties. If the tire is tested when *not* rolling, the force-deflection curve is quite nonlinear, giving an equivalent linearized spring constant which varies with amplitude of motion. However, when the same tire is tested[7] while rolling at about 10 mph, the spring constant becomes very nearly independent of motion, indicating linear behavior (see Fig. 2–11).

[5] J. Lyman, *Magnetic Bearings* (Cambridge, Mass.: Cambridge Thermionic Corp., 1967).

[6] H. W. Knoebel, "The Electric Vacuum Gyro," *Control Engineering* (Feb. 1964), pp. 70–73.

[7] *Dynamic Spring Rate Performance of Rolling Tires*, General Motors Engineering Publication 3610, 1968.

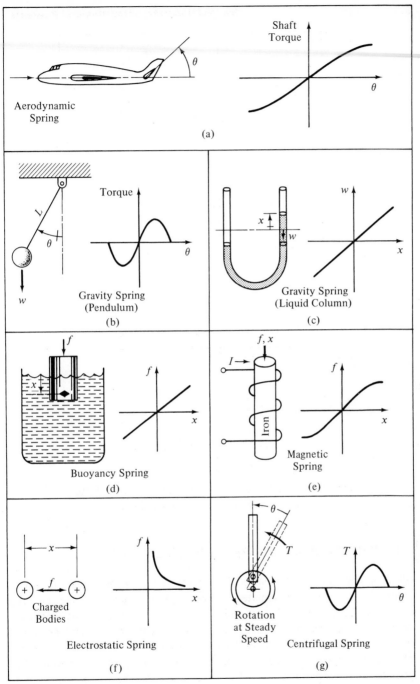

FIGURE 2–10

Some Springlike Effects in Unfamiliar Forms

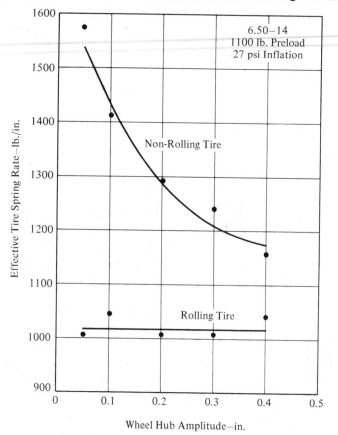

FIGURE 2–11

Effect of Wheel Motion Amplitude on Spring Rate
for Rolling and Non-Rolling Tires

2–4. THE DAMPER OR DASHPOT ELEMENT.

While a pure spring element stores energy with no loss or dissipation, a pure damper element dissipates all the energy supplied to it. Since energy cannot be destroyed, what we mean by dissipation of energy is that it is converted from mechanical to thermal form (heat) which flows away to the surroundings and is thus no longer available for useful work. Various physical mechanisms, usually associated with some form of friction, can provide this dissipative action. The definition of the pure and ideal translational damper element is contained in the input/output relationship (see Fig. 2–12a)

$$f = B \left(\frac{dx_1}{dt} - \frac{dx_2}{dt} \right) = B \frac{dx}{dt} \qquad (2\text{-}18)$$

where

$f \triangleq$ force applied to ends of damper, lb_f

$dx_1/dt \triangleq$ velocity of one end, inch/sec

$dx_2/dt \triangleq$ velocity of other end, inch/sec

$\quad B \triangleq$ damper coefficient, $lb_f/(\text{inch/sec})$

$dx/dt \triangleq dx_1/dt - dx_2/dt \triangleq$ relative velocity of ends, inch/sec

The damper force is thus seen to be directly proportional to the relative *velocity* of its two ends, whereas the spring force is proportional to displacement. For rotational systems we have

$$T = B \left(\frac{d\theta_1}{dt} - \frac{d\theta_2}{dt} \right) = B \frac{d\theta}{dt} \qquad (2\text{-}19)$$

where

$\quad T \triangleq$ torque applied to ends of damper, inch-lb_f

$d\theta_1/dt \triangleq$ angular velocity of one end, radian/sec

$d\theta_2/dt \triangleq$ angular velocity of other end, radian/sec

$\quad B \triangleq$ damper coefficient, inch-$lb_f/(\text{rad/sec})$

$d\theta/dt \triangleq d\theta_1/dt - d\theta_2/dt \triangleq$ relative angular velocity of ends, rad/sec

To draw block diagrams, we again need the concept of the transfer function, and it is now necessary to extend it to include *differential* as well as algebraic equations. This will be facilitated by use of the *operator notation.*

$$Dx \triangleq \frac{dx}{dt} \qquad (2\text{-}20)$$

or

$$D \triangleq \frac{d}{dt} \qquad (2\text{-}21)$$

That is, any quantity found immediately to the right of the differential operator D is to be differentiated with respect to time t. Equations (2–18) and (2–19) can thus be written as

$$f = BDx \qquad (2\text{-}22)$$

and

$$T = BD\theta \qquad (2\text{-}23)$$

We define the *operational transfer function* $(f/x)(D)$ by treating Eq. (2–22) as algebraic and forming the output/input ratio

$$\frac{f}{x}(D) \triangleq BD \qquad (2\text{-}24)$$

This is read "f over x of D equals by definition BD." The notation $(f/x)(D)$ is used since the simpler f/x could be (wrongly) interpreted as an ordinary

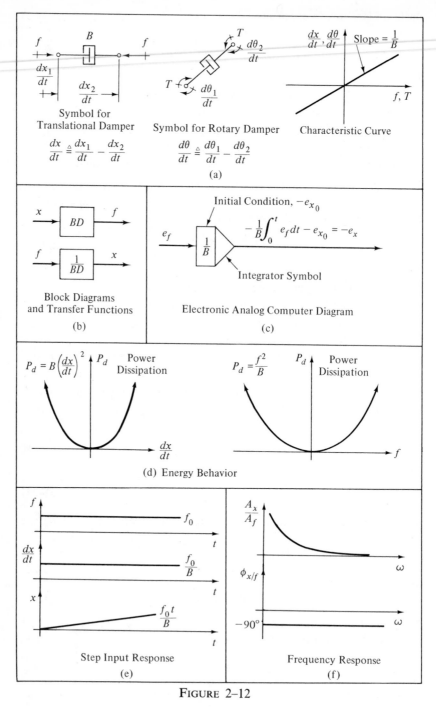

FIGURE 2–12

The Damper Element

instantaneous ratio of the time-varying quantities f and x, *which the transfer function is not.* For example, if $x = 2t^2$, $f = 4Bt$ and $f/x = 2B/t$, while if $x = \sin t$, $f = B \cos t$, and $f/x = B/\tan t$, whereas $(f/x)(D) = BD$ *for any and all forms of input motion x.* That is, the operational transfer function is a means of describing the *general* dynamic behavior of the system, not its *specific* response to a specific input. In block diagrams (see Fig. 2–12b), giving the element's transfer function inside the block gives a *complete* description of the element's dynamics.

If we consider the input to be force f rather than displacement x we get

$$\frac{x}{f}(D) \triangleq \frac{1}{BD} = \left(\frac{1}{B}\right)\left(\frac{1}{D}\right) \qquad (2\text{–}25)$$

The meaning of the operator $1/D$ is revealed by writing from Eq. (2–18)

$$f \, dt = B \, dx \qquad (2\text{–}26)$$

$$\int_0^t f \, dt = B \int_{x_0}^x dx = B(x - x_0) \qquad (2\text{–}27)$$

where x_0 is the *initial value* (value at $t = 0$) of x. Then

$$x - x_0 = \frac{1}{B} \int_0^t f \, dt \qquad (2\text{–}28)$$

which may be compared to Eq. (2–25) rewritten as

$$x = \frac{1}{B} \frac{1}{D}(f) \qquad (2\text{–}29)$$

If we take the initial value $x_0 = 0$, Eq. (2–28) gives

$$x = \frac{1}{B} \int_0^t f \, dt \qquad (2\text{–}30)$$

thus the operator $1/D$ indicates *integration* with respect to t, however the constant of integration x_0 is understood, not expressly stated.

To simulate a damper element on an electronic analog computer, we make use of the *electronic integrators* found on all such machines. (Electronic *differentiation* rarely turns out to be practical due to noise problems, thus it is generally avoided in computers.) The integrator contains a resistor R and capacitor C whose product RC may be set to give the desired value of $1/B$. There is also a mechanism for charging the capacitor to an initial ($t = 0$) voltage corresponding to the initial displacement x_0. Note that again the computer, in addition to performing the desired operation (integration) unavoidably changes the algebraic sign (see Fig. 2–12c). When we use the CSMP digital simulation language for this element, we write the statement X = (1.0/B) * INTGRL (XO, F) if the numerical value of B and initial displacement XO are given elsewhere on a data card. If, say, $1/B = 10.$ and XO $= -1.0$ we could also write

$X = 10.$ *INTGRL $(-1.0, F)$. In either case, if the computer is given force F as a function of TIME, say, $F = 5.0*$ TIME, the statement for X will cause the computer to carry out a step-wise numerical integration (using Simpson's rule, trapezoidal rule, or some other selected method) to find and plot X as a function of time.

Since instantaneous power is the product of instantaneous force and velocity, a damper element completely dissipates the mechanical energy supplied to it at a rate

$$P_d \triangleq \text{power dissipation} = (f)\left(\frac{dx}{dt}\right) = B\left(\frac{dx}{dt}\right)^2 \frac{\text{inch-lb}_f}{\text{sec}} \quad \text{(2--31)}$$

The total energy dissipated over any time interval would be the time integral of the power $\int P_d \, dt = \text{inch-lb}_f$. A constant force f_0, for example, gives an energy dissipation of $f_0^2 t / B$ inch-lb$_f$ for a time interval t. To see that the damper dissipates all the energy supplied, note that any force applied to it produces a velocity in the *same* direction. The source which is supplying the force must thus provide power *to* the damper, since when the force on a device and the velocity have the same sign, the power input to the device is positive. With a damper it is *impossible* for the applied force and resulting velocity to have opposite signs; thus the damper can never supply power to another device. A spring, however, absorbs and stores energy as a force is applied to it (say as in Fig. 2–2), but if the force is gradually relaxed back to zero, the external force and the velocity now have *opposite* signs, showing that the spring is *delivering* power.

A step-input force f_0 instantly (since a pure damper has no inertia) causes a velocity f_0/B which is maintained as long as f_0 is maintained (see Fig. 2–12e). This constant velocity produces a displacement x which increases linearly with time. This linear increase with time is called a *ramp function*. Thus, a step of f causes a step in dx/dt and a ramp in x. To find the frequency response we let $f = f_0 \sin \omega t$. Then

$$x - x_0 = \frac{1}{B}\int_0^t f \, dt = \frac{1}{B}\int_0^t f_0 \sin \omega t \, dt = \frac{f_0}{B\omega}(1 - \cos \omega t) \quad \text{(2--32)}$$

These relations are graphed in Fig. 2–13 for an arbitrarily chosen value of x_0. We see that the "sine" wave representing the oscillation of x has a 90° phase lag with respect to the f sine wave, that is, x "starts" at point s, whereas f started at $t = 0$. When a phase angle represents lagging behavior, it is conventionally given a negative sign, thus $\phi_{x/f} = -90°$. Note that this is true for any frequency ω. The amplitude ratio is clearly

$$\frac{A_x}{A_f} = \frac{f_0/B\omega}{f_0} = \frac{1}{B\omega} \quad \text{(2--33)}$$

giving the frequency response curves of Fig. 2–12f. Note that a sinusoidal

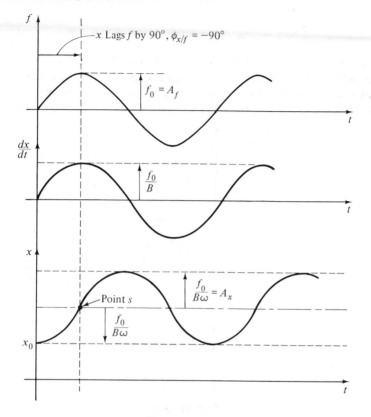

FIGURE 2–13

Damper Frequency Response

force of, say, 5 pound amplitude can produce a very large displacement (approaching infinity as $\omega \to 0$) if applied at a low frequency, but a smaller and smaller displacement (approaching zero as $\omega \to \infty$) as the frequency is raised. Compare this with the behavior of the spring element, which, for a given force, produces exactly the same displacement for every frequency.

The determination of frequency response curves for system elements is relatively quick and simple; however, it becomes much more tedious when complete systems are considered. Fortunately, a shortcut method (which will be derived in a later chapter) called the *sinusoidal transfer function* is available. The sinusoidal transfer function is obtained from the operational transfer function by directly substituting the term $i\omega$ for the D operator wherever it appears, where $i \triangleq \sqrt{-1}$, and ω is the frequency of the sinusoidal input. Thus from Eq. (2–25) we have

$$\frac{x}{f}(i\omega) \overset{\triangle}{=} \frac{1}{i\omega B} \overset{\triangle}{=} \text{sinusoidal transfer function} \qquad \textbf{(2–34)}$$

This is read "x over f of $i\omega$ equals by definition one over $i\omega B$." Since sinusoidal transfer functions are in general complex numbers, we now quickly review some basics from this area. A general complex number $a + ib$ can be plotted in the complex plane as in Fig. 2–14 and can also

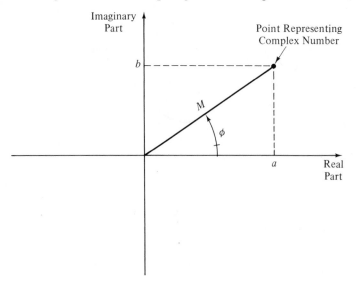

FIGURE 2–14

Complex Number Definitions

be given in the polar form $M\underline{/\phi}$ where $M \overset{\triangle}{=} \sqrt{a^2 + b^2}$ and $\phi \overset{\triangle}{=} \tan^{-1}(b/a)$. In interpreting sinusoidal transfer functions such as Eq. (2–34), we use the polar form $M\underline{/\phi}$. If this is done one can show that the magnitude M is the amplitude ratio A_x/A_f and ϕ is the phase angle $\phi_{x/f}$ by which output x leads input f. (If ϕ turns out *negative*, x lags f.) We thus have

$$\frac{x}{f}(i\omega) = \frac{1}{i\omega B} = \left(\frac{1}{B\omega}\right)\left(\frac{1}{i}\right) = \frac{1}{B\omega}(-i) = \frac{1}{B\omega}\underline{/-90°}$$

$$= M\underline{/\phi} = \frac{A_x}{A_f}\underline{/\phi_{x/f}} \qquad \textbf{(2–35)}$$

which we see agrees with our earlier result.

2–5. REAL DAMPERS.

Just as with springs, a damper element is sometimes used to represent an intentionally introduced device and sometimes for unavoidable "parasitic" effects. Let us consider intentional dampers first. The classic device is perhaps the viscous (piston/cylinder) damper whose configuration is the basis for the standard damper symbol of Fig. 2–12a. A damper of this type, used to damp vibrations of the Ranger spacecraft's solar panels,[8] is shown in Fig. 2–15a. A relative velocity between the cylinder and piston forces the viscous oil through the clearance space h, shearing the fluid and creating the damping force. An analysis in the reference gives

$$B = \frac{6\pi\mu L}{h^3}\left[\left(R_2 - \frac{h}{2}\right)^2 - R_1^2\right]\left[\frac{(R_2^2 - R_1^2)}{R_2 - \frac{h}{2}} - h\right], \frac{\text{lb}_f}{\text{inch/sec}} \quad \text{(2–36)}$$

where $\mu \triangleq$ fluid viscosity, $\text{lb}_f\text{-sec/inch}^2$. For this application the desired value of B was about 7.6 $\text{lb}_f/(\text{inch/sec})$. Twenty-eight such dampers were constructed and experimentally tested to determine B (see Fig. 2–15b). The vertical bands indicate the "scatter" of experimental measurements for an individual damper. Note also the possible variation in B due to unavoidable manufacturing tolerances. Nevertheless, the agreement between predicted and actual values is on the average very good. It should be pointed out that, even using silicone oil (which is the least-temperature-sensitive liquid suitable for dampers) a temperature change from 70°F to 140°F causes about a 50% decrease in viscosity and thus in B; therefore, experimental values must be temperature-corrected to make a fair comparison with theoretical calculations. A sinusoidal test method was used to measure B, and the maximum velocity reached was about 12 inches/sec. Figure 2–15c shows a fairly linear force/velocity relationship within these limits. If tests had been carried to higher forces and velocities, nonlinear behavior would have been revealed. This would be mainly due to the fluid flow processes changing from smooth (called *laminar* or *streamline* flow) at low velocities to *turbulent* at high velocities. For laminar flow, pressure drop and flow velocity have a linear relation, while for turbulent the pressure drop is proportional to about the 1.8 power of velocity.

Figure 2–16a shows a simple form of damper which is easily analyzed using the basic definition of fluid viscosity given in Fig. 2–16b. Here a flat plate of area A floats on a liquid film of thickness t, pulled by a steady force F which causes a constant plate velocity V. The definition of the viscosity is

$$\mu \triangleq \text{fluid viscosity} \triangleq \frac{\text{shearing stress}}{\text{velocity gradient}} = \frac{F/A}{V/t}, \frac{\text{lb}_f\text{-sec}}{\text{inch}^2} \quad \text{(2–37)}$$

The viscosity of fluids is actually measured with an instrument (viscosimeter) based on a rotational version of this scheme in which one measures

---

[8] M. Gayman, *Development of a Point Damper for the Ranger Solar Panels*, Jet Propulsion Lab Rept. 32-793, Cal. Inst. of Tech., 1965.

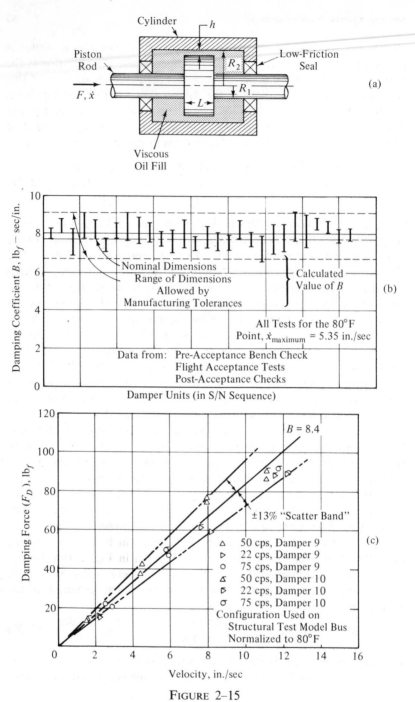

FIGURE 2–15

Damper for Ranger Spacecraft

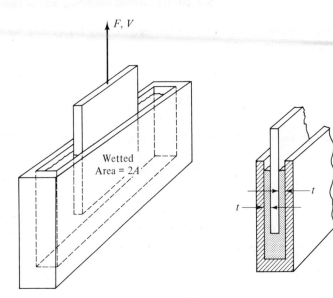

Wetted
Area = 2A

F, V

t

t

(a)

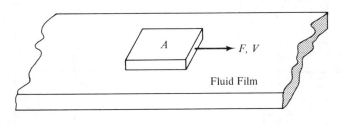

A

F, V

Fluid Film

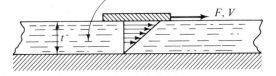

Fluid of Viscosity μ

F, V

t

(b)

FIGURE 2–16
Simple Shear Damper and Viscosity Definition

F, A, V, and t and then calculates the viscosity. Using Eq. (2–37), we get for the damper of Fig. 2–16a.

$$F = \frac{2A\mu}{t} V \qquad (2\text{–}38)$$

$$B = \frac{F}{V} = \frac{2A\mu}{t}, \frac{\text{lb}_f}{\text{inch/sec}} \qquad (2\text{–}39)$$

The rotational versions shown in Fig. 2–17a and b may be similarly analyzed to yield

$$B = \frac{\pi D^3 L \mu}{4t}, \frac{\text{inch-lb}_f}{\text{rad/sec}} \qquad (2\text{–}40)$$

and

$$B = \frac{\pi D_0^4 \mu}{16t}, \frac{\text{inch-lb}_f}{\text{rad/sec}} \qquad (2\text{–}41)$$

where the shear area loss due to D_i has been neglected.

Radial-Gap Damper   Axial-Gap Damper

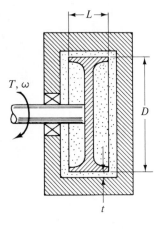
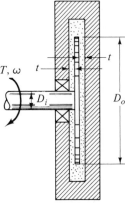

Damping Effects are
Assumed to be Confined
to the Gaps of Width t

(a)   (b)

FIGURE 2–17

Two Types of Rotary Dampers

Gases may also be used as the damping fluid. Since their viscosity is much lower, they do not give as large a value of B; however, they are less temperature-dependent. Also, if the gas used is atmospheric air, there is no leakage or sealing problem. Figure 2–18 shows a small damper of this type available commercially.[9] The graphite piston and glass cylinder are

---

[9] "Airpot," (Norwalk, Conn.: Electric Regulator Corp.).

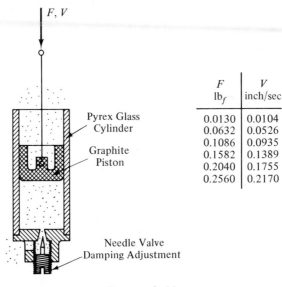

F lb$_f$	V inch/sec
0.0130	0.0104
0.0632	0.0526
0.1086	0.0935
0.1582	0.1389
0.2040	0.1755
0.2560	0.2170

FIGURE 2–18

A Commercial Air Damper

fitted to a tolerance of 0.0001 inch and give practically no rubbing friction since the air forms a thin film between them. The fluid damping action occurs in this air film and also in an adjustable needle valve, which forms a flow restriction between the cylinder and atmosphere. Flow in the air film is laminar (giving a linear damping relation), while that in the needle valve is more nearly turbulent (giving nonlinear damping) unless the valve is almost shut. If the valve is shut tight we get the strongest damping and, since it is now all due to the air film, it is quite linear. The table of Fig. 2–18 is actual data taken with the needle valve shut by applying dead weights to the piston and measuring the resulting velocity. Plot this data to check linearity and also find B for this device.

Certain electrical effects also provide a damping action which closely approximates that of the pure and ideal damper element. The damping forces available in this way are relatively small but are sometimes sufficient for low-power devices such as measuring instruments. Figure 2–19 shows such an *eddy-current* damper.[10] Motion of the conducting cup in the magnetic field generates a voltage

$$V = 10^{-8}B_m\pi DV, \text{ volts} \qquad (2\text{–}42)$$

in the cup where

---

[10] H. K. P. Neubert, "Instrument Transducers" (London: Oxford University Press, Inc.), 1963, p. 47.

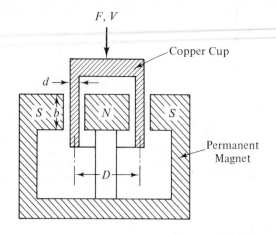

Cross-Section of
Circular Configuration

FIGURE 2–19

Eddy-Current Damper

$B_m \triangleq$ magnetic induction, gauss

$D \triangleq$ cup mean diameter, cm

$V \triangleq$ relative velocity, cm/sec

The resistance of the circular conducting path within the field in the cup is

$$R = \frac{\pi D \rho}{bd}, \text{ ohms} \tag{2–43}$$

where

$\rho \triangleq$ cup resistivity, ohm-cm

$b, d \triangleq$ width and thickness of conducting path, cm

The current in this path is thus $B_m(bdV/10^8\rho)$ and since a current-carrying conductor in a magnetic field experiences a force proportional to the current, we get a force proportional to and opposing the velocity V. The damper coefficient B is found to be

$$B = \frac{B_m{}^2 \pi bd D}{10^9 \rho}, \frac{\text{dyne}}{\text{cm/sec}} \tag{2–44}$$

The dissipated energy shows up as an I^2R heating of the cup. A rotational version is essentially a DC generator with output terminals short-circuited. Eddy-current damping is relatively insensitive to temperature, as shown in Fig. 2–20.

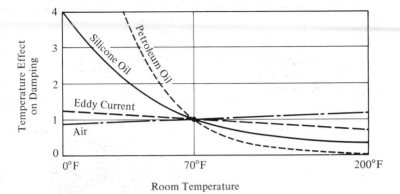

FIGURE 2–20

Temperature Sensitivity of Damping Methods

The use (Fig. 2–21a) of a porous plug as a flow restriction for an air damper has been studied[11] theoretically and experimentally. While the arrangement exhibits some nonlinearity and also a significant air-spring effect, it has been successfully applied in practice. In Fig. 2–21b a capillary tube[12] provides a laminar flow resistance between the ends of a piston/cylinder to give essentially linear damping. Squeeze-film damping,[13] Fig. 2–21c, is quite nonlinear but can provide large forces for small motions. It may employ either gases or liquids. A commercially available[14] rotary damper with adjustable damping is shown in Fig. 2–21d.

Let us now leave the realm of intentionally introduced damping devices and consider briefly the use of the damping element to represent unavoidble "parasitic" energy dissipation effects in mechanical systems. A list of such effects would include

   a. Frictional effects in moving parts of machines
   b. Fluid drag on vehicles such as ships, aircraft, and trains
   c. Windage losses of rotors in machines
   d. Hysteresis losses associated with cyclic stress in materials
   e. Structural damping due to riveted joints, etc.
   f. Air damping of vibrating structural shapes

Frictional effects in machines are usually a complex combination of dry rubbing (so-called Coulomb) friction plus linear and nonlinear fluid

[11] R. L. Peskin and E. Martinez, ASME Papers 65-WA/FE-8 and 65-WA/FE-9, 1965.

[12] H. H. Richardson, *Fluid Control, Components and Systems*, Agardograph 118 (Dec. 1968).

[13] E. A. Sommer, "Squeeze-Film Damping," *Machine Design* (May 26, 1966), p. 163.

[14] EFDYN Corp., 8700 S. Dobson, Chicago.

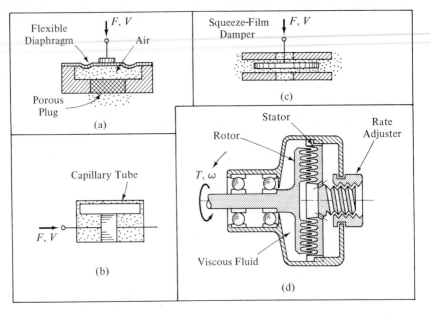

FIGURE 2–21

Some Other Examples of Damper Forms

friction. The friction torque of a hydraulic motor, for example, is due to rubbing, sliding, and rolling of various parts such as pistons, cylinders, ball bearings, plain bearings, seals, and valve plates. An experimentally measured friction characteristic for such a motor might appear as in Fig. 2–22a. Just as in springs, the nonlinear characteristic may be linearized for approximate analysis in the neighborhood of an operating point by taking the slope of the curve as a value for B. Dry or Coulomb friction is generally assumed independent of velocity except for the difference between static and running values of friction coefficient (see Fig. 2–22b). When the operation of a system involves *large* motions or speed changes rather than small variations around an operating point, linearizing schemes other than the local tangent line may be appropriate (see Fig. 2–23). The incentive for linearization is of course the desire to obtain linear differential equations so that rapid and revealing analytical methods may be applied in the early stages of design and analysis. Later, computers are profitably employed, and either analog or digital techniques allow us to include nonlinear damping characteristics "exactly," so that we may check whether our earlier linearizations obscured any essential features of behavior.

The drag force or torque on vehicles or other solid bodies moving in a fluid medium is essentially proportional to velocity for low velocities and becomes proportional to the square of velocity at high speeds. For bodies

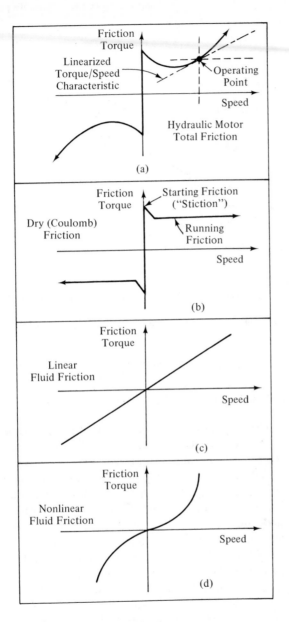

FIGURE 2–22

Hydraulic Motor Friction and Its Components

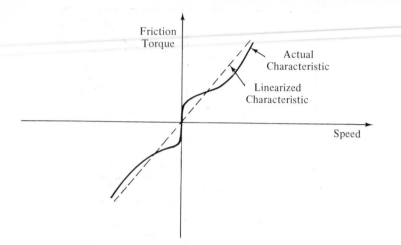

FIGURE 2–23

Linearization Method for Large Speed Range

of simple geometry, theoretical results are available[15,16] for the low-velocity (viscous) range:

$$\text{sphere of radius } r, \quad B = \frac{F}{V} = 6\pi r\mu \quad (2\text{–}45)$$

$$\text{"thin" cylinder of radius } r \text{ and length } \ell, \quad B = \frac{F}{V} = 2\pi \ell \mu \quad (2\text{–}46)$$

For bodies of complex shape, it is necessary to run experiments to determine the force/velocity relationship. Even for simple shapes, high velocities cause turbulent flow, and again experiments are necessary; however, most such results indicate the drag force to be proportional to the *square* of velocity[17] giving a nonlinear damping. The windage torques of rotating electrical machines[18] have a complex nonlinear characteristic.

When structures such as the frames of machine tools and aircraft vibrate at resonance, the stresses and deflections are limited only by the damping provided by the atmospheric air and the metal hysteresis losses. Spacecraft structures are often damped only by hysteresis and rivet-joint friction, since there is no air damping in the vacuum of space. The treat-

---

[15] Horace Lamb, *Hydrodynamics*, 6th ed. (New York: Dover Publications, Inc., 1945).

[16] G. G. Stokes, *On the Effect of Internal Friction of Fluids on the Motion of Pendulums*, Trans. Cambridge Phil. Soc., Vol. IX, pt. II, 1851, pp. 8–106.

[17] D. G. Stephens and M. A. Scavullo, *Investigation of Air Damping of Circular and Rectangular Plates, a Cylinder and a Sphere*, NASA TND-1865 (April 1965).

[18] J. E. Vrancik, *Prediction of Windage Power Losses in Alternators*, NASA TND-4849 (Oct. 1968).

ment of damping due to hysteresis requires a different approach, because one cannot identify an obvious "damping force." Rather, the energy dissipation is occurring at a microscopic level in the metal and is distributed over its volume. The magnitude of the energy loss appears to depend on the local stress raised to some power which, unfortunately, varies over a wide range and must be determined experimentally for each material. Since the stress level also varies over wide ranges and in complicated fashion over the volume of a structure, calculation of total damping is very difficult. Once the structure (or a suitable scale model) has been constructed, however, vibration test measurements allow determination of damping factors associated with each mode of vibration. When the vibration mode of interest is excited, and then allowed to die out freely, the rate of decay of the vibration permits calculation of an equivalent linear damping factor.

2-6. THE INERTIA ELEMENT.

A designer rarely inserts a component into a system for the purpose of adding inertia; thus the mass or inertia element often represents an undesirable effect which, unfortunately, is unavoidable, since all materials (solid, liquid, or gas) possess the property of mass. There are, of course, *some* applications in which mass itself serves a useful function. Figure 2–24a shows an accelerometer, an instrument for measuring acceleration. Every accelerometer must contain a mass, called the proof mass, since the principle of acceleration measurement lies in the measurement of the force required to give the proof mass the measured acceleration. In Fig. 2–24a, the spring element measures the force by deflecting, while the damper element suppresses spurious vibrations. Rotary inertia in the form of *flywheels* is sometimes employed as an energy storage device[19] or as a means of smoothing out speed fluctuations in engines or other machines (see Fig. 2–24b).

The physical law defining the force/motion behavior of inertia elements is Newton's law

$$\sum \text{Forces} = (\text{mass})(\text{acceleration}) \qquad (2\text{–}47)$$

which refers basically to an idealized "point mass" which occupies infinitesimal space. To apply this law directly to practical situations, the concept of a rigid body is introduced. For a purely translatory motion (no rotation), every point in a rigid body has identical motion and thus

---

[19] D. W. Rabenhorst, *Primary Energy Storage and the Super Flywheel*, Johns Hopkins Applied Physics Lab., Rept. TG-1081 (Sept. 1969).

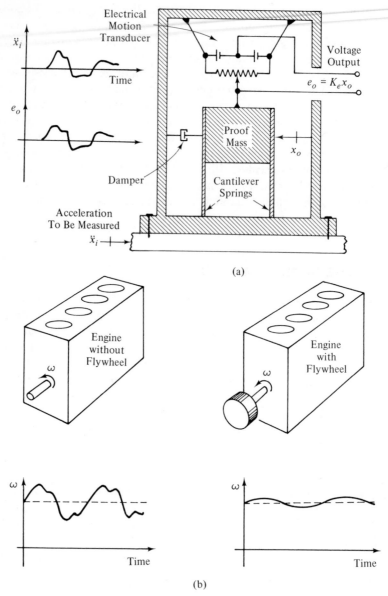

(a)

(b)

FIGURE 2–24

Useful Applications of Inertia

Eq. (2–47) applies to such a body (see Fig. 2–25). Real physical bodies can, of course, never display this ideal rigid behavior when being accelerated, since they experience internal elastic deflections which allow relative motion between points in the body. However, in many practical cases,

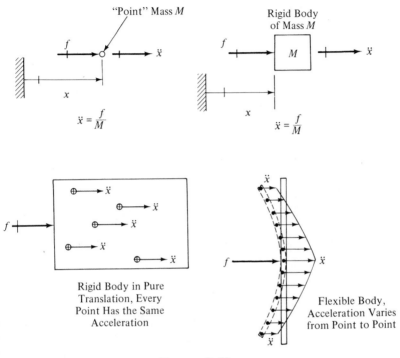

FIGURE 2–25

Definitions and Behavior of Rigid and Flexible Bodies

these internal deflections are very small *relative* to the gross motion of the entire body, and the assumption of an ideal rigid body gives good results.

For bodies undergoing pure rotational motion about a fixed axis of rotation, we have Newton's law in the rotational form

$$\sum \text{torques} = (\text{moment of inertia})(\text{angular acceleration}) \quad \textbf{(2–48)}$$

The concept of moment of inertia also considers the rotating body to be perfectly rigid. The "particles" of the body now do *not* all have the same acceleration but they *do* have accelerations which are intimately related, and in a known way, so that their combined inertia effect (moment of inertia) can be computed using integral calculus. For a homogeneous right circular cylinder (Fig. 2–26), for example, we may apply the basic (trans-

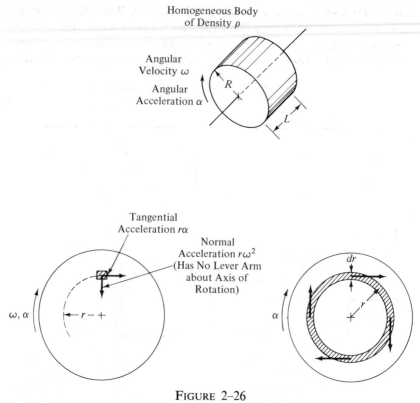

FIGURE 2–26

Rotational Inertia

lational) form of Newton's law to the ring-shaped mass element of infinitesimal width dr at radius r. Every particle in this element has exactly the same tangential acceleration $r\alpha$; thus the tangential force to produce this acceleration must be

$$\text{tangential force} = (\text{mass})(\text{acceleration}) = (2\pi r L dr \rho)(r\alpha) \quad \textbf{(2–49)}$$

The torque associated with this force is simply r times the force and the total torque is obtained by integration

$$\text{Total torque} = \int_0^R 2\pi\rho L\alpha r^3 \, dr = \pi R^2 L\rho \, \frac{R^2}{2} \alpha = \frac{MR^2}{2} \alpha \quad \textbf{(2–50)}$$

Since moment of inertia J is defined by Torque $= J\alpha$, we get

$$\text{Torque} = \frac{MR^2}{2} \alpha = J\alpha \quad \textbf{(2–51)}$$

$$J = \frac{MR^2}{2}, \text{lb}_f\text{-inch-sec}^2 \qquad (2\text{-}52)$$

For geometrically simple and homogeneous bodies, J can be calculated with relative ease, and Fig. 2–27 gives some useful results. For real machine parts, the shapes are usually complex and several materials of different density may be involved, making computation of J difficult and subject to

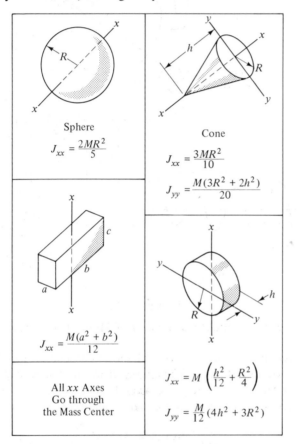

Sphere

$$J_{xx} = \frac{2MR^2}{5}$$

Cone

$$J_{xx} = \frac{3MR^2}{10}$$

$$J_{yy} = \frac{M(3R^2 + 2h^2)}{20}$$

$$J_{xx} = \frac{M(a^2 + b^2)}{12}$$

All xx Axes
Go through
the Mass Center

$$J_{xx} = M\left(\frac{h^2}{12} + \frac{R^2}{4}\right)$$

$$J_{yy} = \frac{M}{12}(4h^2 + 3R^2)$$

FIGURE 2–27

Moments of Inertia for Some Common Shapes

error. However, at the design stage, where the actual part exists only on paper, an estimate of this sort is necessary. Once the part has been constructed, experimental methods of finding J can be applied. One such methods mounts J in "frictionless" bearings so as to define the desired

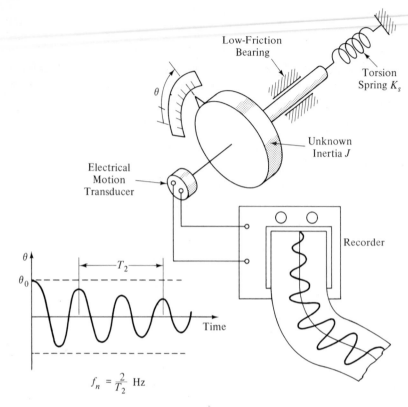

$$f_n = \frac{2}{T_2} \text{ Hz}$$

FIGURE 2–28

Experimental Measurement of Moment of Inertia

axis of rotation. A torsional spring of spring constant K_s inch-lb$_f$/radian is attached to J as in Fig. 2–28, and an electrical motion-measuring device is applied to measure angle θ and record its variation with time. If the inertia is manually deflected by an angle θ_0 and released, we may write

$$\sum \text{Torques} = J\alpha = J\frac{d^2\theta}{dt^2} \qquad (2\text{–}53)$$

$$-K_s\theta = J\frac{d^2\theta}{dt^2} \qquad (2\text{–}54)$$

$$\frac{J}{K_s}\frac{d^2\theta}{dt^2} + \theta = 0 \qquad (2\text{–}55)$$

Solution of this differential equation with initial conditions $\theta = \theta_0$, $\dot{\theta} = 0$ at $t = 0$ gives

$$\theta = \theta_0 \cos \omega_n t, \qquad \text{where } \omega_n^2 \triangleq \frac{K_s}{J} \triangleq \text{ undamped natural frequency} \qquad (2\text{–}56)$$

which indicates a sustained oscillation of frequency $f_n = \omega_n/2\pi = \sqrt{K_s}/(2\pi\sqrt{J})$ Hz. Actually, the oscillation will slowly die out due to unavoidable damping (friction), but generally sufficient oscillations occur to allow measurement of f_n from the recorder chart. Since K_s can be accurately found from a static calibration of the spring with dead weights, the value of J can now be calculated from

$$J = \frac{K_s}{4\pi^2 f_n^2} \qquad (2\text{-}57)$$

Whereas the linear characteristic curves relating force and displacement for spring elements, and force and velocity for damper elements, are closely approximated only by careful choice of materials, clever design and limited range of operation; the linear force/acceleration characteristic of the inertia element is for all practical purposes perfectly realized in those (many) cases where the body is "sufficiently" rigid. That is, Newton's law, while being strictly an empirical relation based on experimental measurements, has been found to hold very closely except for relativistic situations in which the velocity of the mass becomes comparable with the speed of light. Thus real inertias may be impure (have some springiness and damping) but are very close to ideal (linear).

The operational transfer functions for translational inertia elements with force or displacement inputs, and rotational inertia elements with torque or angular displacement inputs, are

$$\left. \begin{array}{ll} \dfrac{x}{f}(D) = \dfrac{1}{MD^2} & \dfrac{\theta}{T}(D) = \dfrac{1}{JD^2} \\[2ex] \dfrac{f}{x}(D) = MD^2 & \dfrac{T}{\theta}(D) = JD^2 \end{array} \right\} \qquad (2\text{-}58)$$

with block diagrams as in Fig. 2–29b and analog computer diagram as in Fig. 2–29c. Using the CSMP digital simulation language, the inertia element could be described by two statements:

$$\text{XD}\overline{\text{O}}\text{T} = \text{XD}\overline{\text{O}}\text{TO} + (1.0/M)* \text{ INTGRL (O.O, F)}$$

and

$$X = \text{INTGRL (XO, XD}\overline{\text{O}}\text{T)}$$

where $\text{XD}\overline{\text{O}}\text{T}$ is the velocity and $\text{XD}\overline{\text{O}}\text{TO}$ is its initial value.

Whereas the spring element stores energy as potential energy of deformation, the inertia element stores it as kinetic energy of motion. A mass M with velocity $\dot{x}$ has kinetic energy $M\dot{x}^2/2$, while a rotary inertia J with angular velocity $\dot{\theta}$ has kinetic energy $J\dot{\theta}^2/2$. Consider a steady force f_0 applied to a mass M initially at rest with displacement $x = 0$. We have

$$f_0 = M\frac{d^2x}{dt^2} = M\frac{dv}{dt} \qquad (2\text{-}59)$$

$$\int_0^t f_0 \, dt = \int_0^v M \, dv \tag{2-60}$$

$$f_0 t = Mv = M\frac{dx}{dt} \tag{2-61}$$

$$\int_0^t f_0 t \, dt = \int_0^x M \, dx \tag{2-62}$$

$$x = \frac{f_0 t^2}{2M} \tag{2-63}$$

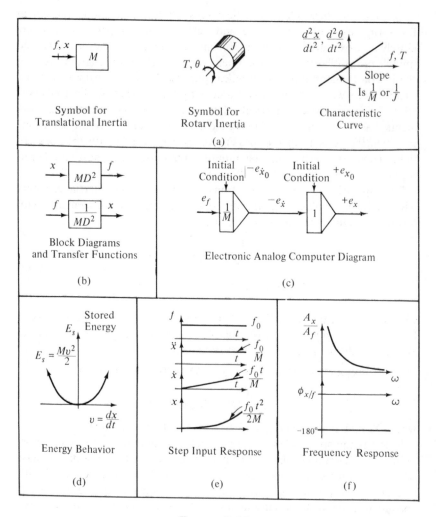

FIGURE 2–29

The Inertia Element

The work done by the constant force f_0 on the mass M in moving it the distance x is $f_0 x = f_0^2 t^2 / 2M = M\dot{x}^2/2$, thus the mass now has this energy stored and can give it up to another body when it is slowed down by this other body.

Equations (2–59) to (2–63) also give the step input response as shown in Fig. 2–29e. The frequency response is obtained from the sinusoidal transfer function

$$\frac{x}{f}(i\omega) = \frac{1}{M(i\omega)^2} = -\frac{1}{M\omega^2} = \frac{1}{M\omega^2} \underline{/-180°} \qquad (2\text{–}64)$$

and is graphed in Fig. 2–29f.

Since the validity of the ideal and pure inertia element as a model for real inertias depends mainly on the rigidity of the real body, it is instructive to investigate this situation for a body of simple shape. The prismatical rod of Fig. 2–30 has its left end driven by a motion input x_i. If the body were perfectly rigid, every particle would have this same motion. While an "exact" analysis requires a distributed-parameter (partial differential equation) treatment, a simple lumped-parameter model reveals the essence of the behavior. The rod is modeled as a single mass M equal to the actual rod mass ρAL and located at the center of mass of the rod. This mass is connected to the left and right ends with massless spring elements having spring constants equal to those of actual pieces of rod of length $L/2$. Newton's law gives us

$$(x_i - x_o)\frac{2AE}{L} = \rho AL\, \frac{d^2 x_o}{dt^2} \qquad (2\text{–}65)$$

$$\frac{\rho L^2}{2E}\frac{d^2 x_o}{dt^2} + x_o = x_i \qquad (2\text{–}66)$$

Using our operator notation and $\omega_n \triangleq \sqrt{2E}/(L\sqrt{\rho})$ we get

$$\left[\frac{D^2}{\omega_n^2} + 1\right] x_o = x_i \qquad (2\text{–}67)$$

the operational transfer function

$$\frac{x_o}{x_i}(D) = \frac{1}{\dfrac{D^2}{\omega_n^2} + 1} \qquad (2\text{–}68)$$

and the sinusoidal transfer function

$$\frac{x_o}{x_i}(i\omega) = \frac{1}{\left(\dfrac{i\omega}{\omega_n}\right)^2 + 1} = \frac{1}{1 - \left(\dfrac{\omega}{\omega_n}\right)^2} \qquad (2\text{–}69)$$

Now for a perfectly rigid body $(x_o/x_i)(i\omega)$ would be identically equal to 1.0 for *every* frequency ω if $x_i = x_{i0} \sin \omega t$. From Eq. (2–69) we see that the real body approaches this as $(\omega/\omega_n) \to 0$, that is, if the forcing frequency ω

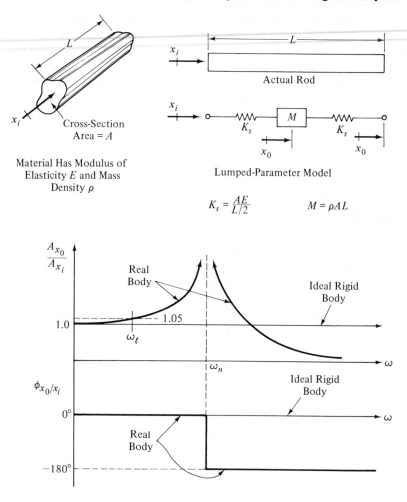

$$K_s = \frac{AE}{L/2} \qquad M = \rho A L$$

FIGURE 2–30

Useful Frequency Range for Rigid Model of Real (Flexible) Body

is small *compared to* ω_n. If we arbitrarily decide that a 5% deviation from perfection is tolerable, we may write

$$\frac{x_o}{x_i}(i\omega) = 1.05 = \frac{1}{1 - \left(\dfrac{\omega_\ell}{\omega_n}\right)^2} \qquad (2\text{–}70)$$

$$\omega_\ell = 0.218\omega_n = \frac{0.308}{L}\sqrt{\frac{E}{\rho}} \qquad (2\text{–}71)$$

where $\omega_\ell \triangleq$ upper limiting frequency for rigid behavior. Notice that as the rod length L approaches zero, $\omega_\ell \to \infty$, since this corresponds to the real mass approaching a "point" mass which would behave rigidly for all frequencies. Also, the limiting frequency is higher for "stiffer" materials (larger modulus of elasticity E) and lighter materials (smaller mass density ρ). To get some feel for actual numbers, consider a steel rod of length 6 inches.

$$\omega_\ell = \frac{0.308}{6} \sqrt{\frac{3 \times 10^7}{0.3/386}} = 10060 \frac{\text{rad}}{\text{sec}} \qquad (2\text{--}72)$$

$$f_\ell = \frac{\omega_\ell}{2\pi} = 1605 \text{ Hz} = 96200 \text{ cycles/minute} \qquad (2\text{--}73)$$

We see that this body will act essentially like a rigid body for oscillatory motions up to a frequency of about 96200 cycles/minute. Since no reciprocating and very few rotating machines run at such high speeds, this body could be modeled as a pure inertia element in many practical problems.

2-7. REFERRAL OF ELEMENTS ACROSS MOTION TRANSFORMERS.

Mechanical systems often include mechanisms such as levers, gears, linkages, cams, chains, and belts (see Fig. 2–31). While the named devices differ considerably in form, they all serve a common basic function, the transformation of the motion of an input member into the kinematically related motion of an output member. While the analysis of systems containing such motion transformers does not require any new elements or methods, it may be simplified in many cases by reducing the actual system to a fictitious but dynamically equivalent one. This is accomplished by a process of "referring" all the elements (masses, springs, dampers) and driving inputs to a *single* location, either the input or output, and writing a single equation for this equivalent system, rather than two (or more) equations for the real system.

We will illustrate the procedure by carrying it through for the lever system of Fig. 2–32a. Note that in this example there are *three* related motions, x_1, x_2, and θ, and that if one specifies any *one* of these the other two are immediately known, since they are *kinematically*, rather than dynamically, related. Suppose we decide to write our equations in terms of motion x_1, perhaps because x_1 might be a point where this system couples to another one. All elements and inputs will thus be "referred" to the x_1 location and we will define a fictitious equivalent system whose motion

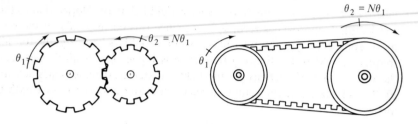

Gears

Belts, Chains

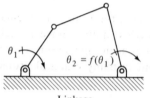

$\theta_2 = f(\theta_1)$

Linkage

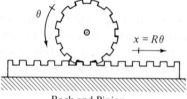

$x = R\theta$

Rack and Pinion

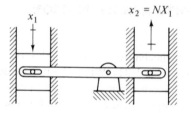

$x_2 = NX_1$

Lever

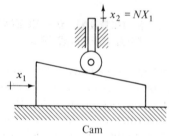

$x_2 = NX_1$

Cam

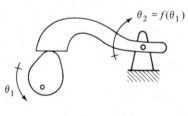

$\theta_2 = f(\theta_1)$

Cam

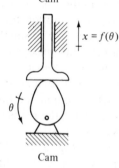

$x = f(\theta)$

Cam

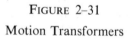

FIGURE 2–31

Motion Transformers

80

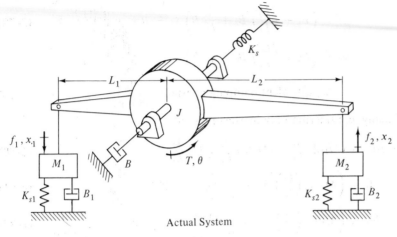

Actual System

(a)

Small Motions
Assumed
$\left(\theta \approx \dfrac{x_1}{L_1} = \dfrac{x_2}{L_2}\right)$

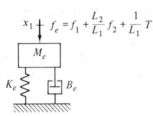

$x_1 \!\uparrow\! f_e = f_1 + \dfrac{L_2}{L_1} f_2 + \dfrac{1}{L_1} T$

Equivalent System Referred to x_1

(b)

FIGURE 2–32

Translational Equivalent for Complex System

will be the same as x_1. To find the equivalent spring element, apply a static load f_1 at x_1. We have then

$$f_1 L_1 = x_1 K_{s1} L_1 + \frac{L_2^2}{L_1} x_1 K_{s2} + \frac{x_1}{L_1} K_s \qquad (2\text{-}74)$$

$$f_1 = \left(K_{s1} + \left(\frac{L_2}{L_1}\right)^2 K_{s2} + \frac{1}{L_1^2} K_s \right) x_1 = K_{se} x_1 \qquad (2\text{-}75)$$

where $K_{se} \triangleq$ equivalent spring constant, referred to x_1.

$$K_{se} \triangleq \left(K_{s1} + \left(\frac{L_2}{L_1}\right)^2 K_{s2} + \frac{1}{L_1^2} K_s \right), \frac{\text{lb}_f}{\text{inch}} \qquad (2\text{-}76)$$

To find the equivalent damper, "mentally" remove the inertias and springs and again apply a force f_1 at x_1 to get

$$f_1 L_1 = \dot{x}_1 B_1 L_1 + \dot{x}_2 B_2 L_2 + B\dot{\theta} = \dot{x}_1 B_1 L_1 + \frac{L_2^2}{L_1} \dot{x}_1 B_2 + \frac{\dot{x}_1}{L_1} B \qquad (2\text{-}77)$$

$$f_1 = \left(B_1 + \left(\frac{L_2}{L_1}\right)^2 B_2 + \frac{1}{L_1^2} B \right) \dot{x}_1 = B_e \dot{x}_1 \qquad (2\text{--}78)$$

$$B_e \triangleq \left(B_1 + \left(\frac{L_2}{L_1}\right)^2 B_2 + \frac{1}{L_1^2} B \right)$$

$$\triangleq \text{ equivalent damping constant, } \frac{\text{lb}_f}{\text{inch/sec}} \qquad (2\text{--}79)$$

Finally, considering only inertias present

$$f_1 L_1 = (M_1 L_1^2)\frac{\ddot{x}_1}{L_1} + (M_2 L_2^2)\frac{\ddot{x}_1}{L_1} + (J)\frac{\ddot{x}_1}{L_1} \qquad (2\text{--}80)$$

$$f_1 = \left(M_1 + \left(\frac{L_2}{L_1}\right)^2 M_2 + \frac{1}{L_1^2} J \right) \ddot{x}_1 = M_e \ddot{x}_1 \qquad (2\text{--}81)$$

$$M_e \triangleq \left(M_1 + \left(\frac{L_2}{L_1}\right)^2 M_2 + \frac{1}{L_1^2} J \right) \triangleq \text{ equivalent inertia, } \frac{\text{lb}_f\text{-sec}^2}{\text{inch}} \qquad (2\text{--}82)$$

To refer the driving inputs to the x_1 location, we note that a torque T is equivalent to a force T/L_1 at the x_1 location, and a force f_2 is equivalent to a force $(L_2/L_1) f_2$ at x_1. We can now show the dynamically equivalent fictitious system as in Fig. 2–32b.

The "rules" (Eqs. (2–76), (2–79), and (2–82)) for calculating the equivalent elements without deriving them "from scratch" each time may be summarized as follows:

1. When referring a translational element (spring, damper, or mass) from location A to location B, where A's motion is N times B's, multiply the element's value by N^2. (This is also true for rotational elements coupled by motion transformers such as gears, belts, and chains, although we have not shown it here.)

2. When referring a rotational element to a translational location, multiply the rotational element by $1/R^2$ where the relation between translation x and rotation θ is $x = R\theta$. For the reverse procedure (referring a translational element to a rotational location) multiply the translational element by R^2.

3. When referring a force at A to get the equivalent force at B, multiply by N. (Holds also for torques.) Multiply a torque at θ by $1/R$ to refer it to x as a force. A force at x is multiplied by R to refer it as a torque to θ.

These rules actually apply to *any* mechanism, no matter what its form, so long as the motions at the two locations are *linearly* related. For mechanisms with nonlinear input/output relations (most cams and linkages) these results are good approximations for small motions in the neighborhood of an operating point at which the motion-transformation ratio is N.

2–8. MECHANICAL IMPEDANCE.

In the theoretical and experimental study of complex mechanical systems, particularly when trying to predict the behavior of an *assemblage* of subsystems from their calculated or measured individual behavior, the use of so-called impedance methods[20] may have advantages. We cannot here pursue this deeply, but merely introduce some definitions. *Mechanical impedance* is defined as a transfer function (can be either operational or sinusoidal) in which force is the numerator term and velocity the denominator. The simplest impedances are those of the spring, damper, and inertia elements themselves.

$$\text{mechanical impedance of spring} \triangleq Z_s(D) \triangleq \frac{f}{v}(D) = \frac{K_s}{D} \quad \text{(2–83)}$$

$$\text{mechanical impedance of damper} \triangleq Z_B(D) \triangleq \frac{f}{v}(D) = B \quad \text{(2–84)}$$

$$\text{mechanical impedance of inertia} \triangleq Z_M(D) \triangleq \frac{f}{v}(D) = MD \quad \text{(2–85)}$$

Figure 2–33 shows the frequency response curves for the sinusoidal

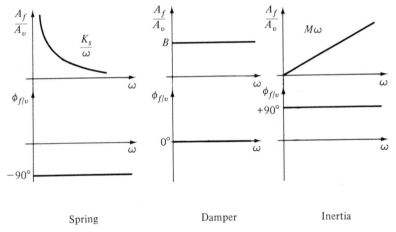

Spring   Damper   Inertia

FIGURE 2–33

Mechanical Impedance for Basic Elements

transfer function versions of these impedances. When mechanical impedance is determined experimentally, sinusoidal testing is almost always used. The machine or structure under test is driven by hydraulic or electrodynamic "shakers," and careful measurements of force and velocity are

---

[20] R. Plunkett, ed., *Mechanical Impedance Methods for Mechanical Vibrations*, ASME, 1958.

made over wide ranges of frequency. When much work of this sort is done, automatic systems for sweeping the frequency and plotting the amplitude ratio and phase angle of the impedance are desirable and commercially available.[21]

2–9. FORCE AND MOTION SOURCES.

We have shown the various elements being driven by input forces or motions, but have said little about how these inputs actually arise in practice. Let us first be clear that, to be precise, the ultimate driving agency of any mechanical system is always a *force*, not a motion. This follows from the cause-and-effect relation stated in Newton's laws, i.e., the force *causes* the acceleration, the acceleration does *not* cause the force. Thus, while there are problems in which the concept of a motion input is preferable to a force input, we should not delude ourselves that a motion occurs without a force occurring *first*. Perhaps the proper point of view is simply to ask, at the input of the system, what is it that is *known*, force or motion? If, for example, the motion is known, the fact that this motion is *caused* by some (perhaps unknown) force should not stop us from postulating a problem with a motion input. It may be appropriate to recall at this point that the forces available to drive mechanical systems can be put into two classes: forces associated with physical contact between two bodies and the "mysterious" action-at-a-distance forces, namely gravitational, magnetic, and electrostatic forces. When using Newton's law, Σ forces = (mass) (acceleration), the terms entered into the force summation must arise either from physical contact or else from magnetic, gravitational, or electrostatic origin; *there are no other kinds of forces*. If the reader has encountered D'Alembert's method of dynamic analysis, he may be thinking at this point about the so-called *inertia force*, but should recall that this is a *fictitious* force mentally added to the system strictly for analysis purposes, and is *not* a real force capable of causing a body initially at rest to move.

Some examples will help to establish a physical feeling for the distinction between force and motion sources. The design of multistory buildings may require consideration of stresses due to both wind and, in some regions of the world, earthquake effects.[22] Figure 2–34 shows an idealized model of a multistory building made up of mass and spring elements. The effect of wind is distributed over the surface of the building, and of course varies

---

[21] Spectral Dynamics Corp., San Diego, Cal.
[22] J. A. Blume, *Design of Multistory Reinforced Concrete Buildings for Earthquake Motions*, Portland Cement Assoc., 1961.

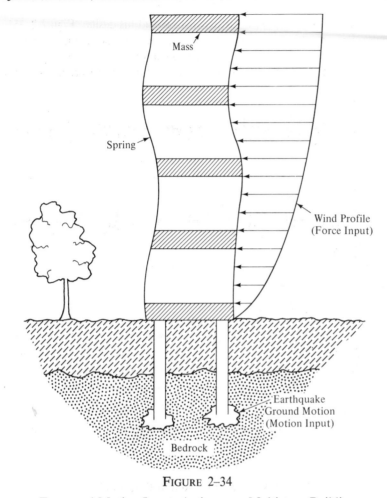

FIGURE 2–34

Force and Motion Inputs Acting on a Multistory Building

with time in a random fashion, but would generally be modeled as a force (or pressure) source or input. Earthquake-resistant design is a most complex field still based greatly on experience; however, attempts to put it on a rational basis generally consider the structure to be excited by the so-called *ground motion*, that is, a motion rather than a force input. The assumption is that the portion of the earth's crust to which the building is "fastened" is so massive relative to the building that the presence of the building has no effect on this "ground motion" and thus the base of the building is constrained to move with it. The ground motion caused at a particular location by a particular earthquake is of course impossible to predict, as is the occurrence of the earthquake itself. The engineer must thus rely on measurements of *past* disturbances to aid his design. Figure 2–35[23] shows seismographic recordings of the horizontal component in the north-south direction of an actual earthquake. Such a record can be used as a motion input to a mathematical model of the building to predict stresses

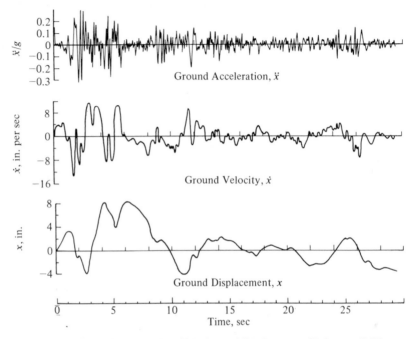

Ground Acceleration, Velocity, and Displacement, El Centro, Calif.,
Earthquake of May 18, 1940, N-S Component

FIGURE 2–35

Ground Motion of an Actual Earthquake

and deflections. Since these records are easily obtained on magnetic tape and thus can be reproduced as voltages, one can "play" this input into an electronic (analog computer) model of the building and read out stresses as voltages. To further reduce the need for simplifying assumptions, a *physical* scale model of the building could be constructed, and the base forced to move according to the earthquake ground motion by playing the magnetic tape into an *electro-hydraulic shaker.*[24] This is a machine which faithfully reproduces as a motion any electrical voltage-time variation

---

[23] *Ibid.*
[24] *The Minivib Dynamic Test System*, Brochure 268/20M/3-70 (New Haven, Conn.: MB Electronics).

which is applied to its input terminals. Such testing machines allow us to accurately apply predetermined motion inputs of many different forms to systems which we wish to study experimentally. Laboratory simulation testing of this sort has certain advantages over actual field testing, and is widely used in many industries[25] to qualify equipment for service in severe environments.

Many types of complicated machinery (packaging and printing equipment, computer peripherals such as card readers, etc.) are driven by a power source (often an electric motor) which runs at essentially constant speed. This steady rotation may be due to an inherent characteristic of the power source (a synchronous motor, for example), the flywheel action of a large inertia, or a feedback control system for speed regulation. In any case, the motions of the functional parts of the machine (motions which are generally complex and *not* of uniform velocity) are usually calculated assuming a known motion (the uniform rotation) at the input of the mechanism; thus we have another example of a motion input. Inertial navigation systems for submarines must take into account the motion caused by the vessel being carried along by the earth. That part of the vessel's total motion caused by the earth's motions is a motion input, since the earth's motions are very accurately known and, for all practical purposes, totally unaffected by the presence or absence of a ship. The operation of the gyroscopes and accelerometers used in such navigation systems is also analytically studied by assuming these instruments (which are fastened to the ship and thus are constrained to follow its motions) to be driven by the ship's rotary and translational motions as inputs. Again, the ship is so massive (relative to the instruments) that its motion is unaffected by their presence. Finally, the study of suspension systems for land vehicles generally assumes the vehicle to be driven over a terrain of a certain profile (perhaps random), thus causing the wheels to have a prescribed vertical motion. This motion is the input for a mass/spring damper mathematical model of the vehicle, and causes motions of the frame and body which can be analytically calculated. If an actual vehicle is available, the electro-hydraulic shakers mentioned earlier can be used (one under each wheel) to simulate driving over an actual terrain whose profile has been measured and tape recorded.

Force inputs that excite vibratory motions often arise from rotational and/or reciprocating unbalance. A rotating rigid body whose mass center and center of rotation do not coincide is said to be unbalanced and, if rotating at a constant speed ω rad/sec will produce a radial force of magnitude $MR\omega^2$ where M is the mass and R the distance between the

---

[25] E. R. Betz, *Studying Structure Dynamics with the Cadillac Road Simulator*, SAE Paper 660101, SAE Trans., Vol. 75, 1967.

mass center and center of rotation. The vertical component of this force would be $MR\omega^2 \sin \omega t$, an oscillatory force which would act on the bearings, and thereby be transmitted into the machine frame and possibly cause the machine to vibrate excessively. If the machine is suitably modeled with masses, springs, and dampers, the unbalance force serves as a force input and allows calculation of vibration of other machine parts (see Fig. 2–36). When force inputs are to be intentionally produced for testing

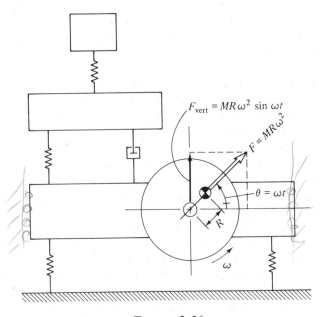

FIGURE 2–36

Rotating Unbalance as a Force Input

purposes in the laboratory, the *electrodynamic shaker* is often used (see Fig. 2–37). Here, current is passed through a coil suspended in a magnetic field. For a fixed field, a magnetic force directly and instantaneously proportional to the current is produced. A body attached to the coil will, of course, experience this same force. Such shakers are available from small units one may hold in his hand and producing a maximum force of 1 or 2 pounds to huge machines weighing 35,000 pounds and producing 25,000 pounds of driving force. Oscillating forces of several thousand cycles per second may be produced. The steering of air, space, and water vehicles is generally accomplished by the manipulation of force or torque inputs. A ship or airplane is maneuvered by deflecting control surfaces (rudders, ailerons, diving planes, etc.) into the relative wind or current. The pressure of the fluid against the control surface produces forces and moments

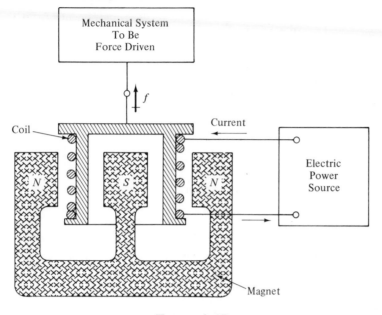

FIGURE 2-37

Electrodynamic Shaker as a Force Source

which act on the vehicle to change its attitude and direction of travel. For space vehicles, no fluid medium exists to provide the control surface pressures; thus the vehicle employs reaction jets, or else swivels the propulsion engine to position the thrust vector so as to cause a turning moment. In addition to such *control* inputs, an aircraft is also subject to *disturbing* force inputs in the form of wind gusts, the so-called atmospheric turbulence. A motion source may sometimes be converted into a satisfactory force source by interposing a "soft" spring between the motion source and the system force input point. Figure 2-38 shows an example of such an arrangement using a Scotch yoke mechanism (a sinusoidal motion source) to produce a sinusoidal force input of selected amplitude and frequency. If the motion x_{si} at the input point of the driven system is small compared to the Scotch yoke motion x, then the spring force applied to the system is determined almost entirely by x, whose amplitude and frequency can be set at desired values by adjusting R and ω. The requirement $x_{si} \ll R$ can be satisfied by choosing K_s sufficiently small (a "soft" spring) *relative* to the "stiffness" of the driven system at its input point. Since small K_s also results in a small driving force, this technique may not be usable when large forces are needed.

A little reflection on the above examples of motion and force inputs should suggest to the reader that, just as in the choice of a model to

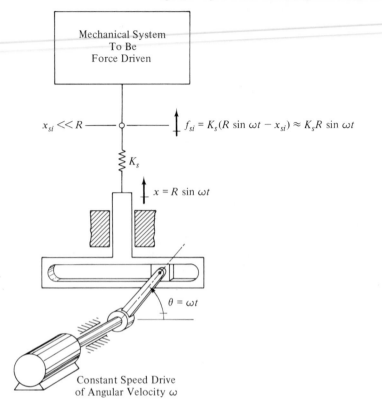

FIGURE 2–38

Force Source Constructed from Motion Source and Soft Spring

represent a component or system, the choice of the input form to be applied to the system also requires careful consideration and is subject to some interpretation. Let us conclude our discussion of sources with a brief look at energy considerations. We should first note that a system can be caused to respond only by the source supplying some energy to it. Thus, while we have chosen to speak of motion and force sources, whichever of these we employ, an interchange of energy must occur between source and system. That is, if we postulate a force source, there will also be an associated *motion* occurring at the force input point. We can compute the instantaneous power being transmitted through this *energy port*, as it is called, as the product of instantaneous force and velocity. If the force applied by the source and the velocity caused by it are in the same direction, power is supplied by the source to the system. If the force and velocity are opposed, the system is returning power to the source. The concept of mechanical impedance mentioned briefly earlier is of some help here. The transfer

function relating force and velocity at the input port of a system is called
the driving-point impedance Z_{dp}.

$$Z_{dp}(D) \triangleq \frac{f}{v}(D) \qquad (2\text{-}86)$$

$$Z_{dp}(i\omega) \triangleq \frac{f}{v}(i\omega) \qquad (2\text{-}87)$$

Now since power $P = fv$ and $v = f/Z_{dp}$

$$P = fv = (f)\frac{f}{Z_{dp}} = \frac{f^2}{Z_{dp}} \qquad (2\text{-}88)$$

Thus if we apply a force source to a system with a high value of driving-
point impedance, not much power will be taken from the source, since the
force produces only a small velocity. The extreme case of this would be the
application of a force to a perfectly rigid wall (driving-point impedance is
infinite since no motion is produced no matter how large the force). In
this case the source would not supply *any* energy. The higher the driving-
point impedance, the more a real force source behaves like an ideal force
source. The lower the driving-point impedance, the more a real motion
source behaves like an ideal motion source. More comprehensive studies
show that real sources may be described accurately as combinations of
ideal sources and an impedance (called the output impedance) characteristic of the physical device. A complete description of the situation thus
requires knowledge of two impedances; the output impedance of the real
source and the driving-point impedance of the system to be driven.
Fortunately, many practical problems do not require these advanced
considerations.

BIBLIOGRAPHY

1. Faires, V. M., *Design of Machine Elements*. New York: The Macmillan
   Company, 1955.

2. Roark, R. J., *Formulas for Stress and Strain*. New York: McGraw-Hill Book
   Company, 1954.

3. Spotts, M. F., *Design of Machine Elements*. Englewood Cliffs, N. J.: Prentice-
   Hall, Inc., 1961.

4. Wahl, A. M., *Mechanical Springs*. Cleveland: Penton Publishing Company,
   1944.

PROBLEMS

2–1. A nonlinear spring has $f = 100x + 20x^3$, lb$_f$, where x is in inches. Find its linearized spring constant for operating points $x = 0, 1, 2,$ and 5 inches.

2–2. Using the graph of Fig. 2–8, estimate the range of linearized spring constant to be expected over the full range of this air spring.

2–3. A tension rod spring as in Fig. 2–7 is made of steel with a working stress of 100,000 psi; the applied stress is f/A. At a load corresponding to the working stress, the energy storage is to be 10,000 inch-lb$_f$. Find all combinations of A and L which meet this requirement. What is the energy storage per cubic inch of material?

2–4. Find an expression for the spring constant of the buoyant spring of Fig. 2–10d in terms of dimensions and material properties. Discuss the effect of a float which is not cylindrical.

2–5. Get an expression for the gravity spring torque on the pendulum of Fig. 2–10b and then linearize it for oscillations around $\theta = 0$.

2–6. Derive expressions for the equivalent spring constant of
    a. Two springs "in series," Fig. P2–1.

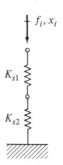

Fig. P2–1

    b. Two springs "in parallel," Fig. P2–2.

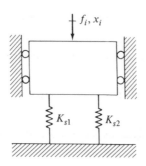

Fig. P2–2

2-7. For the damper of Fig. 2–15a, find the viscosity needed to give $B = 10$ lb$_f$/(inch/sec) if

$$h = 0.05 \text{ inch} \qquad R_1 = 0.2 \text{ inch}$$
$$L = 1 \text{ inch} \qquad R_2 = 1.0 \text{ inch}$$

If the operating temperature is estimated to be 100°F, choose a suitable fluid from Appendix B. If this unit is cycled with a displacement $x = 1 \sin 10t$ inches (t in seconds), what is the average rate of heat generation?

2-8. Derive Eq. (2–40).

2-9. Derive Eq. (2–41).

2-10. In a coil spring (Fig. 2–7a) the stress tending to cause failure is given by $16f_i R/\pi d^3$. Using a spring steel with working stress of 60,000 psi, $R = 0.5$ inch, find d needed to sustain a load of 40 lb$_f$. If K_s is to be 40 lb$_f$/inch, find the number of coils needed.

2-11. Repeat Prob. 2–6, except use dampers in place of springs.

2-12. Using the experimental data of Fig. 2–9, find the spring constant of the structure shown there.

2-13. In Fig. P2–3 a combination of springs from Fig. 2–7b, c and i is shown.

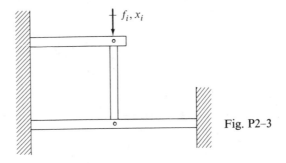

Fig. P2–3

Find the equivalent spring constant for this assemblage. What would this simplify to if the rod connecting the two beams may be considered rigid?

2-14. In Fig. 2–7, for all the springs which might be made of metal, what is the effect on K_s of changing from steel to aluminum?

2-15. In Fig. 2–7b, c, e, and i, assume all parameters fixed except L. Sketch a graph showing the variation of spring compliance $1/K_s$ versus L.

2-16. Extend the results of Prob. 2–6 to *any* number of springs in series or parallel.

2-17. A rotational damper with one end fixed has a torque given by $(3 - 3t + 62.8 \sin 62.8t)$ applied to the free end. Find expression for:

  a.   The angular velocity of the free end.
  b.   The angular displacement of the free end.

Sketch the shape of the angular displacement versus time curve for t going from 0 to 3.

2–18. Using the silicone fluids of Appendix B, design for operation at a temperature of 100°F a damper as in Fig. 2–17a. It is to have $B = 1.5$ inch-lb$_f$/(rad/sec), t no smaller than 0.050 in. (to ease manufacturing tolerances), and must fit in a cylindrical space 2 inches deep and 3 inches in diameter. When you have decided on appropriate dimensions and viscosity, estimate how much B changes if the temperature actually turns out to be 200°F.

2–19. A damper with $B = 5$ lb$_f$/(inch/sec) has one end moving with displacement ($3t + 4 \sin 10t$) inches (t in seconds), while the other end has displacement $1 \sin 10t$. Find the force in the damper.

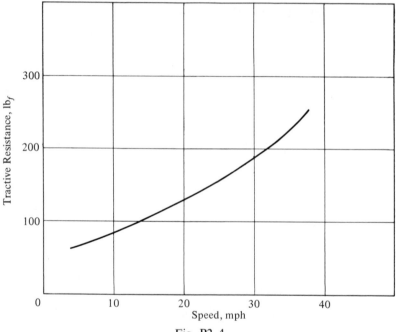

Fig. P2–4

2–20. Figure P2–4 shows the experimentally measured tractive resistance force of an 8000-pound truck. This represents all the power losses except air drag. The air drag force may be estimated as $0.04V^2$ lb$_f$, where V is the vehicle speed in ft/sec. Find a linearized damping coefficient B to represent the total energy dissipation in the neighborhood of speeds of 10 mph, 20 mph, and 30 mph (3 different values).

2–21. An empirical formula used to estimate the total resistance to motion of railway passenger cars gives (for a certain type of car) the force as ($130 + 1.5V + 0.034V^2$) lb$_f$, where V is car speed in mph. What *kind* of friction is represented by each term in this formula? Sketch a graph of each term versus V for speeds up to 100 mph. Get a linearized damping coefficient B for the total force if V is near 50 mph.

2–22. For a damper with $B = 2.0$ lb$_f$/(inch/sec), calculate and sketch the frequency response curves (Fig. 2–12f). If the amplitude of the sinusoidal driving force is 10 lb$_f$, find the frequency at which the displacement amplitude is 0.01 inch.

2–23. Check the correctness of solution Eq. (2–56) by substituting it into (2–55).

2–24. Obtain the result of Eq. (2–64) without the use of the sinusoidal transfer function.

2–25. In Fig. 2–30 the rod acts as an inertia element for sufficiently low frequencies. At low frequencies, how does this same rod behave if its right-hand end is held fast ("built-into a wall")?

2–26. In the system of Fig. 2–30, why does the rod cross-sectional area not influence the final results?

2–27. For the gears of Fig. 2–31 let $J_1 = J_2 = 0.01$ lb$_f$-inch-sec^2 and gear ratio $N = 8$. Find equivalent inertia of the whole system referred to the θ_1 shaft; find it referred to the θ_2 shaft. If shaft 1 has a damper with $B = 3$ inch-lb$_f$/(rad/sec) find the total damping referred to the number 2 shaft.

2–28. In Fig. 2–32 let $L_2 = 3$ in., $L_1 = 1$ in., $K_{s1} = K_{s2} = 20$ lb$_f$/in., $K_s = 10$ inch-lb$_f$/rad, $B = 0$, $B_1 = B_2 = 3$ lb$_f$/(inch/sec), $J = 0.25$ lb$_f$-sec^2/in., $W_1 = W_2 = 10$ lb$_f$, $f_1 = 3$ lb$_f$, $f_2 = -2t$ lb$_f$, $T = 5 \sin 10t$ inch-lb$_f$. Find the equivalent system, referred to x_1.

2–29. Repeat Prob. 2–28, but refer everything to x_2.

2–30. Repeat Prob. 2–28, but refer everything to θ.

2–31. For what frequency will the magnitude of the sinusoidal mechanical impedance of a spring with $K_s = 1000$ lb$_f$/in. be the same as that of a damper with $B = 10$ lb$_f$/(in./sec)? What must the weight of a mass M be, such that its impedance will be 10 at this same frequency?

2–32. In section 2–9, the use of laboratory simulation testing as opposed to field testing was discussed. Speculate on the possible relative advantages of simulation. Does it have any disadvantages? Explain.

2–33. Considering a road profile as a motion input to an automobile suspension system, how does the motion input change when the car is driven over the road at various constant speeds? If, at 20 mph, the motion input contains frequencies from 0 to 5 Hz, what will the frequency content be at 60 mph? What frequency is caused by driving over expansion joints spaced every 30 feet at 60 mph?

2–34. In the earthquake record of Fig. 2–35, why are the velocity and displacement traces "smoother" than the acceleration?

2–35. A mechanical engineer wishes to build a rotating vibration exciter for lab testing, using the principle of Fig. 2–36; however, he wants *only* a vertical oscillating force. Can you come up with a modification of Fig. 2–36 which gives no horizontal force?

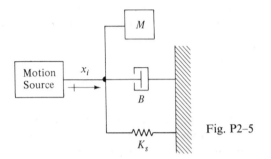

Fig. P2–5

2–36. In Fig. P2–5 we have $K_s = 10$ lb$_f$/inch, $B = 10$ lb$_f$/(inch/sec), $M = 10$ lb$_f$-sec^2/in., and $x_i = \sin \omega t$ inches. Find the total force that must be provided by the motion source. Plot as separate curves the individual contributions of the spring, damper, and mass to the total force, considering amplitude only, not phase angle. Use a logarithmic force scale going from 0.001 to 1000 lb$_f$ and a logarithmic frequency scale going from $\omega = 0.01$ to 10. Compare the relative importance of the three forces at $\omega = 0.01$, 1.0, and 10 rad/sec.

2–37. In the accelerometer of Fig. 2–24a the motion transducer is designed for full-scale motion of 0.10 inch. If the full-scale acceleration $\ddot{x}_i$ is 100 g's (3217 ft/sec^2) find an expression giving all the possible combinations of M (proof mass) and K_s (cantilever spring constant) which will give this. If space requirements allow 0.5 cubic inches of space for a steel mass (specific weight 0.3 lb$_f$/in^3), what K_s must be used?

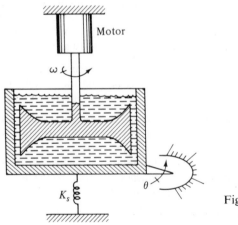

Fig. P2–6

2-38. Figure P2–6 shows a viscosimeter (viscosity measuring device) based on the damper configuration of Fig. 2–17a. Derive a formula showing how viscosity may be calculated from the known dimensions of the device, motor speed, and angular deflection θ.

2-39. In the cam of Fig. 2–31, over its specified range of motion, $\theta_2 = 0.5\theta_1 + 0.2\theta_1^2 + 0.05\theta_1^3$. If the θ_1 shaft has a torsion spring with $K_s = 10$ inch-lb_f/rad attached to it, refer this spring to the θ_2 shaft for small motions in the neighborhood of $\theta_1 = 0.2$ radians.

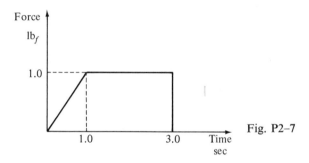

Fig. P2–7

2-40. The force of Fig. P2–7 is applied to a pure mass M initially at rest at $x = 0$. Calculate and sketch the time history of

  a. Acceleration
  b. Velocity
  c. Displacement
  d. Stored energy

2-41. Repeat Prob. 2–40, substituting a spring K_s for the mass.

2-42. Repeat Prob. 2–40, substituting a damper B for the mass, deleting the plot of stored energy, and adding plots for dissipated power and dissipated energy.

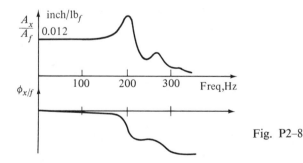

Fig. P2–8

2-43. A machine part of complex shape was frequency-response tested with results as in Fig. P2–8. Over what range of frequencies could it be modeled as a pure spring element? What is K_s?

2–44. Using CSMP or whatever digital simulation language is available to you, find and plot (computer printer-plot) the velocity and displacement of a mass weighing 10 lb$_f$ whose initial velocity is -1 ft/sec, initial displacement is $+3.5$ ft, and which is subjected to a force given by $(3 + 0.3t + 1000 \sin 3.5t)$, lb$_f$, where t is in seconds and runs from 0 to 10.

3

SYSTEM ELEMENTS, ELECTRICAL

3–1. INTRODUCTION.

It is difficult to think of any physical, man-made systems of medium or large scale which do not involve electricity in *some* way. If, however, we restrict our consideration to cases in which system functions are performed *largely* by electrical means the following examples might come to mind:

    a. Electrical power transmission and distribution.
    b. Communications—telegraph, telephone, radio, television, microwave.
    c. General-purpose computers: digital, analog, and hybrid.

Application areas which, in general, are electromechanical but which might in specific cases be largely electrical include:

    a. Measurement systems.
    b. Control systems.

The electrical components or elements which are used in systems can be described in terms of the relationship between currents and voltages existing in the devices, just as mechanical elements are described by their force/motion relations. Note that an electric motor is *not* an electrical device but rather an *electromechanical* one, since its description requires specification of not just voltage/current but also force/motion relations. Similarly, a phototransistor (transistor sensitive to light) is an *electro-optical* device, since its description involves voltage, current, and light flux. A loudspeaker would be an electro-mechanical-acoustic device, while a microphone would be acoustic-mechanical-electrical. Such "mixed-media" devices are extremely important in many systems, but since they are not strictly electrical we will not include them in this chapter.

Even with the restriction to strictly electrical devices, the scope is still too large for the purposes of this text and this chapter. Some attempt at classification will help us narrow down to the desired scope. We choose here to classify according to

a.   Network versus field concept.
b.   Passive versus active device.
c.   Linear (proportional) versus digital (on-off) device.

The network-versus-field classification is essentially that of lumped versus distributed parameters and is based on a wavelength/physical size criterion. *If the physical size of the device is small compared to the wavelength associated with signal propagation, the device may be considered lumped and a network model employed.* Wavelengths may be calculated from the known frequency range of a given system and the wave-propagation law

$$\text{wavelength } \lambda = \frac{\text{velocity of wave propagation } V}{\text{frequency } f} \qquad (3\text{--}1)$$

The velocity of propagation for electrical waves in free space is 186,000 miles per second. As an example, consider the electrical portion of a high-fidelity music reproduction system. Such systems deal with frequencies in the range 20 to 20,000 Hz, the so-called audio range. The shortest wavelength is associated with the highest frequency, in this case 20,000 cycles/sec, so we get (using the simplifying assumption of free space conditions)

$$\lambda = \frac{186,000 \text{ miles/sec}}{20,000 \text{ cycles/sec}} = 9.3 \text{ miles per cycle} \qquad (3\text{--}2)$$

Since a typical resistor or capacitor is less than 1 inch long, it is clear that audio electrical systems can be (and are) treated by the simpler lumped-parameter (network) approach rather than the more general field approach. It is interesting to note that the wavelength/size concept is applicable to any physical system which exhibits wave propagation, such as mechanical vibrating systems and acoustic systems. For acoustic systems in air, the velocity of propagation V is the speed of sound, 1100 ft/sec. Thus to check whether the acoustical portions (loudspeaker cones, microphone diaphragms, etc.) of a hi-fi system may be treated as lumped or distributed, we calculate the shortest wavelength as $\lambda = 1100/20000 = 0.055$ ft. $= 0.66$ inch. Since a speaker for high frequencies (a "tweeter") may be several inches in diameter and a microphone diaphragm 0.25 to 1.0 inch, we see that the acoustical system is "right on the ragged edge" for validity of a lumped model at the 20000-Hz frequency. At lower frequencies the lumped model would get better and better. In treating electrical elements we will take strictly the lumped or network approach and thus eliminate consideration of high-frequency phenomena associated with devices such as radar and microwave antennas, waveguides, etc. This restriction fortunately is not a severe one, since a majority of practical systems can be and are treated by the lumped approach.

The distinction between passive and active devices is based on energy considerations. From physics the reader will recall that a resistor, capacitor, or inductor is not a *source* of energy in the sense of a battery or generator. A charged capacitor or current-carrying inductor *does* store energy which can be supplied to another device, but the capacitor did *not* charge itself, nor did the inductor establish its current itself; rather some energy source was needed for this. A resistor does not even temporarily store energy; it dissipates into heat all the electrical energy supplied to it. Thus, the three basic circuit elements—resistance, capacitance, and inductance—are called *passive elements*, since they contain no energy sources.

The basic *active elements* in electric circuits are energy sources such as batteries (electrochemical source), generators (electromechanical source), solar cells (electro-optical source), and thermocouples (thermoelectric source). When these basic sources are suitably combined with the two basic power modulators, the vacuum tube and the transistor, we obtain active devices called *controlled sources* whose outstanding characteristic is the capability of *power amplification*. These controlled sources will accept as input a low-power voltage or current signal and accurately reproduce it, but at a much higher power level at the output of the device. The vacuum tube or transistor does not *itself* supply the power difference between output and input, it simply modulates, in a precise and controlled fashion, the power taken from the *basic* source (battery, etc.) and delivered to the output. These combinations of vacuum tubes or transistors with their power supplies are generally called active devices and because of their amplification capability are, in a sense, the fundamental base of all electronic systems. The basic principles of tube and solid-state amplification are generally introduced in physics courses and, for non-electrical engineers, extended to the point of practical application in a later electronics course. We will not here duplicate this material but rather emphasize the practical use of what is perhaps the single most useful active device, the *operational amplifier*. Through integrated circuit techniques, this device has been reduced in size and cost to the point where it is comparable in some cases to a single resistor or capacitor. Its function as a basic building block for many different types of useful circuits is enhanced by its ease of application. That is, careful and expert circuit design is needed to produce the operational amplifier (commonly called op-amp) itself, but once such a device is available, its further application to design of instrumentation amplifiers, analog computers, filters, etc., can be successfully accomplished by those with somewhat less electronic expertise. While an op-amp is not strictly an *element*, since it contains resistors, transistors, etc. (which are conventionally considered to be the elements), it is coming to be treated more and more like a component or element and we choose to include it in this chapter.

Since their development in the 1940s, electronic digital computers have had an increasing impact on many aspects of technology and society. Their future influence, while difficult (and sometimes frightening) to predict, will undoubtedly be great and widespread. Two complementary (sometimes competitive) concepts forecast to play large roles in this development are the *time-sharing* among many users of very large and fast machines (in the nature of a public utility like gas, electricity, telephone) and the individual application of the small and cheap (though potent) *minicomputer*. We mention these computers here since they are perhaps the most significant applications of *digital electronic devices*. We do not intend to give here a comprehensive description of such devices. Suffice it to say that they essentially perform on-off type switching functions needed to implement the logic operations required in digital computation. For example, a two-input AND gate produces an output signal only if *both* of the input signals are simultaneously "on." The "on" state for both input and output signals can fall in a wide voltage range, thus the devices are tolerant of noise voltages and need not be individually very "accurate" even though the overall computer is extremely accurate. While high "accuracy" is not needed, these devices must be very small, cheap, and fast, since so many of them are needed to make a computer. In contrasting these digital devices with linear (proportional) devices such as the op-amp, we note that in linear devices the specific waveform of input and output signals is of prime importance, while in digital devices it is simply the presence (logical 1) or absence (logical 0) of a voltage within some wide range which matters; the *precise* value of the signal is of no consequence. Just as for the op-amp, these digital devices are not really elements in the accepted sense because they contain resistors, transistors, and diodes. However, integrated circuit technology produces them in such small sizes and low costs that logic system designers treat them as basic building blocks. Since a properly functioning digital system operates in the realm of arithmetic rather than differential equations, its modeling, analysis, and design do not fit the pattern of linear system dynamics and we thus do not here treat these digital devices.

Having tried to give some picture of those types of basic electrical devices which we will *not* treat, it may be appropriate to summarize briefly what *will* be covered: the three basic passive elements (resistance, capacitance, and inductance), energy sources (current and voltage), and the active device the operational amplifier.

3–2. THE RESISTANCE ELEMENT.

The pure and ideal resistance element rigorously follows Ohm's law, which gives the current/voltage relation as

$$i = \frac{e}{R} \tag{3–3}$$

where

$i \triangleq$ current through the resistor, amperes

$e \triangleq$ voltage across the resistor, volts

$R \triangleq$ resistance of the resistor, ohms

The main features of this element are the strict linearity between e and i, the instantaneous response of i to e (or e to i), and the fact that all the electrical energy supplied is dissipated into heat. Real resistors are always somewhat "impure" (they exhibit some capacitance and inductance) and non-ideal (the i/e characteristic curve is not exactly a straight line). Capacitive and inductive effects make themselves known only when current and voltage are *changing* with time, thus the "impurity" of a resistor will not be revealed by a steady-state experiment which establishes the i/e curve by measurements with a voltmeter and an ammeter. Such an experiment will, of course, reveal departures from ideal (straight line) behavior. Many practical resistors are found to be very close to ideal (less than 1% nonlinearity), and this allows us to make resistance measurements with a rather simple instrument, the ohmmeter. In the ohmmeter a known and fixed current (from a current source) is passed through the unknown resistor, causing a voltage drop across the resistor. This voltage is measured with a voltmeter and since (assuming Ohm's law to hold) it is directly proportional to resistance ($e = iR$) the scale of the voltmeter can be marked off in *ohms*, rather than volts, thus giving us a direct reading of resistance. For ideal resistors we can thus think of *defining R* in terms of e and i as

$$\text{resistance } R \triangleq \frac{e}{i}, \text{ ohms} \qquad (3\text{-}4)$$

Sometimes the reciprocal of resistance (the conductance) is used

$$\text{conductance } G \triangleq \frac{i}{e}, \text{ mhos} \qquad (3\text{-}5)$$

Note in Fig. 3–1a that a resistor has associated with it only one current i, but two potentials e_1 and e_2 at its terminals. The current is, however, determined (in both magnitude and direction) by the potential *difference* (voltage) $e_1 - e_2$ which we call simply e. That is, in an ideal resistor, the same current is caused if we apply $e_1 = 10001$ volts, $e_2 = 10000$ volts, as by $e_1 = 1$ volt, $e_2 = 0$ volts. (A *real* resistor, if not carefully designed for high voltage, might be destroyed by the first situation.) It is necessary to establish *algebraic sign conventions* for current i and voltage e and Fig. 3–1a shows the accepted form. When e_1 is greater than e_2, e is a positive number. We choose the polarity shown for positive e. If e is negative, the *actual* polarity would be reversed. The fixed $+$ and $-$ signs shown in Fig. 3–1a are thus a *sign convention* for the variable e, *not* an indication of the actual instantaneous polarity, which in dynamics problems often changes with time. When we are solving system problems, the voltage e would be an *unknown*. If the solution came out $e = +6.2$ volts or $e = -3.7$ volts, if we had not *at the beginning* decided on a sign convention we would not know what actual polarity was meant by $+6.2$ or -3.7. Since from physics we know that a positive e in Fig. 3–1a causes a current i to flow from left to right, we take the positive direction for i ($+\!\!\longrightarrow$) to the right. Having *chosen* the e sign convention as above, we *must* now choose i positive as shown, or else Ohm's law would be violated. That is, $i = +e/R$, not $i = -e/R$, and R is defined as a positive number. Summarizing, we may choose (from the two possibilities available) either the positive direction for i or e first, but having made this choice, the other sign convention must conform to Ohm's law; that is, a positive voltage *must* cause a positive current.

The block diagrams and transfer functions of Fig. 3–1b should be self-evident as should the analog computer diagrams of Fig. 3–1c. To determine the energy behavior, we recall from physics that the instantaneous electric power supplied to a device is the product of the instantaneous current through the device and the instantaneous voltage across it. If the power has a negative sign it means the device is *supplying* power rather than using it up. For the resistance element we have

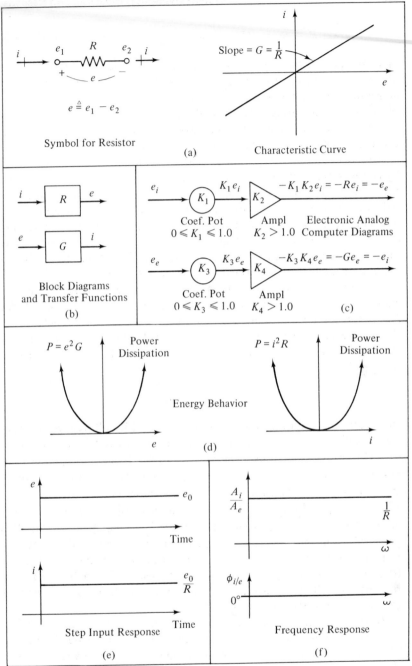

FIGURE 3–1

The Resistance Element

$$\text{power } P \triangleq ie = i(iR) = i^2R = e\frac{e}{R} = e^2/R = e^2G \text{ watts} \quad \text{(3-6)}$$

Since the power is always positive irrespective of the polarity of e or the direction of i, the resistor always takes power from the source. If a resistor at room temperature is suddenly connected to a voltage source, it instantly starts generating heat internally at a rate e^2G watts $= e^2G$ Newton-meters/sec $= e^2G$ Joules/sec $= 0.000948e^2G$ Btu/sec. This internal heat generation causes the temperature of the resistor to rise. As soon as resistor temperature is higher than that of its surroundings, heat transfer by conduction, convection, and radiation causes heat to flow away from the resistor. When the resistor gets hot enough, this heat transfer rate just balances the e^2G heat generation and the resistor achieves an equilibrium temperature above room temperature. In a real resistor this temperature cannot be allowed to get too high, or else the R value changes excessively or the resistor may actually burn out. Resistors used in electronic equipment are usually thermally rated at 1 or 2 watts or less.

The instantaneous dynamic response of the pure resistance element is shown in the step and frequency response graphs of Fig. 3–1e and f. Since $i = e/R$ is an algebraic equation, changes in e of any form whatever are *instantly* reflected in proportional changes in i. A random voltage e will, for example, produce a random i of exactly the same shape. The sinusoidal transfer function

$$\frac{i}{e}(i\omega) = \frac{1}{R}\underline{/0°} \quad \text{(3-7)}$$

shows that the amplitude ratio is constant at $1/R$ for all frequencies from zero to infinity and the phase shift between e and i is zero for all frequencies. (In electrical analysis some authors prefer to use the symbol j for $\sqrt{-1}$, since i might be confused with current. Since system dynamics treats all kinds of systems, not just electrical, we choose to retain i for $\sqrt{-1}$ since it is universally so used in mathematics.) Real resistors are always impure (contain some capacitance and inductance), and this prevents the instantaneous step response and the perfectly flat frequency response. Since practical systems always deal with a limited range of frequencies (*not* zero to infinity), if a real resistor has a flat frequency response over its necessary range, the fact that it deviates from flatness elsewhere is of little consequence.

Resistance elements can be pure while not being ideal (linear). Some examples of practically useful nonlinear resistors are vacuum tube and semiconductor diodes and the Varistor. Figure 3–2a shows a vacuum tube diode used for rectification of alternating current and for various computing functions in analog computers. The i/e characteristic curve shows this

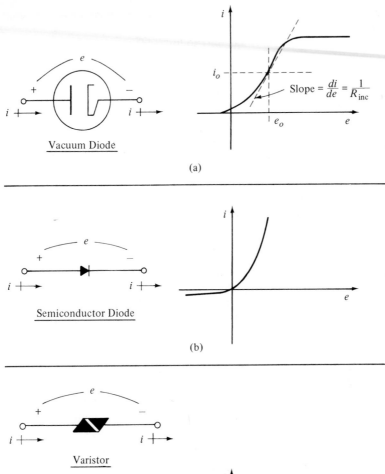

Vacuum Diode

Slope $= \dfrac{di}{de} = \dfrac{1}{R_{inc}}$

(a)

Semiconductor Diode

(b)

Varistor

e	i
Volt	Amp
0.0	0.0000
2.0	0.0001
4.0	0.0004
6.0	0.0012
8.0	0.0031
10.0	0.0068
12.0	0.0135
14.0	0.0246

(c)

FIGURE 3–2

Some Nonlinear Resistances

107

device allows positive currents only to flow, giving it the ability to convert AC voltages to pulsating, but unidirectional, DC. This is a resistive device since the current is an instantaneous function of the voltage; however, due to the nonlinearity of the i/e curve, there is some question as to just what its resistance might be, as a numerical value. The most general viewpoint might be that the purpose of defining resistance is to specify the e/i relationships. For a linear resistance this requires only a statement of a single number; for nonlinear resistances, a function, graph, or table of values is necessary; a single-number description is basically inapplicable. In some applications an approximate linearized treatment based on the local slope of the e/i curve is feasible. Such a resistance is called an incremental or small-signal resistance and is defined by

$$\text{incremental resistance } R \triangleq \frac{de}{di} \qquad (3\text{–}8)$$

where the slope de/di is measured at the operating point as in Fig. 3–2a. If the device is operated in the neighborhood of e_0, i_0, such a definition gives a single-number relation between e and i adequate for many purposes. The semiconductor diode of Fig. 3–2 has similar applications, but is smaller and requires no heater power. It, however, allows a small negative or reverse current flow. The Varistor[1] is a semiconductor element with a symmetrical e/i relation of approximate fourth power shape, $i = Ke^4$. The table of Fig. 3–2c gives actual measured e, i values; check how closely they conform to the fourth power law. Varistor applications include meter overload protection, signal limiting, and low-voltage regulation.

3–3. THE CAPACITANCE ELEMENT.

Two conductors separated by a nonconducting medium (insulator or dielectric) form a capacitor whose capacitance is defined by

$$C \triangleq \frac{q}{e} \text{ farads} \qquad (3\text{–}9)$$

The process of charging a capacitor consists of removing charge from one conductor and placing an equal amount on the other. The *net* charge of a capacitor is thus always zero and we understand the term "charge on the capacitor" to mean the magnitude of the charge on *either* conductor. Since C is defined to be a positive number, the algebraic sign of charge q is the same as that of the voltage e across the capacitor. In Eq. (3–9) then

---

[1] Victory Engineering Corp., Springfield, N. J., Bulletins MSCV 121 and MSV 111.

$q \triangleq$ charge on the capacitor, coulombs

$e \triangleq$ voltage across the capacitor, volts

In the pure and ideal capacitance element, the numerical value of C is absolutely constant for all values of q or e. Real capacitors exhibit some nonlinearity (C as defined by Eq. (3–9) varies with q) and are "contaminated" by the presence of resistance and inductance; however, the approximation is quite satisfactory in many cases.

Since a voltage-current rather than a voltage-charge relation is often more desirable, we recall the definition of current i

$$i \triangleq \frac{dq}{dt} \tag{3–10}$$

Equation (3–9) then gives

$$e = \frac{1}{C}q \tag{3–11}$$

$$\frac{de}{dt} = \frac{1}{C}\frac{dq}{dt} = \frac{i}{C} \tag{3–12}$$

Using the D operator we may write

$$i = CDe \tag{3–13}$$

giving the operational transfer function for a voltage input as

$$\frac{i}{e}(D) = CD \tag{3–14}$$

Alternatively,

$$de = \frac{1}{C}i\,dt \tag{3–15}$$

$$\int_{e_0}^{e} de = \frac{1}{C}\int_{0}^{t} i\,dt \tag{3–16}$$

$$e - e_0 = \frac{1}{C}\int_{0}^{t} i\,dt \tag{3–17}$$

where e_0 is the voltage across the capacitor which existed at time equal to zero. If e_0 were zero (capacitor initially uncharged), we would have

$$e = \frac{1}{C}\int_{0}^{t} i\,dt \tag{3–18}$$

The operational transfer function for a current input is

$$\frac{e}{i}(D) = \frac{1}{CD} \tag{3–19}$$

The pure and ideal capacitance element stores in its electric field all the electrical energy supplied to it during a charging process and will give up all this energy when discharged. For example, if we applied a constant current, i_0, to an initially uncharged capacitor, Eq. (3–18) indicates the voltage would rise as a ramp function

$$e = \frac{1}{C} \int_0^t i_0 \, dt = \frac{i_0 t}{C} \qquad (3\text{–}20)$$

The instantaneous power into the capacitor is $P = ei = i_0^2 t / C$; thus the total energy supplied up to time t is

$$\text{energy} = \int P \, dt = \int_0^t \frac{i_0^2 t}{C} \, dt = \frac{i_0^2 t^2}{2C} = \frac{Ce^2}{2} = \frac{q^2}{2C} \qquad (3\text{–}21)$$

Actually, the energy stored by a charged capacitor is $Ce^2/2 = q^2/2C$, irrespective of how the final voltage e or charge q was built up; the constant current i_0 used above was just an example. This can be shown by recalling that the work done to transfer a charge dq through a potential difference e is edq. Since for a capacitor $e = q/c$ we have

$$\text{total energy} = \int_0^q e \, dq = \int_0^q \frac{q}{C} \, dq = \frac{q^2}{2C} = \frac{Ce^2}{2} \qquad (3\text{–}22)$$

The energy supplied to the capacitor during the charging process is all stored in the capacitor and can be recovered by connecting the charged capacitor to some energy-using device (like a resistor) and letting the capacitor discharge into it. In this process the voltage polarity remains the same as during charging, but now the current is *reversed* (giving it a minus sign) and thus the "power *into* the capacitor" is now negative, which is the same as saying that power is being taken from the capacitor.

We should note at this point that when we speak of the "current *through* a capacitor," the current does not really pass through the dielectric material between the capacitor plates. Rather, an equal amount of charge is taken from one plate and supplied to the other by way of the circuit *external* to the capacitor. This flow of charge is, of course, a current, but it does not go *through* the capacitor the same way as it would go through a resistor or inductor. However, as a matter of common usage, we will continue to speak of the current through a capacitor and rely on the reader to recall the true physical situation. Figure 3–3a shows the standard symbol and sign conventions for current, voltage, and charge. Again, the voltage (and charge) $+$ $-$ sign convention can be chosen either of two ways, but once this choice is made the direction of positive current must conform to Eq. (3–18); that is, a positive current *must* cause a positive change in e. The measurement of actual capacitors to obtain a numerical value of C cannot easily be accomplished from the defining Eq. (3–9) as it

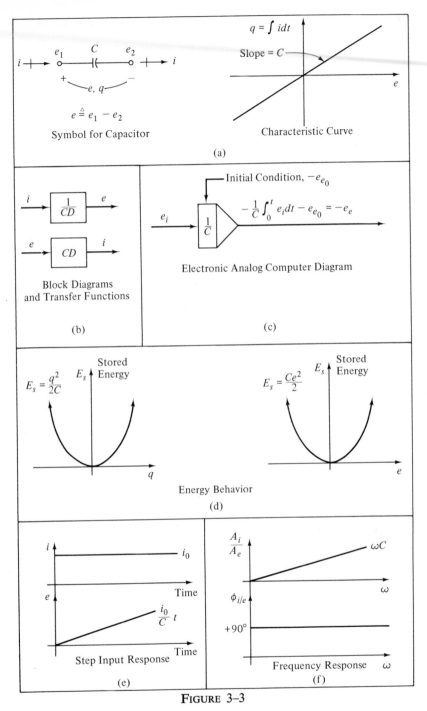

FIGURE 3–3

The Capacitance Element

could for resistors, because instruments for accurately measuring charge q are not practical. Various methods are available; they usually employ application of a sinusoidal voltage of known frequency. For such an AC signal, both current and voltage can be measured, and C computed from the sinusoidal transfer function as follows:

$$\frac{i}{e}(i\omega) = i\omega C = \omega C \; \underline{/90°} \tag{3-23}$$

Since the amplitudes I and E of the sine waves of current and voltage can be measured, and frequency ω is known, we compute C from

$$C = \frac{I}{\omega E} \tag{3-24}$$

Actually, commercial capacitance meters generally use an AC bridge method to compare the unknown capacitor with a known standard. (The thoughtful reader may wish to speculate on how the standard capacitor's capacitance came to be known.)

Figure 3–3e shows the voltage response to a step input of current, which presents no mathematical difficulties. If instead we apply a step input of voltage, we get

$$i = C\frac{de}{dt} = C\frac{d}{dt} \text{ (step function)} \tag{3-25}$$

Application of the classical definition of the derivative to the step function shows the derivative to be zero everywhere but at the location of the step, and there it is undefined. In Fig. 3–4, we approximate the step function as a terminated ramp function. By letting the rise time get smaller and smaller, we can approach the perfect step function as closely as we please. As we do this we see that the magnitude of de/dt will approach infinity, and its duration will approach zero, but the area under it will always be e_0. The "function" defined by this limiting process is called the *impulse function* of strength (area) e_0; if e_0 is 1.0 (a *unit* step function) its derivative is a *unit impulse function;* that is, its area is one unit. From Eq. (3–25) we see that a step input voltage produces a capacitor current of infinite magnitude and infinitesimal time duration. Since real physical quantities are limited to finite values, these events cannot, of course, occur in the real world. First, a true (instant-rising) step voltage cannot be achieved, and secondly, a real capacitor has parasitic resistance and inductance which limit current and its rate of change. Thus, a real capacitor will exhibit a short-lived (but not infinitesimal) and a large (but not infinite) current spike. We will find in later chapters that such a spike (which *does* have a definite area) may many times be treated as a perfect impulse of the same area with good accuracy.

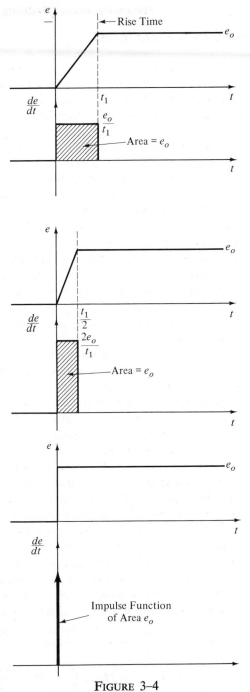

FIGURE 3–4

Approximate and Exact Impulse Functions

3–4. THE INDUCTANCE ELEMENT.

An electric current (motion of charge) always creates an associated magnetic field. If a coil or other circuit lies within this field, and if the field changes with time, an electromotive force (voltage) is induced in the circuit. The magnitude of the induced voltage is proportional to the rate of change of flux linking the circuit $d\Phi/dt$, weber/sec and its polarity is such as to oppose the cause producing it. If no ferromagnetic materials (such as iron) are present, the rate of change of flux is proportional to the rate of change of current di/dt which is producing the magnetic field. The proportionality factor relating the induced emf to the rate of change of current is called the *inductance*. The presence of ferromagnetic materials greatly increases the strength of the effects, but also makes them significantly nonlinear, since now the flux produced by the current is not proportional to the current. Thus, iron can be used to get a large value of inductance, but the value will be different for different current levels. For small changes in current about some operating point, one can define an incremental inductance for a linearized analysis. Large current swings require nonlinear treatment using computers to model the experimentally measured magnetization curve of the particular material.

The pure inductance element has induced voltage e instantaneously related to di/dt, but the relation can be nonlinear. The pure and ideal element has e directly proportional to di/dt; that is, it is linear and free from capacitance and resistance. While it is possible to make resistors and capacitors which are very close to pure and ideal from DC up to rather high frequencies, a real inductor always has considerable resistance, which means that at DC and low frequencies (where di/dt is zero or small) it will behave like a resistor, not an inductor. However, for some intermediate range of frequencies (not including zero), a real inductor may behave quite like the ideal model—particularly if the required inductance is small enough to be achievable without use of magnetic materials.

Inductive effects can be classified as *self-inductance* and *mutual inductance*. Self-inductance is a property of a single coil due to the fact that the magnetic field set up by the coil current links the coil itself. Mutual inductance causes a changing current in one circuit to induce a voltage in *another* circuit. Figure 3–5 shows a configuration illustrating these concepts. Considering voltages induced into circuit A we would have:

$$e_A = e_{A1} + e_{A2} = L_1 \frac{di_A}{dt} \pm M_{B/A1} \frac{di_B}{dt} \pm M_{A2/A1} \frac{di_A}{dt}$$

$$+ L_2 \frac{di_A}{dt} \pm M_{B/A2} \frac{di_B}{dt} \pm M_{A1/A2} \frac{di_A}{dt} \qquad (3\text{–}26)$$

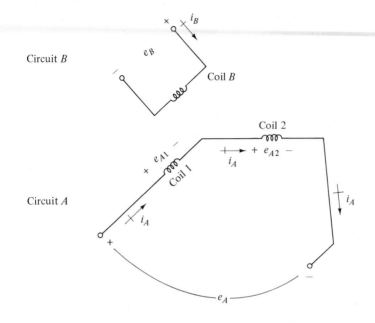

For Any Pair of Coils X and Y, $M_{X/Y} \lesseqgtr \sqrt{L_X L_Y}$

FIGURE 3–5

Self-Inductance and Mutual Inductance

$$e_A = (L_1 + L_2 \pm M_{A2/A1} \pm M_{A1/A2}) \frac{di_A}{dt} + (\pm M_{B/A1} \pm M_{B/A2}) \frac{di_B}{dt} \quad (3\text{--}27)$$

where

$$L_1 \triangleq \text{self-inductance of coil 1, henries} \quad (3\text{--}28)$$

$$L_2 \triangleq \text{self-inductance of coil 2, henries} \quad (3\text{--}29)$$

$$M_{A2/A1} = M_{A1/A2} \triangleq \text{mutual inductance of coils 1 and 2, henries} \quad (3\text{--}30)$$

$$M_{B/A1} \triangleq \text{mutual inductance of coils } B \text{ and } A_1, \text{ henries} \quad (3\text{--}31)$$

$$M_{B/A2} \triangleq \text{mutual inductance of coils } B \text{ and } A_2, \text{ henries}$$

(all e's are in volts, i's in amps, t in seconds)

Note that mutual inductance is symmetrical ($M_{A2/A1} = M_{A1/A2}$). That is, a current changing with a certain di/dt in coil 1 induces the same voltage in coil 2 as would be induced in coil 1 by the same di/dt current change in coil 2. This also holds for *separate* circuits such as A and B; the voltage induced in coil B by di_A/dt in coil 1 would be $\pm M_{A1/B}(di_A/dt)$ where $M_{A1/B} = M_{B/A1}$. The $\pm$ signs used on the mutual inductance terms in the above equations require explanation. For any *specific* fixed physical orientation of the coils all these signs would each be definitely + or definitely −, and not the ambiguous ±. That is, the induced voltage in circuit *A* due to current change in *B* can either add to or subtract from the self-induced voltage in *A*. Since a two-dimensional circuit drawing does not show the actual geometry clearly enough, unless additional information is provided we cannot decide whether a given mutual term should be given a + or a − sign. A common convention for providing this polarity information on a drawing (once it has been reasoned out from the physical arrangement or by experimental test) is to place a dot on one end of each coil of a mutual pair. The dots are placed so that the following rule holds:

> If both the assumed positive directions for the two currents are toward the dots (or both away from the dots), then the sign of the *M* term will be the same as the sign of the *L* term.

The possibility of inductive effects opposing one another is made use of in the manufacture of wire-wound resistors, to have a minimum of parasitic self-inductance. Figure 3–6 shows two such methods in practical use. Note in each case that the currents are directed such that their magnetic fields tend to cancel, thus reducing inductive effects and making the resistor behave more like a pure resistance element at high frequencies.

Figure 3–7 displays the behavior of the pure and ideal self-inductance element *L*. The defining equation is

$$e = L \frac{di}{dt} \qquad (3\text{–}32)$$

with a consistent set of sign conventions as shown in Fig. 3–7a. If we choose current positive to the right as shown, and if di/dt is positive, we would get a drop in voltage from e_1 to e_2; thus we must take the sign convention on *e* as shown to conform to Eq. (3–32). That is, *L* is defined to be a positive number and Eq. (3–32) says that a positive di/dt corresponds to a positive *e*. For nonlinear inductances (such as those containing iron) the current/voltage relation can be given as

$$e = \left(N \frac{d\Phi}{di} \right) \frac{di}{dt} \qquad (3\text{–}33)$$

where

$N \triangleq$ number of turns on coil

$\dfrac{d\Phi}{di} \triangleq$ rate of change of flux with respect to current

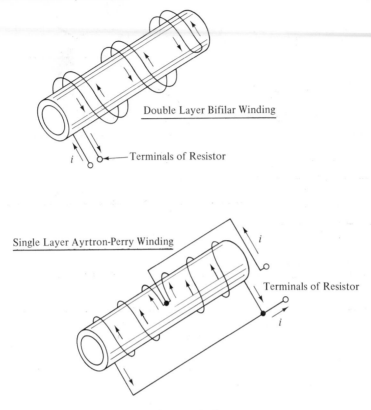

Double Layer Bifilar Winding

Terminals of Resistor

i

Single Layer Ayrtron-Perry Winding

Terminals of Resistor

i

i

FIGURE 3–6

Resistors Wound to Minimize Inductance

The term $d\Phi/di$ can be obtained from experimentally measured curves of Φ versus i; that is, $d\Phi/di$ is a function of i. For linear inductances $d\Phi/di$ is a *constant* and then Eq. (3–33) becomes (3–32) with $N(d\Phi/di) \triangleq L$.

If we apply a constant voltage e_0 to an inductance element, we cause current to increase at a constant rate $di/dt = e_0/L$. Since this current and voltage are both positive, the inductor is absorbing energy at a rate $e_0 i$. Let us assume the inductor carried zero current at time zero when the voltage was applied. Then

$$\text{power} = P = e_0 i = (e_0)\left(\frac{e_0}{L}\right) t \qquad (3\text{–}34)$$

and

$$\text{energy} = \int_0^t P\, dt = \int_0^t \frac{e_0^2}{L} t\, dt = \frac{e_0^2 t^2}{2L} = \frac{i^2 L}{2} \qquad (3\text{–}35)$$

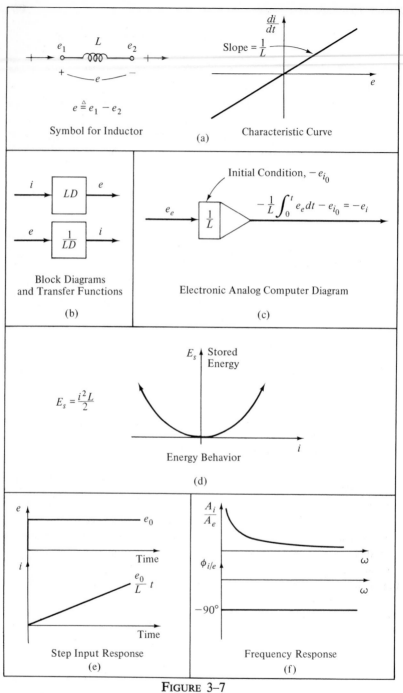

FIGURE 3–7

The Inductance Element

Thus at an instant t when the current is i the inductor has received energy in the amount $i^2L/2$. This energy is actually stored in the magnetic field since we find that if we connect a current-carrying inductor to an energy-using device (a resistor, for example) the inductor will *supply* energy in an amount $i^2L/2$ as its current decays from i to zero. During this decay process i (if originally positive) stays positive, but di/dt becomes negative and thus e is negative. The product ei is thus negative, showing that the inductor is supplying power to the external circuit. The energy storage $i^2L/2$ is correct irrespective of how the current i was achieved, as we can see from

$$\text{power} = ei = L\frac{di}{dt}\,i \tag{3-36}$$

$$\text{energy} = \int_0^t iL\frac{di}{dt}\,dt = \int_0^i Li\,di = \frac{i^2L}{2} \tag{3-37}$$

We have seen that a step input of voltage e_0 causes a ramp current $i = e_0 t/L$. A step input of current i_0 gives rise to a voltage impulse of strength (area) Li_0. The frequency response of the inductance element is obtained from the sinusoidal transfer function as follows:

$$e = L\frac{di}{dt} = L\,Di \qquad \frac{i}{e}(D) = \frac{1}{LD} \tag{3-38}$$

$$\frac{i}{e}(i\omega) = \frac{1}{i\omega L} = \frac{1}{\omega L}(-i) = \frac{1}{\omega L}\underline{/-90°} = \frac{A_i}{A_e}\underline{/\phi_{i/e}} \tag{3-39}$$

Note that at very low frequencies ($\omega \to 0$) a small voltage amplitude can produce a very large (approaching ∞ as $\omega \to 0$) current. Thus an inductance is sometimes said to approach a short circuit for low frequencies and under such conditions could be replaced, in a circuit diagram, with just a piece of "connecting wire." (Recall that in circuit diagrams one shows R's, C's, and L's connected by pieces of perfectly conducting (no voltage drop) "wire"). At high frequencies ($\omega \to \infty$) note that the current produced by any finite voltage approaches zero. Thus we often say an inductor approaches an *open circuit* at high frequencies, and could thus just be "cut out" of a circuit diagram under such conditions. For a capacitance, just the reverse frequency behavior was observed; the capacitance approaches a short circuit at high frequencies and an open circuit at low frequencies.

3-5. REAL RESISTORS, CAPACITORS, AND INDUCTORS.

Just as we saw in mechanical elements, electrical elements are sometimes specifically designed into a system and other times they appear unintentionally and perhaps undesirably. In control systems which use DC motors, for example, the inductance of the field must often be included in the model, even though it slows the response and is thus undesirable. This inductance is not something the designer "wired into" his system, it is simply an effect present in motor fields that must be taken into account. The cables and wires used to interconnect electrical components are (ideally) perfect conductors devoid of all resistance, capacitance, and inductance; however, at high frequencies they exhibit all three properties and must be so modeled to properly predict system behavior. Vacuum tubes are basically intended as instantaneous power modulators; however, their basic construction is such that the electrodes (metal conductors) are separated by a vacuum (dielectric), and thus inter-electrode capacitances are formed which influence the dynamic response. These capacitances were not deliberately designed into the tube; they were an unavoidable by-product of the basic tube configuration. Such parasitic or unintentional electrical elements are usually difficult to calculate from theory, but generally can be measured once the equipment has been built.

Turning now to "intentional" circuit elements, let us first consider resistors. While an ideal and pure resistance element is completely described for all purposes of circuit analysis by a single number R, the practical choice of a resistor involves many complex factors such as:[2]

1.	physical size	9.	temperature coefficient of R
2.	wattage rating	10.	voltage coefficient
3.	stability	11.	solderability or weldability
4.	manufacturing tolerance	12.	humidity tolerance
5.	maximum temperature	13.	shock and vibration tolerance
6.	maximum voltage	14.	shelf life
7.	frequency range	15.	load life
8.	electrical noise	16.	reliability

A number of manufacturing processes[3] are used to fabricate commercial resistors; we here describe briefly only three of these. The most obvious method is that of winding a coil of fine wire on an insulating form, the so-called wire-wound resistor, Fig. 3–8a. The resistance of a length L of wire of cross-sectional area A is given by

$$R = \frac{\rho L}{A} \tag{3–40}$$

where

$\rho \triangleq$ material resistivity, ohm-meters

---

[2] G. W. A. Dummer, *Modern Electronic Components* (London: Sir Isaac Pitman & Sons Ltd., 1966), p. 57.
[3] *Ibid.*

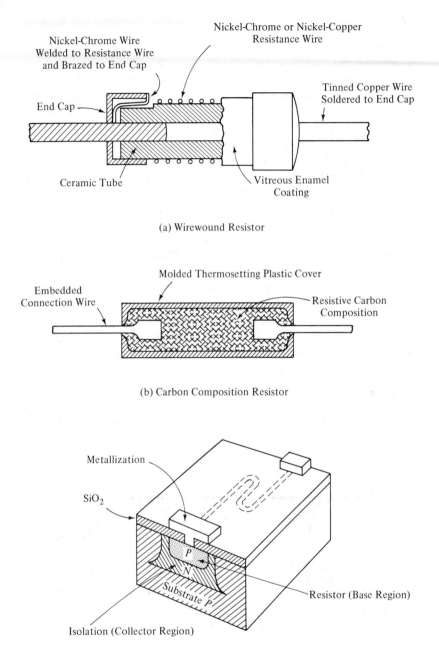

Nickel-Chrome or Nickel-Copper
Resistance Wire

Nickel-Chrome Wire
Welded to Resistance Wire
and Brazed to End Cap

Tinned Copper Wire
Soldered to End Cap

End Cap

Ceramic Tube

Vitreous Enamel
Coating

(a) Wirewound Resistor

Molded Thermosetting Plastic Cover

Embedded
Connection Wire

Resistive Carbon
Composition

(b) Carbon Composition Resistor

Metallization

SiO$_2$

P

N

Substrate P

Resistor (Base Region)

Isolation (Collector Region)

(c) Diffused Microcircuit Resistor

FIGURE 3–8

Discrete Component and Integrated-Circuit Resistors

For commercial annealed copper $10^8\rho = 1.72$, for Constantan (Cu 60, Ni 40) 49.0, for Nichrome 100, and for Manganin (Cu 84, Mn 12, Ni 4) 44.0, all at room temperature. The carbon-composition resistor of Fig. 3–8b is molded from a powder made up of carbon black, a resin binder, and a refractory filling. Lead wires may be attached by several methods; Fig. 3–8b shows wires with enlarged ends molded directly into the carbon rod. In monolithic integrated circuits,[4] the tiny resistors needed are produced from semiconductor materials by a diffusion process at the same time that the transistors, diodes, and capacitors are formed (see Fig. 3–8c). An extremely thin and very narrow strip of doped material is diffused into a substrate of oppositely doped silicon. A 100,000-ohm resistor can be put onto a square chip 0.020 inch on a side and 0.006 inch thick.

Resistors vary in size from the microscopic integrated circuit units mentioned above, which can dissipate a few milliwatts of power, to wire-wound power resistors rated at 300 watts and measuring 1 by 10 inches. *Stability* of a resistor refers to constancy of the R value under shelf life or working conditions. Precision, hermetically sealed wire-wound resistors may have a stability as good as 0.01%. Carbon composition stability is in the order of 5% for ordinary conditions, but may be as bad as 25% for severe environments. The *manufacturing tolerance* ranges from 20% for carbon composition to 0.01% for precision wire-wound. *Maximum allowable temperatures* go from 70°C to 600°C, while *maximum voltages* go up to about 3000 volts.

The *useful frequency range* of a resistor is determined by the magnitude of the parasitic capacitance and/or inductance effects associated with its construction. Figure 3–9 shows typical behavior for precision metal and

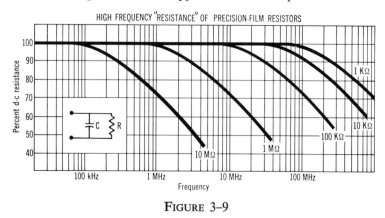

FIGURE 3–9

High Frequency Behavior of Film Resistors

---

[4] L. Stern, *Fundamentals of Integrated Circuits* (New York: Hayden Book Co., 1968), p. 73.

carbon film resistors.[5] Note that the model for the resistor includes a parallel capacitance representing an unavoidable parasitic effect. The trend that high-value resistors deviate from the pure resistance model at lower frequencies than low-value is typical of most types of construction. Wire-wound resistors may exhibit sufficient inductance to require its inclusion in the device model. Figure 3–10a shows the model employed. For R values about 100 ohms or less, the inductance predominates with a gradual shift to capacitive dominance at higher resistance values. Typical ranges of L and C are shown in Fig. 3–10b, and 3–10c shows step responses for a class of special "high speed" resistors.[6] Since the method of mounting the resistor influences the dynamic behavior, tests should be run with mounting methods similar to those expected in the actual application.

Electrical noise refers to random voltages caused internally in the resistor by thermal agitation (Johnson noise) and so-called current noise. Johnson noise is present in all resistors, and contains a very wide range of frequencies. At room temperature a 1-megohm resistor will exhibit about 100 μV of noise voltage over a 1-MHz bandwidth of frequencies. Current noise is peculiar to carbon composition resistors and other non-metallic films, and increases with current through the resistor. It can be as great as several thousand microvolts. The main interest in resistor noise concerns the use of the resistor in amplifying circuits for small voltage signals. The presence of noise limits the smallest signals which can be detected and amplified.

Temperature coefficient of resistance refers to the fact that all resistance materials change resistance with temperature. For small temperature changes near an operating point, the effect is essentially linear and one can define the temperature coefficient of resistance as dR/dT. Typical values are: carbon composition $\pm 0.12\%$ per C°, carbon film -0.02 to -0.1% per C°, precision wire-wound $+0.002\%$ per C°. *Voltage coefficient* is most important for carbon composition resistors and refers to an immediate change in resistance (usually a decrease) following application of a DC voltage. This effect is distinct from the temperature coefficient, and can be as much as 0.02% per volt. Wire-wound resistors do not show this effect, and carbon film have it only to the extent of 0.002% per volt or less.

While resistors themselves have quite good *reliability*, due to the large numbers used in a typical electronic equipment they usually are the greatest contributor to equipment failure. Typically, for each transistor or vacuum tube one finds five to ten resistors. The average failure rate for *all* components ranges from 0.0004% per 1000 hours in undersea cable amplifiers (which must be very reliable) to about 2% per 1000 hours in commer-

---

[5] Texas Instruments, Brochure 3-66(20M), 1954.
[6] RCL Electronics, Inc., Catalog 678A.

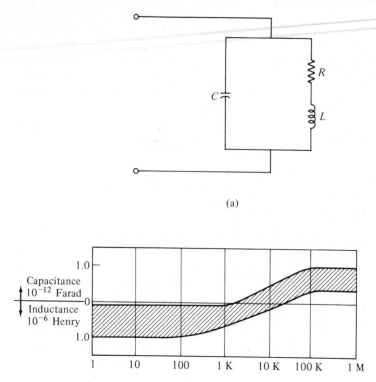

(a)

(b)

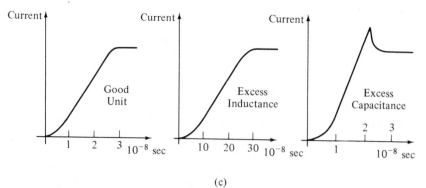

(c)

FIGURE 3–10

Dynamic Behavior of "High-Speed" Wire-Wound Resistors

cial radio and television service. This wide range is due to variations both in component quality and in severity of the environment of use. For resistors used in general purpose ground-based electronics, average failure rates[7] range from 0.01% per 1000 hours for oxide film types to 0.2% per 1000 hours for wire-wound. Soldered connections, the second greatest contributor to failure, have a rate of 0.01, whereas transistors range from 0.01 to 0.1.

Turning now to capacitors, Fig. 3–11 shows formulas for computing capacitance of some common configurations. Most intentional capacitors are based on the parallel flat plate arrangement for which

$$C = \frac{\epsilon A}{d}, \qquad \text{farad} \qquad \qquad \textbf{(3–41)}$$

$\epsilon \triangleq$ permittivity of the dielectric, F/meter

$A \triangleq$ area of plates, meter^2

$d \triangleq$ distance between plates, meters

For a vacuum between the plates, the permittivity is the smallest possible, 8.85×10^{-12}, while for air at atmospheric pressure it is $K(8.85 \times 10^{-12})$, where $K = 1.00059$. The constant K of a dielectric material is called the dielectric coefficient, and is simply the ratio of the permittivity of the material to that of a vacuum. For a spacing d of one millimeter, a 1-F flat plate air capacitor would be a square 6.5 *miles* on a side, showing that the farad is usually an inconvenient unit of measurement, although there are 1-farad commercial capacitors which can be held in one hand.[8] For most applications the microfarad (μF), which is 10^{-6} farad, and the picofarad (pF), 10^{-12} farad, are more convenient, with a 1-μF capacitor (other than the electrolytic type) being considered quite large. Common dielectrics in use include impregnated paper ($K = 4$ to 6), glass and mica (4 to 7), polystyrene (2.3), Mylar (2 to 5), and ceramics (6 to 3000). Permittivity for electrolytic capacitors is not generally quoted, because the dielectric film is somewhat indefinite in thickness; however, a K of about 3 appears to be the right order of magnitude. Electrolytic capacitors achieve very large C values in small space, because the dielectric film is much thinner ($d \approx 10^{-5}$ cm) than it is practical to make a sheet of paper or plastic film (about 10^{-3} cm minimum).

---

[7] G. W. A. Dummer, *Modern Electronic Components* (London: Sir Isaac Pitman & Sons Ltd., 1966), p. 456.

[8] "The Measurement of Electrolytic Capacitors," *The Experimenter* (Vol. 40, No. 6, June 1966), General Radio Corp.

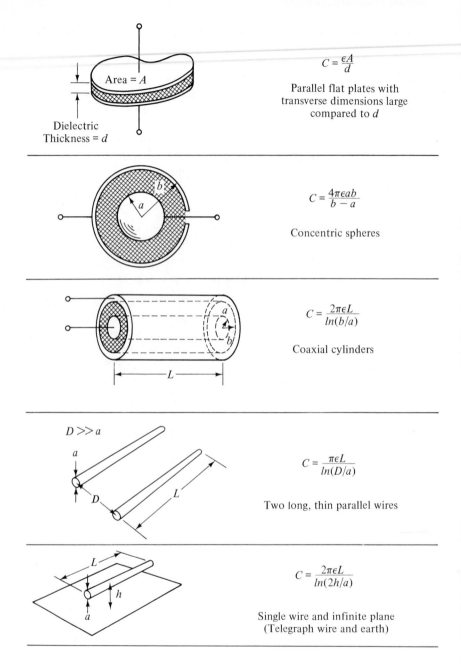

$$C = \frac{\epsilon A}{d}$$

Parallel flat plates with transverse dimensions large compared to d

$$C = \frac{4\pi\epsilon ab}{b - a}$$

Concentric spheres

$$C = \frac{2\pi\epsilon L}{\ln(b/a)}$$

Coaxial cylinders

$$C = \frac{\pi\epsilon L}{\ln(D/a)}$$

Two long, thin parallel wires

$$C = \frac{2\pi\epsilon L}{\ln(2h/a)}$$

Single wire and infinite plane
(Telegraph wire and earth)

FIGURE 3–11

Capacitance Formulas

Capacitors may be constructed in several ways, using various materials.[9] One basic method is shown in Fig. 3–12a where 0.00025-inch aluminum foil (the electrodes) is sandwiched between 0.0005-inch oil-impregnated paper or plastic film (the dielectric), rolled up into a cylinder and provided with terminals and an outer protective case. Metallized paper, a variation on this technique, deposits or sprays a thin metal film onto the paper dielectric to reduce the volume of the finished capacitor. Figure 3–12b shows a tubular ceramic capacitor; the ceramic dielectrics used include steatite, titanium dioxide, and barium titanate. A typical electrolytic capacitor construction is shown in Fig. 3–12c. This type of capacitor has a unique principle of operation which puts it somewhat in a class apart from all other types. The sketch shows sandwiched metal foil and paper separators which appear quite similar to a "conventional" capacitor; however, the paper is *not* the dielectric in this case. The paper is saturated with an electrolyte paste of glycol and ammonium tetraborate. One of the 0.002-inch aluminum foils has had a very thin layer of aluminum oxide (the dielectric) formed on it by applying a constant voltage while it was immersed in an ammonium borate solution. This foil will be the anode, or positive terminal, of the capacitor, and *must never have a negative voltage applied to it.* That is, the electrolytic capacitor is a polarized device and is not usable where currents and voltages actually reverse, whereas all other types of capacitors are completely "reversible." (Actually, reversible electrolytics can be and are made by preforming the dielectric film on *both* of the aluminum foils and making suitable connections; however, most electrolytics are of the polarized type and must be used with this in mind.) Air-dielectric capacitors, Fig. 3–12d, are used mainly as laboratory standards and are made in parallel-plate and concentric-cylinder configurations. They are extremely accurate and stable, varying the order of 0.01 to 0.04% over several years and having a temperature coefficient of capacitance of less than 10 parts per million per Celsius degree. Diffused capacitors for monolithic integrated circuits are produced by techniques similar to those explained for the resistor of Fig. 3–8c. Thin-film capacitors are tiny discrete components used in microelectronic circuits when capacitors of higher quality than can be produced by the diffusion process are needed. A chip 0.0024 inch thick and 0.032 inch square accommodates a 3000-pF ±5% capacitor. Tiny "beam-leads" 0.006 inch long, 0.004 inch wide, and 0.0004 inch thick are used to attach the capacitors to other circuit elements. Figure 3–12e shows a cross section of a typical [10] unit using tantalum oxide as the dielectric.

---

[9] G. W. A. Dummer, *Modern Electronic Components* (London: Sir Isaac Pitman & Sons Ltd., 1966), Chapters 7 and 8.

[10] W. E. Wesolowski and M. Tierman, "Beam-Leaded, Thin-Film Capacitors," *Electronic Capabilities* (Winter 1969–70), p. 34.

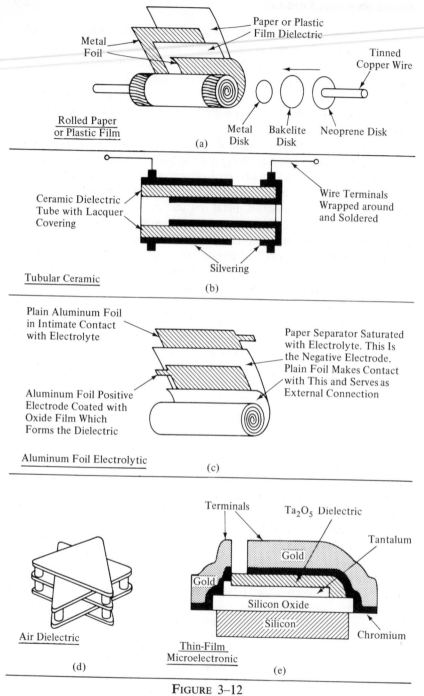

Figure 3–12

Capacitor Constructions

While a pure capacitor stores and can then release all the energy supplied to it, real capacitors exhibit losses for various reasons. The *power factor* of a capacitor is defined as the ratio of energy wasted per cycle of AC voltage, divided by the energy stored per cycle. It may vary with frequency, and values measured at 1000 Hz range from 0.00001 for precision air-dielectric types through 0.0005 for polystyrene film to 0.05 for some electrolytics. This wasted energy shows up as heat in a real capacitor, whereas a pure capacitor experiences no temperature rise whatever. An equivalent circuit sometimes used to model a real capacitor is shown in Fig. 3–13. The above-mentioned power losses are due to the resisitive elements in this model. The presence of the parallel resistance element R_p, called the *leakage resistance*, in the model shows that a real capacitor will allow a DC current to flow. For a given type of capacitor, the leakage resistance is very nearly inversely proportional to capacitance; that is, CR_p is a constant. Due to R_p, a charged capacitor will not hold its voltage indefinitely; it will decay exponentially to zero. The time for this decay varies from several days for Teflon and polystyrene types to a few seconds

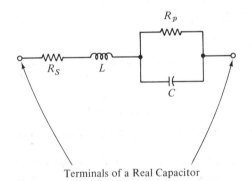

Terminals of a Real Capacitor

FIGURE 3–13

Model of a Real Capacitor

for some electrolytics. For polystyrene, for example, the product CR_p at room temperature is about 10^6, where R_p is in megohms and C in microfarads. A 1–μF unit would thus be expected to have an R_p of about 10^{12} ohms. *Dielectric absorption* refers to the reappearance of a voltage after a charged capacitor has been discharged by short-circuiting and is then open-circuited. As an example, for a 200-volt initial charge applied for one minute and a two-second short-circuit discharge, the voltage reappearing after one minute is 0 for an air capacitor, 0.02% (40 millivolts) for polystyrene, and 2.0% for oil-impregnated paper. Together with other effects, the above-mentioned deviations from perfection place frequency restrictions on the application of various capacitor types as shown in Fig. 3–14. The temperature coefficient of capacitance in parts per million per Celsius degree varies from $+10$ for precision air capacitors through about ±200 for paper and plastic films to $+1000$ to 2000 for many electrolytics. Failure rates in percent per 1000 hr are in the range of 0.005 for polystyrene to 0.5 for aluminum foil electrolytics.

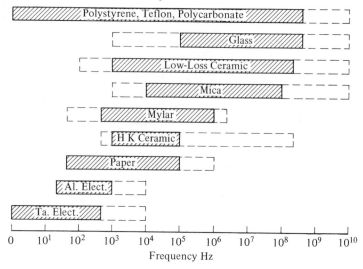

FIGURE 3–14

Useful Frequency Range of Various Capacitor Types

Commercial inductors are not generally available in as wide a selection of sizes and types as are resistors and capacitors. In fact many modern circuit synthesis methods endeavor to achieve the required performance entirely by the use of R and C because of the practical difficulties associated with the construction and use of L. We shall see later that the use of active (operational amplifier) network methods allows one to achieve circuit behavior previously requiring use of R, C, and L with R and C alone. A further disadvantage of inductance is the present unavailability of any method for constructing microcircuit inductance elements for use in integrated circuits. The theoretical calculation of the inductance of configurations of most practical shapes can be carried out with high accuracy so long as no magnetic materials are used.[11] When magnetic materials are used

---

[11] F. W. Grover, *Inductance Calculations* (New York: D. Van Nostrand, 1946).

(to get large L values) only rough calculations are possible and the inductance varies with current level. Figure 3–15 shows formulas[12] for some common configurations of non-magnetic inductances. These values may be multiplied by the relative permeability of the magnetic material to estimate inductance for magnetic core inductors for small-signal operation near a given operating point.

For real inductors the *quality factor* Q plays a role similar to that of the power factor in capacitors; that is it gives an indication of the energy losses due to the presence of resistance. Since most inductors are coils of many turns of wire, they must of necessity have considerable resistance. At low and intermediate frequencies a real inductor may be modeled as in Fig. 3–16a. For such a model Q is defined as

$$Q \triangleq \frac{\omega L}{R} \qquad (3\text{–}42)$$

$$\omega \triangleq \text{frequency of AC voltage, rad/sec}$$

and is proportional to the ratio of stored to dissipated energy. We see that for a pure inductor ($R \equiv 0$) the quality factor would be infinity irrespective of the frequency; thus high Q indicates a more nearly pure inductor than does low Q. Note this trend is opposite to that for the power factor in capacitors; there one wants a *small* power factor. For real inductors $R \neq 0$ and Q approaches zero for low frequencies. If the model of Fig. 3–16a held for all frequencies, Q would approach infinity as ω approached infinity; however, at intermediate frequencies core-loss effects appear which change the model to Fig. 3–16b. Core losses are energy losses due to eddy currents and hysteresis in the core and are represented by an equivalent shunt resistance. For this new circuit $Q = \omega L / [R_c + ((R_c + R_e)/R_e^2)\omega^2 L^2]$, which approaches zero for both very low and very high frequencies.

This explains the shape of the Q versus frequency curve actually measured on a commercial inductor[13] made as a magnetic core toroid and shown in Fig. 3–16c. An inductance of 60 henries is achieved in a unit measuring 7/16 by 31/64 by 9/16 inch and weighing 0.2 ounce. The DC resistance is 5160 ohms. Because magnetic materials are used, the inductance varies with current; to keep this variation below 5% the DC current must be limited to 0.2 mA or less. At a given frequency the inductance may be quite constant over a fairly wide range of AC voltage amplitudes so long as the instantaneous AC current remains at or below the DC limiting

---

[12] *Ibid.*, p. 26.
[13] Miniductor ML-10, United Transformer Co., New York.

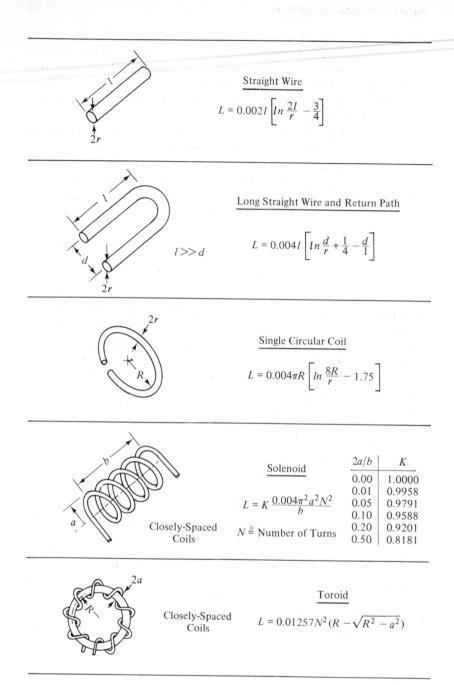

Straight Wire

$$L = 0.002l \left[\ln \frac{2l}{r} - \frac{3}{4} \right]$$

Long Straight Wire and Return Path

$l \gg d$

$$L = 0.004l \left[\ln \frac{d}{r} + \frac{1}{4} - \frac{d}{l} \right]$$

Single Circular Coil

$$L = 0.004\pi R \left[\ln \frac{8R}{r} - 1.75 \right]$$

Solenoid

$$L = K \frac{0.004\pi^2 a^2 N^2}{b}$$

Closely-Spaced Coils

$N \triangleq$ Number of Turns

$2a/b$	K
0.00	1.0000
0.01	0.9958
0.05	0.9791
0.10	0.9588
0.20	0.9201
0.50	0.8181

Toroid

Closely-Spaced Coils

$$L = 0.01257 N^2 (R - \sqrt{R^2 - a^2})$$

FIGURE 3–15

Inductance Formulas (All lengths in cm, inductance in μH)

132

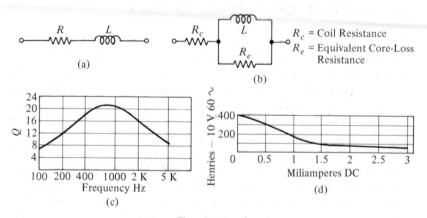

FIGURE 3–16

Models of Real Inductors

value just mentioned. A 2-henry unit with DC resistance of 130 Ω and DC current limit of 8 mA shows only a 3% change in inductance for a 400-Hz voltage of 0.1 to 35 volts. A typical curve showing variation of inductance (measured at 60 Hz, 10 volts AC) with DC current is given in Fig. 3–16d. The leveling off of L at high currents is due to saturation of the magnetic core which reduces the permeability of the magnetic material. Lest the large L values (60 H and 2 H) and low frequency ranges mislead the reader, we should indicate that many practical inductors of millihenry and micro-henry value work in high kilohertz and megahertz frequency ranges. A 0.15-μH inductor is about 0.1 inch in diameter and 0.4 inch long, with a DC resistance of 0.02 ohms and a Q of about 65 at 25 MHz, the frequency at which L is measured. If we take the simple definition of Q as $\omega L/R$ in this example, we get $Q = (25)(6.28)(10^6)(0.15)(10^{-6})/0.02 = 1180$, which is much higher than the measured value of 65. One explanation of this lies in the so-called "skin-effect" which makes the resistance much higher than its DC value when we operate at very high frequencies. At DC and low frequencies the current is uniformly distributed over the cross section of a conductor while at high frequencies the current is "crowded" toward the surface, making the effective cross section smaller and thus raising the resistance. This is caused by self-induced emf's set up by variations in the internal flux in the conductor. In our numerical example, it appears that the effective resistance at 25 MHz is about $1180/65 = 18$ times the DC value. Part of this resistance could also be caused by the core losses men-tioned earlier.

3-6. CURRENT AND VOLTAGE SOURCES.

The energy sources which drive electrical systems may conveniently be classified as voltage sources or current sources. An ideal voltage source supplies the intended voltage to the circuit no matter how much current (and thus power) this might require; conversely for a current source. That is, a 10-volt source applied to a 10-ohm resistor produces a 1-amp current and draws 10 watts. The same 10-volt source applied to a 0.001-ohm resistor produces 10,000 amps and draws 100,000 watts. No *real* voltage source, such as a battery, can behave in this ideal manner; however, over restricted ranges we may approach the ideal. Furthermore, the behavior of real sources may be modeled by a combination of an ideal source and a passive element, or elements, such as a resistor. Just as in mechanical systems, where the force source is fundamental because of the cause-and-effect relation given by Newton's law (a force *causes* a motion—not the reverse), we might say that voltage sources are fundamental since it is the electromotive force (voltage) which *causes* a current to flow. However, in practice, use of a current source as a model may be quite correct in some systems, since real devices which behave very much like current sources are available.

Batteries of various types are widely used as electrical power sources.[14] Experimentally measured performance curves of a modern rechargeable lead-acid unit[15] are shown in Fig. 3–17a. By plotting voltage versus current at time = 1 second from these discharge curves, we get Fig. 3–17b. We there see that this battery is not an ideal voltage source, since voltage drops off as more current is drawn. However, the curve is very nearly a straight line of slope -0.1 volt/amp, suggesting that we can model this real battery as an ideal 12.6-volt source in series with a 0.1-ohm resistor, as in Fig. 3–17c. The 0.1-ohm resistor is called the internal impedance of the real source; an ideal voltage source has zero internal impedance, thus real voltage sources approach perfection as their internal impedance approaches zero.

The vast majority of the electric power used by man is produced by the rotating mechanical-electrical energy converters called generators, usually of the AC type. These are basically voltage sources which again exhibit an internal impedance so that their terminal voltage drops off as more current is drawn. Since equipment in homes and industry is not generally connected directly to the output terminals of a generator, we might be

---

[14] "Batteries," *Machine Design* (April 11, 1963), p. 189.
[15] Delco Energette, Brochure DR-9647, Delco-Remy Div.

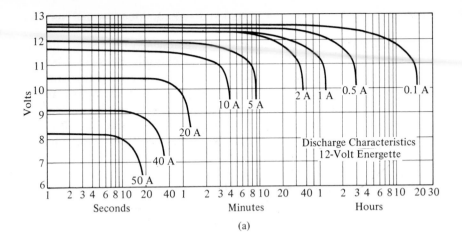

Volts

2 A 1 A   0.5 A   0.1 A

10 A 5 A

20 A

Discharge Characteristics
12-Volt Energette

40 A

50 A

Seconds   Minutes   Hours

(a)

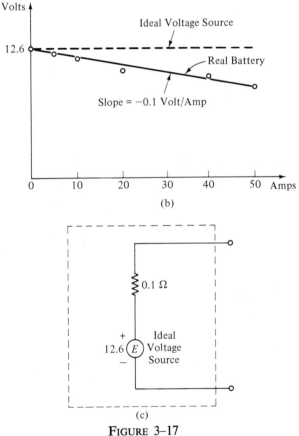

Volts

Ideal Voltage Source

12.6

Real Battery

Slope = −0.1 Volt/Amp

0   10   20   30   40   50   Amps

(b)

0.1 Ω

+
12.6 E Ideal
−   Voltage
Source

(c)

FIGURE 3–17

Modeling of a Real Battery

135

more interested in modeling the source represented by, say, a 110-volt, 60-Hz wall plug. We will not pursue this analytically, but the reader may wish to speculate on how an experiment might be set up to gather information needed to develop such a model. Direct-current generators are still used as power amplifiers in some servomechanisms (motion control systems) such as are employed in automatic machine tools. A 5000-watt generator[16] designed for such applications is driven by a 7.5-hp induction motor at 3450 rpm. The peak output voltage is ± 100 volts and the peak current is ± 50 amps. The output power is controlled by manipulating the field current with a transistor amplifier. When the field current is zero, the generator output is zero even though it is turning at 3450 rpm. Applying field current in one direction produces output of one polarity; reversing the field current reverses the output voltage. If this generator is run at constant speed with a fixed field current, it produces a definite voltage at its output terminals, if these terminals are open-circuit. If we now connect a load to these terminals and draw current, we find that the terminal voltage drops because of the internal impedance of the generator armature. In the machine being considered, this was measured as 0.31 ohms. Thus if we have, say, 10 volts open-circuit output voltage and then connect a load which draws 1 amp, the voltage will drop to 9.69 volts. Dynamic tests reveal the armature to also have 2.2 millihenries of inductance; thus if the load is changed suddenly (say by switching a resistor across the output terminals) the generator behaves as a source with dynamic characteristics. Figure 3–18 illustrates some of these concepts.

Electronically regulated DC power supplies, rather than batteries, are used as power sources for much electrical equipment. These power supplies take 110-volt/60-Hz AC power from the "wall plug" and provide regulated DC voltage or current at their output terminals. Many such regulators use feedback control principles and provide performance closely approaching the ideal voltage or current source. Details of operation of such supplies are beyond the scope of this text, but may be found in manufacturers' catalogs.[17] Operating specifications of a typical 15-volt supply are: current range 0–1.2 amp, voltage change for 0–1.2-amp current change $<0.05\%$, voltage change for 105–125-volt AC power line change $<0.05\%$, temperature effect on voltage $<0.05\%$ per Celsius degree, time drift $<0.05\%$ in 8 hours. The internal impedance is quoted as 0.008 ohms for DC to 100 Hz, 0.02 ohms for 100 Hz to 1000 Hz, and 0.1 ohms $+ 1~\mu\text{H}$ of inductance for 1 kHz to 100 kHz. Since all such power supplies produce DC by rectifying and filtering the input AC power, there is always a little AC left "riding on top" of the DC, since no filter is perfect. For the unit

---

[16] *Ro-Tran Power Control Units*, Brochure SGM 968 10M, Inland Motor Corp., Radford, Va.

[17] Kepco, Inc., Flushing, N. Y., Catalog #146-1255.

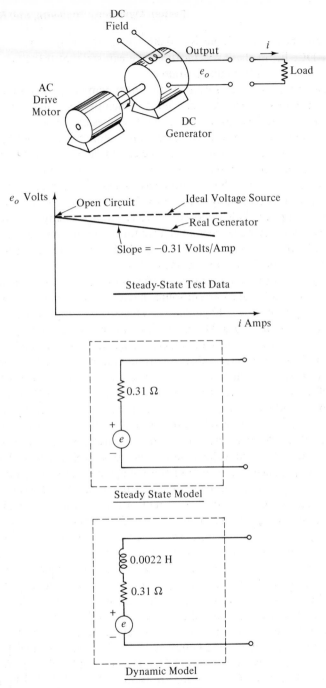

FIGURE 3–18

Modeling of DC Generator Power Source

described above, this ripple is <0.0005 volts which is <0.003% of the 15-volt DC. From these specifications we can see that this device comes quite close to the ideal voltage source.

While voltage sources are by far the most common, some applications require current sources. Voltage sources which use feedback control principles can usually be converted to current sources by connecting a current-sensing resistor in series with the load, and feeding back the voltage across this resistor to the regulator circuit. The circuit now tries to keep the voltage across the sensing resistor constant, which is the same as keeping the load current constant; thus we now have a current source. Specifications for a unit of this type designed to provide 0 to 0.5 amp at 0–40 volts are: current change for full rated (0–40 volt) voltage change <0.005% or 2 μA, whichever is more; current change for AC input voltage change from 105 to 125 volt <0.0005% or 0.2 μA, whichever is greater; 8-hour time drift <0.02% or 2 μA; temperature effect <0.01% per C°; and ripple <0.02%. The output impedance is 800,000 ohms plus 0.1 μF of shunt capacitance. Note that a current source has a very high internal impedance, just the opposite of a voltage source. This 800,000 ohms does *not* correspond to an actual 800,000-ohm resistor in the power supply; rather it represents the ratio of the load voltage change necessary to cause a current deviation from the desired value. A current source is loaded by connecting it to a device (load) which requires a certain voltage to force the set current through it. For example, if we set the current control at 0.5 amp in the unit above and short circuit its output (attach a load of zero resistance), the current drawn will be 0.5 amp and the load voltage will be zero. If we now attach an 80-ohm resistance load, the current should ideally stay at 0.5 amp and the load voltage should go to 40 volts. The 800,000-ohm internal impedance is the *ratio* of the voltage change (0–40 volts) to the current change from the ideal value. In this case, $40/800,000 = 0.0005$ amp, which is 0.01% of the set current of 0.5 amp, and ±0.005% if we take a ±20-volt variation around 20 volts. This 0.005% corresponds to the specification given earlier on current regulation. Figure 3–19 shows how a real current source with a known internal impedance can be modeled as an ideal current source in parallel with this impedance. In Fig. 3–19a the load is a short circuit, so no current goes through the internal resistance, the voltage e_{ab} is zero, and all the current from the source goes into the load. By adding an 80.008-ohm load in 3–19b, we cause e_{ab} to be 40 volts and the source current i now splits between the load and the internal resistance such that $i_{IR} = 0.00005$ and $i_L = 0.49995$. The current deviation from the set value is thus 0.00005, in agreement with our earlier calculation. Note that if the internal resistance R_{IR} were infinite (an open circuit), then *all* the source current goes to the load irrespective of the load resistance and we have a perfect current source. Thus a real current source approaches perfection as its internal impedance approaches infinity.

The practical voltage and current sources mentioned so far have provided either constant (DC) or sinusoidally varying (AC) voltages or currents. In analyzing electrical systems we can, of course, assign any time

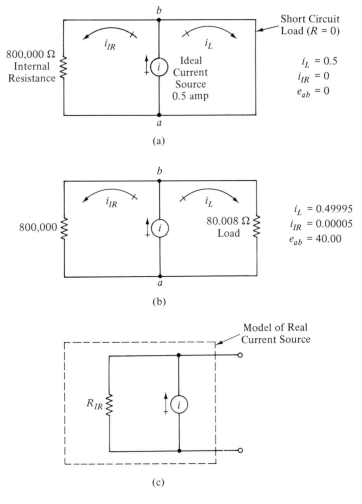

(a)

(b)

(c)

FIGURE 3–19

Modeling of a Real Current Source

variation we wish to our sources, that is, $e(t)$ and $i(t)$ may have any form whatever to suit the physical situation. In experimental work, versatile electronic *signal generators* (usually voltage sources) are available which provide a wide range of useful forms of $e(t)$. Figure 3–20 shows the voltage signals available from a typical [18] unit of this type. The internal resistance of this unit is 50 ohms, so the voltage drops about 0.5 volts for each 0.01 amp of current drawn by the load. At maximum amplitude, and resistance loads of less than 50 ohms, the current drawn is too great (≈ 0.1 amp), the internal circuitry is overloaded, and the waveforms become distorted. By turning down the amplitude dial, the proper waveform is restored, only at a lower level. In this way, loads as small as, say, 0.1 Ω can be driven, but with a voltage amplitude of only a few millivolts. The single-pulse waveforms shown can be triggered by a manual switch or an electrical input signal. The latter capability allows one to produce pulse trains of any desired repetition rate and pulse duration. Signal generators which provide random voltages are also available. These are described in terms of the probability density function of voltage amplitude (probability of the instantaneous voltage being near a certain level) and frequency content (rapidity of variation of the voltage). Because of its mathematical convenience and wide occurrence in natural phenomena, the Gaussian ("normal") probability density function is provided in most random signal generators. Frequency content can be adjusted from 0–0.00015 Hz to 0–50,000 Hz in one commercial unit.[19]

3–7. ELECTRICAL IMPEDANCE AND ELECTROMECHANICAL ANALOGIES.

Electrical impedance is a generalization of the simple voltage/current ratio (called the resistance), and associated with resistors, to include inductors and capacitors and, in fact, entire circuits. Two common forms of impedance, the *operational impedance* and the *sinusoidal impedance*, are in common use. These impedances are defined, respectively, as the operational transfer function

$$Z(D) \triangleq \frac{e}{i}(D) \qquad\qquad (3\text{–}43)$$

and the sinusoidal transfer function

$$Z(i\omega) \triangleq \frac{e}{i}(i\omega) \qquad\qquad (3\text{–}44)$$

---

[18] Wavetek Model 114, Wavetek, San Diego, California.
[19] Hewlett-Packard 3722A, Hewlett-Packard, Palo Alto, Cal.

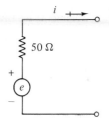

Maximum $e = \pm 16.3$ volts
Maximum $i \approx \pm 0.10$ amp

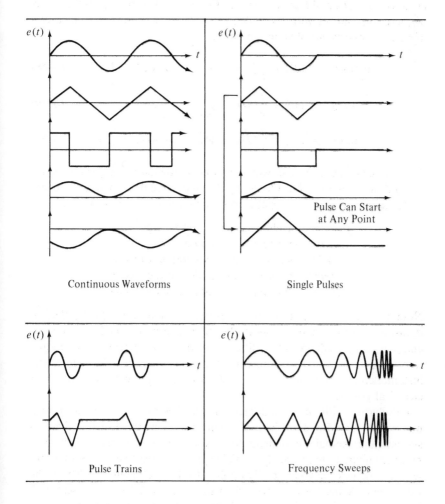

FIGURE 3–20

Electronic Signal Generator Voltage Waveforms

141

For the basic resistance, capacitance, and inductance elements we thus have

$$Z_R(D) = R \qquad Z_R(i\omega) = R \qquad (3\text{-}45)$$

$$Z_C(D) = \frac{1}{CD} \qquad Z_C(i\omega) = \frac{1}{i\omega C} \qquad (3\text{-}46)$$

$$Z_L(D) = LD \qquad Z_L(i\omega) = i\omega L \qquad (3\text{-}47)$$

Since impedances are always voltage/current ratios, their dimensions (whether operational or sinusoidal) are always ohms. The sinusoidal impedances of R, C, and L are graphed in Fig. 3–21a. If we compare the forms of these impedances with those defined in Chapter 2 for mechanical elements, we observe an obvious similarity:

$$\text{damper} \quad Z_B(D) \triangleq \frac{f}{v}(D) = B \qquad Z_B(i\omega) = B \qquad (3\text{-}48)$$

$$\text{spring} \quad Z_s(D) \triangleq \frac{f}{v}(D) = \frac{1}{C_s D} \qquad Z_s(i\omega) = \frac{1}{i\omega C_s} \qquad (3\text{-}49)$$

$$\text{inertia} \quad Z_M(D) \triangleq \frac{f}{v}(D) = MD \qquad Z_M(i\omega) = i\omega M \qquad (3\text{-}50)$$

A signal, element, or system which exhibits mathematical behavior identical to that of another, but physically different, signal, element, or system, is called an *analog*. Based on the above equations we may state that:

> *force* is a mechanical analog of voltage
> *velocity* is a mechanical analog of current
> a *dashpot* is a mechanical analog of a resistor
> a *spring* is a mechanical analog of a capacitor
> a *mass* is a mechanical analog of an inductor

Note also that a dashpot dissipates all the mechanical energy supplied to it, just as a resistor dissipates all the electrical energy supplied to it. Springs and masses *store* energy in two different ways, just as capacitors and inductors store energy in two different ways. The product $(f)(v)$ represents instantaneous mechanical power just as $(e)(i)$ represents instantaneous electrical power.

Familiarity with these analogies may be helpful to an engineer in a number of ways. For a mechanical engineer working on electromechanical systems, the analogies may make the electrical aspects of the system seem more familiar. Often, already-available mathematical solutions to problems in one field may be directly applied to the analogous problem in the other field, saving the time needed to solve the problem "from scratch." Operating principles of successful hardware in one area may be carried

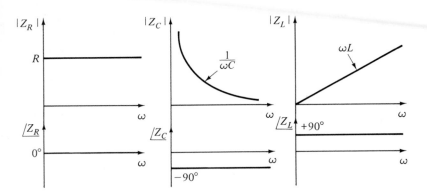

Impedances of Electrical Elements

(a)

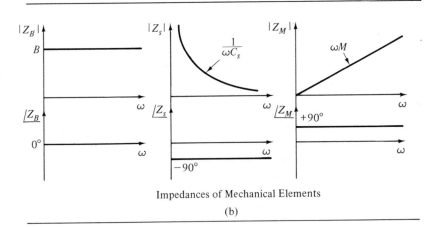

Impedances of Mechanical Elements

(b)

FIGURE 3–21

Analogous Behavior of Electrical and Mechanical Elements

over to "invent" an analogous or similar useful device in another area. It is often possible to reduce an electromechanical system to an all-electric or all-mechanical model using analogies, and electrical engineers even sometimes prefer to study systems which are *entirely* mechanical using electrical models. This is somewhat a matter of personal preference and may in some cases be the best way to proceed; however, it is not recommended as a general rule. It is felt better to model the system *directly*, since many of the parasitic and nonlinear effects in real systems may be overlooked or incorrectly treated if one insists on converting everything to, say, all-electrical or all-mechanical elements. While a system engineer may be academically trained in a particular discipline and feel most at home there, he should be willing to develop, when the need arises, sufficient understanding of "foreign" topics to treat them adequately in his system studies. At one time, use of electrical models based on the above or similar analogies could be justified since these R, L, C models were actually built and used as computers to study, say, mechanical vibrating systems. Modern analog computers do not, however, use this approach; they simulate the *mathematical* operations and work *directly* from the system equations —thus no electrical models are needed. Furthermore, much system analysis is now done by digital simulation, which again works directly from the equations.

Returning to the topic of electrical impedance we should point out that it can be and is used, not just for elements but for complete systems of arbitrary size. In Fig. 3–22a, for example, one can derive a formula for the impedance (e/i) (D) by proper application of physical laws. For devices which are too complex to analyze accurately by theory, experimental measurements can be made to determine the impedance. This is usually done by the frequency response method, where a sinusoidal e is applied at various frequencies and the amplitude ratio A_e/A_i and phase angle $\phi_{e/i}$ are measured and plotted. Such measurements are sufficiently important that instruments specially designed for this purpose are available.[20] The two instruments discussed in the reference together cover the frequency range of 5 Hz to 108 MHz and provide meter readings which directly give the magnitude in ohms and phase angle in degrees at each selected frequency. Figure 3–22b shows measurements made with one of these instruments on two inductors of similar nominal value, but of different physical construction. Note that at high frequencies the inductors exhibit parasitic effects which are clearly different for the two units. Such information is very valuable in the selection and application of components, since it completely documents their dynamic behavior and shows over what frequency ranges they behave essentially like the pure element intended.

While impedance is useful in characterizing the dynamic behavior of components or systems, it also finds application in the solution of AC circuit problems. At any given frequency the sinusoidal impedance of any circuit can be given as a real part R and an imaginary part X:

$$Z = R + iX \tag{3–51}$$

where

---

[20] "Methods of Measuring Impedance," *Hewlett-Packard Jour.*, Vol. 18, No. 5 (Jan., 1967).

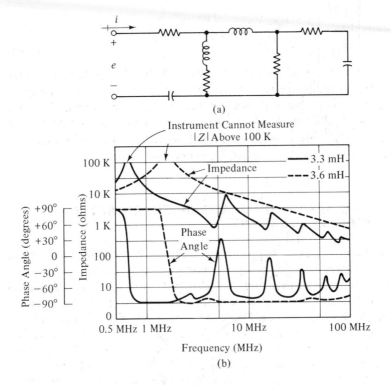

(a)

(b)

FIGURE 3–22

Impedance of Complex Circuits

$$R \triangleq \text{resistive impedance, ohms} \qquad \textbf{(3–52)}$$

$$X \triangleq \text{reactive impedance, ohms} \qquad \textbf{(3–53)}$$

If X is a positive number, the reactive impedance is behaving like an inductor and is called *inductive reactance;* if negative, it is called *capacitive reactance*. Given R and X, one can always compute the magnitude $M = \sqrt{R^2 + X^2}$ and phase angle $\phi = \tan^{-1}(X/R)$ of the impedance. Since sinusoidal impedance gives the amplitude ratio and phase angle of voltage with respect to current, if the impedance of any circuit is known and either the voltage or the current is given, the other can be calculated. For example, a 0.1-μF capacitor at 60 Hz has an impedance $1/(3.77)(10^{-5})i = 26,600$ $\Omega\ \underline{/-90°}$. If we apply a zero phase angle 110-volt 60-Hz voltage to this capacitor, what current will flow?

$$Z = \frac{\text{voltage}}{\text{current}} = 26,600\ \underline{/-90°} \qquad \textbf{(3–54)}$$

$$\text{current} = \frac{\text{voltage}}{26,600 \; \underline{/-90°}} = \frac{110. \; \underline{/0°}}{26,600 \; \underline{/-90°}} = 0.00413 \; \underline{/+90°} \text{ amp}$$

$$(3\text{-}55)$$

This current could be written as $i = 0.00413 \sin(377t + 90°)$ amp. If we apply this same voltage to a "black box" with an unknown assemblage of R, L, and C inside, but with two terminals at which a 60-Hz impedance of, say, $52.4 \; \underline{/+62°}$ ohms has been measured, the current would be $2.10 \sin (377t \; \underline{/-62°})$. These examples should illustrate the convenience which the impedance concept brings to AC circuit analysis.

3-8. THE OPERATIONAL AMPLIFIER, AN ACTIVE CIRCUIT "ELEMENT."

As mentioned earlier, the operational amplifier is not strictly an element in the accepted sense (because it *contains* elements such as resistors and transistors), but we include it in this chapter because of its great usefulness and the fact that, in size and cost, some models of op-amps approach a single capacitor or resistor. While the wide range of sizes, cost, and performance of op-amps prevents one from showing a "typical" unit, Fig. 3–23 will give some idea of the size and internal circuitry of a recent integrated circuit model [21] selling for about $4.50 in quantities of 100 or more. Depending on performance, op-amps are available in price ranges from about $1 to $200. Although the circuitry of Fig. 3–23 looks quite complex, the *user* of op-amps can fortunately neglect this and get by with a rather simple model for most applications. The description and use of this model, which can be found in various texts and manufacturers' literature,[22] will now be given.

In Fig. 3–24a the input voltages e_{i1} and e_{i2} are applied to the amplifier input terminals and produce an amplified signal $A(e_{i2} - e_{i1})$ where A is the gain (amplification) of the amplifier. The configuration shown is called a *differential-input* amplifier; if e_{i2} is connected to ground (allowing input only through e_{i1}), it is called *single-ended*. The impedances Z_i and Z_o are respectively the input and output impedances. To get the simplest model, but one which is still practically useful in many cases, we make the following assumptions:

1. The amplifier gain A is infinite.
2. Z_i is infinite, thus no current is drawn at the input terminals.

---

[21] Model AD502, Analog Devices, Cambridge, Massachusetts.
[22] *Handbook and Catalog of Operational Amplifiers*, Burr-Brown Research Corp., Tucson, Arizona.

Schematic Diagram

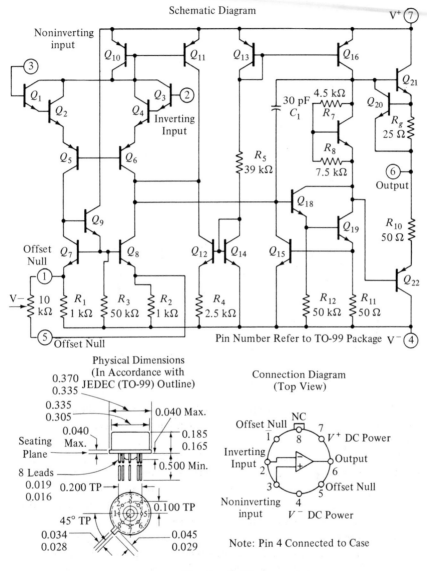

FIGURE 3–23

Integrated Circuit Operational Amplifier

3. Z_o is zero, thus $e_o = A(e_{i2} - e_{i1})$.
4. The time response is instantaneous.
5. The output voltage has a definite design range, usually about ± 5 to ± 100 volts.

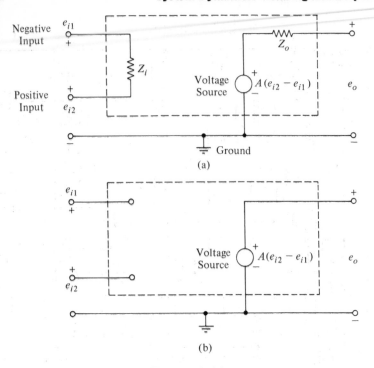

FIGURE 3–24

Models of Op-Amp Input and Output

Figure 3–24b shows the simplified model corresponding to these assumptions. (We will later show the effects of a real amplifier deviating from this ideal behavior.)

We will now demonstrate the use of this model by analyzing the three basic linear analog computing devices: the *coefficient multiplier*, the *summer*, and the *integrator*. Each of these is formed by a particular combination of resistors, capacitors, and the op-amp. Figure 3–25a shows the coefficient multiplier, whose function is to take a computer voltage e_1 and produce an output voltage e_o which is Ke_1, where K is an adjustable constant. The positive input e_{i2} has been connected to ground and two resistors, R_i (the input resistor) and R_{fb} (the feedback resistor), connected to the op-amp. We wish to find the relation between e_1 and e_o. At the point called the *summing junction*, since the current into the amplifier is zero and current cannot accumulate at a point, we see that $i_{R_i} = i_{R_{fb}}$. Since the voltage across R_i is $e_1 - e_{i1}$ and the voltage across R_{fb} is $e_{i1} - e_o$ we have

$$i_{R_i} = \frac{e_1 - e_{i1}}{R_i} = i_{R_{fb}} = \frac{e_{i1} - e_o}{R_{fb}} \qquad (3\text{–}56)$$

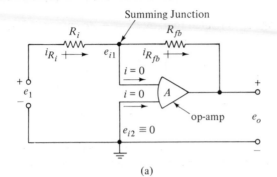

(a)

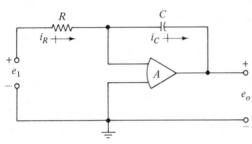

(b)

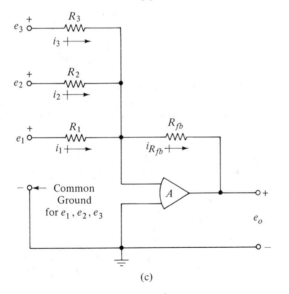

(c)

FIGURE 3–25

Coefficient Multiplier, Integrator, and Summer

Now the voltage e_{i1} is related to e_o by $e_o = -Ae_{i1}$ where A is the amplifier gain; thus $e_{i1} = -e_o/A$. Then Eq. (3–56) becomes

$$\frac{e_1 + (e_o/A)}{R_i} = \frac{(-e_o/A) - e_o}{R_{fb}} \tag{3–57}$$

and if $A = \infty$,

$$e_o = -\frac{R_{fb}}{R_i}e_1 \tag{3–58}$$

Of course, A cannot be infinite, but it can be, say, 10^6 and then the terms neglected in Eq. (3–57) become one millionth of the terms retained, which, while not perfect, is a very good approximation. Since the voltage e_{i1} is always nearly zero in op-amp applications, the summing junction is sometimes called a "virtual ground." We will assume its potential to be exactly zero in subsequent analyses. From Eq. (3–58) we see that the proportionality factor between e_o and e_1 is $-R_{fb}/R_i$; thus if we make $R_{fb} = 10,000$ Ω and $R_i = 1000$, the output voltage will be 10 times the input, but of opposite polarity. Thus, in an analog computer, if we need to multiply a voltage by an adjustable constant, this arrangement would be used.

Using the assumptions just employed above, Fig. 3–25b may be quickly analyzed by writing

$$i_R = \frac{e_1}{R} = i_c = -C\frac{de_o}{dt} = -CDe_o \tag{3–59}$$

$$e_o = -\frac{1}{RCD}e_1 = -\frac{1}{RC}\int e_1\,dt \tag{3–60}$$

We see that this configuration is the integrator which we have employed in symbolic form (without understanding its inner workings) to show analog computer diagrams for dampers, masses, capacitors, and inductors. If we take, say, $R = 10^6\,\Omega$ and $C = 1\,\mu F$, we have $e_o = -\int e_1\,dt$, and if e_1 is a step input of $+2$ volt, e_o will be $-2t$ volts, where t is in seconds. In Fig. 3–25c we show a device for adding voltages, the summer. We have

$$i_1 + i_2 + i_3 = i_{R_{fb}} \tag{3–61}$$

$$\frac{e_1}{R_1} + \frac{e_2}{R_2} + \frac{e_3}{R_3} = -\frac{e_o}{R_{fb}} \tag{3–62}$$

$$e_o = -\left(\frac{R_{fb}}{R_1}e_1 + \frac{R_{fb}}{R_2}e_2 + \frac{R_{fb}}{R_3}e_3\right) \tag{3–63}$$

If we take $R_{fb} = R_1 = R_2 = R_3$,

$$e_o = -(e_1 + e_2 + e_3) \tag{3–64}$$

showing e_o to be the negative sum of the input voltages. If one has enough coefficient multipliers, integrators, and summers, we shall show in a later chapter that he can connect together an analog computer circuit which will "solve" any linear differential equation with constant coefficients. Furthermore, addition of only two basic nonlinear components, the variable multiplier and the arbitrary function generator, allows us to tackle a wide variety of nonlinear differential equations, most of which cannot be solved analytically.

A useful general result for op-amp circuits may be derived from Fig. 3-26a. There Z_i and Z_{fb} represent arbitrary impedances—that is, any combination of R, C, and L exhibiting two terminals, such as Fig. 3–22a. From the definition of impedance,

$$i_i = \frac{e_1}{Z_i(D)} = i_{fb} = -\frac{e_o}{Z_{fb}(D)} \tag{3–65}$$

$$\frac{e_o}{e_1}(D) = -\frac{Z_{fb}(D)}{Z_i(D)} \tag{3–66}$$

If the impedances are as in Fig. 3–26b, for example,

$$\frac{e_o}{e_1}(D) = \frac{-RCD}{(RCD + 1)(RCD + 1)} \tag{3–67}$$

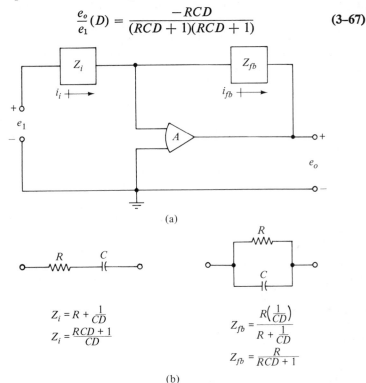

(a)

$$Z_i = R + \frac{1}{CD}$$
$$Z_i = \frac{RCD + 1}{CD}$$

$$Z_{fb} = \frac{R\left(\frac{1}{CD}\right)}{R + \frac{1}{CD}}$$

$$Z_{fb} = \frac{R}{RCD + 1}$$

(b)

FIGURE 3–26

Op-Amp with General Impedances

This transfer function represents a band-pass filter, a useful device in many practical systems. Its frequency response $(e_o/e_1)(i\omega)$ reveals that it allows a band of frequencies to pass through it from e_1 to e_o, but rejects (filters out) any frequencies above or below this band.

We conclude this section with a brief look at the behavior of *real* op-amps as compared to the idealized model used up to this point. Consider first the effect of non-infinite gain A on the circuit of Fig. 3–25a. Equation 3–57 may be manipulated to give

$$e_o = -\frac{R_{fb}}{R_i} e_1 \left[1 + \underbrace{\frac{1}{A} + \frac{R_{fb}}{AR_i}}_{\text{error}} \right] \qquad (3\text{–}68)$$

The gain A (called open-loop gain) may be in the range 10^4 to 10^8, while R_{fb}/R_i rarely exceeds 10^3; thus, the error upper limit is from about 10^{-5} (0.001 %) to 0.1 (10%). The meaning of this error is that if one selects precision resistors for R_i and R_{fb} so as to get a precise e_o/e_1 ratio, and if the gain A is too low, the ratio will be inaccurate. Of course, if A is known and *fixed*, we could select R_i and R_{fb} to compensate for the error due to low A. However, A may drift in a random fashion due to temperature, age, etc., reducing the effectiveness of the compensation. The next errors we consider are those due to *voltage offset* and *bias current*. Voltage offset refers to the fact that if e_1 in Fig. 3–25a is made zero by grounding it, e_o will *not* be exactly zero, due to imperfections in the amplifier. This offset can be trimmed out at a given instant, but temperature drift will cause it to reappear. Figure 3–27a shows a model for computing the error due to this effect. Analysis gives

$$e_o = e_{os} \left(1 + \frac{R_{fb}}{R_i} \right) \qquad (3\text{–}69)$$

Since e_o should be zero, this is the error voltage. The best values of offset voltage e_{os} are the order of 30 μV over a temperature range of -25 to $+85°$C, with a temperature coefficient of 0.2 μV/C°. Cheap integrated circuit models may have 10 mV and 15 μV/C°. Taking an e_o full scale voltage of 10 volts, for $R_{fb}/R_i = 1000$, the maximum error as a percentage of full scale could range from 0.3 % to 100% if no attempt to trim the error were made. *Bias current* is the small current that flows in the amplifier input leads, even when no input voltage is applied. The model in Fig. 3–27b leads to the result

$$e_0 = -i_{b1} R_{fb} \qquad (3\text{–}70)$$

Values of i_{b1} range from 100 pA (1 pA $= 10^{-12}$ A) over the range -25 to $+85°$C, with temperature coefficient 1.0 pA/C° to 1.0 mA and 0.5 nA/C°.

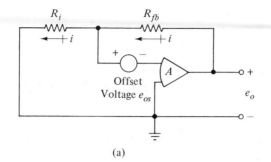

(a)

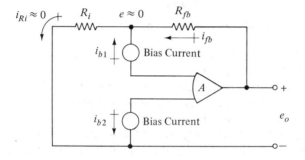

(b)

FIGURE 3–27

Error Models for Offset Voltage and Bias Current

The next deviation of real amplifiers considered here is the non-infinite input impedance and non-zero output impedance. Figure 3–28 shows a model for these effects; note that a load resistance R_L is also shown to represent the input resistance of any device which would be connected to the op-amp output. Analysis shows that the effect of non-infinite input resistance is equivalent to a loss of open-loop gain A, the effective value being given by

$$A_{eff} = \frac{A}{1 + \dfrac{R_i R_{fb}}{R_1(R_i R_{fb})}} \qquad (3\text{–}71)$$

A similar effect is produced by non-zero output impedance,

$$A_{eff} = \frac{A}{1 + \dfrac{R_2}{R_{fb}} + \dfrac{R_2}{R_L}} \qquad (3\text{–}72)$$

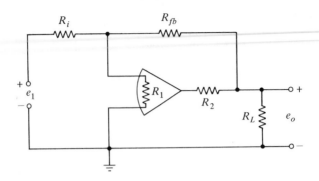

FIGURE 3–28

Error Model for Input and Output Impedance

Input resistances vary from 10^{11} Ω to 3×10^5 Ω, while output resistances are in the range 10 to 30,000 Ω.

The speed of response of op-amps is usually quoted as the range of frequencies for which the rated output and current may be attained without distortion of the shape of the sine wave. This is called the *full power response bandwidth* and ranges from 5 kHz to 20 MHz. Sometimes the *settling time* after a step input is also quoted. High-speed op-amps will settle to within 0.1% of final value in 1 or 2 μsec. The *slew rate* is the maximum rate of change of output voltage achievable and ranges from 0.1 to 500 volts/μsec.

BIBLIOGRAPHY

1. Cowan, J. D. and H. S. Kirschbaum, *Introduction to Circuit Analysis.* Columbus: Charles E. Merrill Publishing Co., 1961.

2. D'Azzo, J. J. and C. J. Houpis, *Principles of Electrical Engineering.* Columbus: Charles E. Merrill Publishing Co., 1968.

3. Fitzgerald, A. E., A. Grabel, and D. E. Higginbotham, *Basic Electrical Engineering.* New York: McGraw-Hill Book Company, 1967.

4. Smith, R. J., *Circuits, Devices and Systems.* New York: John Wiley & Sons, Inc., 1966.

PROBLEMS

3–1. Calculate the wavelengths of a television signal of frequency 80 MHz and a radar signal of 2000 MHz. Estimate the size at which a field-type treatment of components would become advisable.

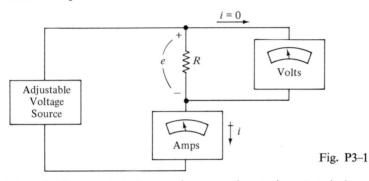

Fig. P3–1

3–2. Figure P3–1 shows a test setup for measuring static current/voltage relationships such as that of Fig. 3–2c. Plot this data and check to see how well it conforms to $i = Ke^4$. Also find the linearized incremental resistance for an operating point of 10 volts. Plot the power dissipated versus both i and e.

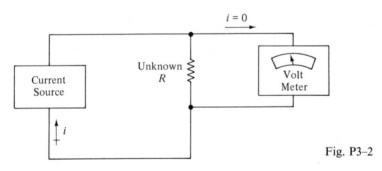

Fig. P3–2

3–3. In the ohmmeter of Fig. P3–2, the current is fixed at 0.0001 amp. If we wish to measure resistances from 1 to 10,000 ohms, what full-scale voltage ranges are necessary in the voltmeter? What is the maximum power dissipation in the resistors being measured? A more rugged voltmeter (higher voltage range) would be used if the current were increased. Is there any disadvantage to this? When an ohmmeter is used to measure a nonlinear resistance, what does the reading mean?

3–4. A particular $\frac{1}{4}$-watt resistor can dissipate heat at the rate 1.75×10^{-6} Btu/sec for each degree Fahrenheit of temperature rise. When the temperature rises, the resistance changes by 0.05% for each degree Fahrenheit. How much does the resistance change for a load change from 0 to $\frac{1}{4}$ watt?

3–5. What is the maximum steady voltage which may be applied to a 1000-ohm, $\frac{1}{2}$-watt resistor? If the voltage is a pulse waveform, as in Fig. P3–3, what is the maximum allowable? (*Hint:* Use the *average* power as the criterion.)

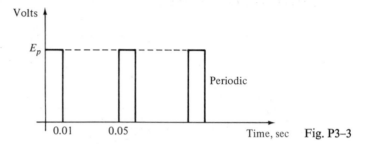

Fig. P3–3

3–6. A 1000-ohm resistor has a voltage $e = -3 + 4t + 5e^{-t} + 6 \sin 10t$ across it, e in volts, t in seconds. Using CSMP or similar computer methods, find and plot versus t both the instantaneous power dissipation and also the total energy dissipated for t from 0 to 1 second.

3–7. In x-y plotters, time-sweeps are generated by applying a ramp voltage to the servo drive system of one of the axes. The ramp voltages are made by applying a constant current to a capacitor. If the servo system sensitivity is 0.10 inches/volt and a current of 0.00001 amp is used, what capacitance is needed to give a sweep of 1 inch/sec?

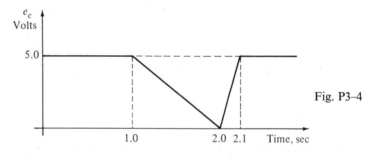

Fig. P3–4

3–8. Find and sketch the current needed to produce the voltage of Fig. P3–4 across a 1-μF capacitor. Also calculate and sketch the stored energy.

3–9. Find the current through a 1-μF capacitor if the voltage is $(1 \sin t + 1 \sin 10t + 1 \sin 1000t)$ volts, t in seconds.

3–10. Find and sketch the voltage across a 1-μF capacitor if the current is as shown in Fig. P3–5.

3–11. In Fig. P3–6 the mutual inductance between coils 1 and 3 and between coils 1 and 2 is negligible, while that between 2 and 3 is 0.01 henry. Find e_1 if $i_1 = -3 + 1000t$ amps, and $i_2 = 10 - 3000t$ amps, t in seconds.

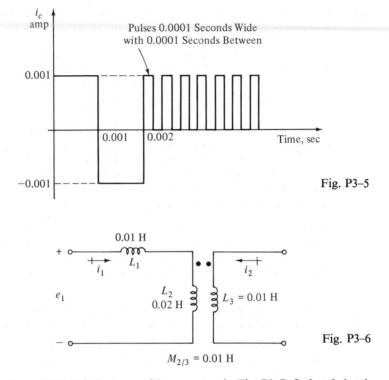

Pulses 0.0001 Seconds Wide
with 0.0001 Seconds Between

Fig. P3-5

Fig. P3-6

$M_{2/3} = 0.01$ H

3-12. For a 0.001-H inductance with current as in Fig. P3-7, find and sketch the voltage.

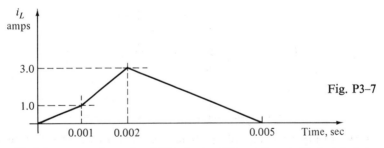

Fig. P3-7

3-13. Find the current through a 1-H inductor if the voltage is $(1 \sin t + 1 \sin 10t + 1 \sin 1000t)$ volts, t in seconds.

3-14. It is desired to use a 50,000-ohm film resistor of the type in Fig. 3-9. What is the highest frequency at which it behaves like a pure resistor?

3-15. What would be an appropriate circuit model and range of numerical values for a 500,000-ohm resistor of the type shown in Fig. 3-10b? How would the model change if the resistance were 100 ohms?

3–16. Estimate the capacitance of 10 feet of coaxial cable with $a = 0.01$ inch, $b = 0.03$ inch, and insulation with dielectric coefficient K of 5.

3–17. For a capacitor constructed as in Fig. 3–12a, with electrodes 0.00025-inch thick, and Mylar dielectric 0.0005-inch thick, estimate the total area needed for a 1-μF unit. If it is 2 inches long, estimate the diameter.

3–18. In Fig. 3–13, what are the values of R_s, R_p, and L in a pure capacitor?

3–19. For a solenoid as in Fig. 3–15, with $a = 1$ inch and $b = 4$ inches, find the number of turns needed to get $L = 0.25$ H. If the wire used has a diameter of 0.01 inch and a resistance of 100 ohms/1000 ft, what will the resistance of this inductor be? Using Eq. (3–42), compute Q for $\omega = 100,000$ rad/sec. If, instead of an air core, a magnetic core with relative permeability of 1000 is used, what would L be?

3–20. For the inductor of Fig. 3–16c ($L = 60$ H, $R = 5160$ ohms) use Eq. (3–42) to estimate Q and compare with the experimental result. Comment on the discrepancy at high frequency. Could R_e of Fig. 3–16b be estimated from the data given and the formula for Q given in the text? Explain how.

3–21. What voltage is produced by an open-circuited current source? What current is produced by a short-circuited voltage source? Discuss.

3–22. How would you set up and run an experiment to determine a model for the electrical source represented by the 110-volt, 60-Hz wall plug in your laboratory?

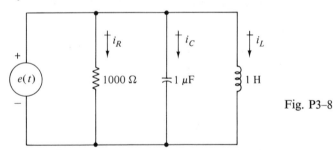

Fig. P3–8

3–23. In Fig. P3–8, the voltage source provides $10 \sin \omega t$ volts (t in seconds). Calculate and sketch the current amplitude versus frequency for R, L, and C separately and also the total current.

3–24. A complex circuit's impedance was measured at 60 Hz and found to be $2827 \underline{/-38°}$ ohms. If we apply a voltage $53.4 \sin 377t$, what will be the current? If the voltage were $53.4 \sin (377t + 10°)$ volts, what would the current be?

3–25. If R_{fb} is 10^6 ohms, design an op-amp summing circuit to implement $e_o = -4.2e_1 - 6.7e_2 - 10e_3$.

3-26. An op-amp integrator for a fast-time-scale analog computer requires
$e_o = -2500 \int e_1 \, dt$. If R is 10^6 ohms, find C.

3-27. A major problem in op-amp integrators is drift, wherein the output voltage gradually (or sometimes rapidly) drifts away from zero, even if the input voltage is exactly zero. This drift is due to the amplifier's offset voltage e_{os} and bias current i_b. Analysis gives the result $|de_o/dt| = |e_{os}|/RC + |i_b|/C$. Find the drift rate in volts/sec for an integrator with $RC = 1$ sec, $C = 1$ μF, $e_{os} = 60$ μV, and $i_b = 110$ pA. If the full-scale output voltage is 10 volts, how long may this integrator be run before the drift becomes 1% of full-scale?

3-28. Using the scheme of Fig. 3-26a, design an op-amp circuit to implement $(e_o/e_1)(D) = -D/(D + 1)$.

3-29. In Eq. (3-68), compare the error caused by a 10% drift in A if A were initially 10^4 with the error caused by a 10% drift in A if it is initially 10^8.

4

SYSTEM ELEMENTS, FLUID AND THERMAL

4-1. INTRODUCTION.

While the linear, lumped-parameter approach of system dynamics has been successfully applied to the design and analysis of many different physical types of systems, it is clear that certain areas lend themselves to such a treatment more than others. Many electrical systems are originally conceived by a designer thinking in terms of putting together a combination of R, L, C, op-amps, etc., to achieve a new device with some useful function. A worker in these areas finds a wide selection of such components readily available to implement his circuit concepts. Mechanical systems, on the other hand, are rarely initially conceived in terms of some sort of connection of B, K_s, and M. Rather, the designer draws on his knowledge of basic mechanisms (cams, gears, linkages, etc.), power sources (hydraulics, pneumatics, electric motors, etc.), sensing instruments, and control schemes to create a system which will, at least nominally, perform the desired functions. To check the details of performance of this proposed system, a dynamic model must often be formulated and analyzed, and at this point lumped-parameter system analysis may be very useful. Fluid and thermal systems follow a somewhat similar pattern in that system dynamics aspects may receive relatively light conscious emphasis during the early conceptual phases. Furthermore, due to the generally less well-

160

defined shapes of bodies of fluid (as compared to solid bodies) and the
fact that heat flow rarely is confined to such simple and obvious paths as
is current in a circuit, these types of systems are often less easily and accurately modeled with linear, lumped elements. Our somewhat abbreviated
study of fluid and thermal elements (relative to the more comprehensive
treatment accorded mechanical and electrical) should thus be viewed more
as a recognition of the limitations of these analysis methods than as an
indication of the relative practical importance of the respective fields, since
it is clear that fluid and thermal systems are of great practical importance.

Since fluid and thermal systems are generally less familiar than mechanical and electrical (mass/spring oscillators and simple electric networks are
usually introduced in beginning physics courses), it may be helpful to show
a few practical examples. Figure 4–1 shows the propellant feed system for

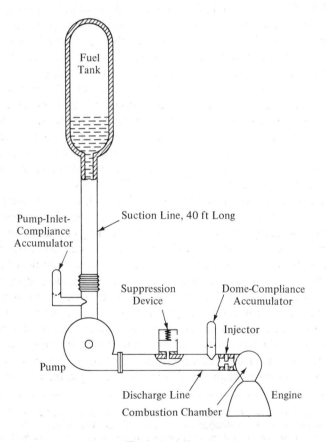

FIGURE 4–1

Rocket Fuel Feed System

a liquid-fueled rocket engine.[1] Such rocket engines may exhibit undesired unstable oscillatory behavior of several types; the one called *chugging* refers to a relatively low frequency (100 to 500 Hz) combustion-chamber pressure oscillation. A dynamic model of this system is needed to allow analytical prediction of those sets of system parameters which result in stable operation and those which will be unstable. Also, it is desired to investigate the feasibility of adding oscillation-suppression devices to unstable systems to stabilize them. The model used in this study had to take into account the inertia, springiness and fluid friction of the fluid in the suction and discharge pipes, the springiness (fluid compliance) at the turbopump inlet due to cavitation and at the combustion chamber injector dome, the combustion chamber pressure/flow dynamics, and the pump pressure/flow relation.

In Fig. 4–2 we see a fluid flow model of the cooling water system of the Scattergood steam power plant of the city of Los Angeles.[2] Sea water is drawn into the system from the ocean, then proceeds through various valves and chambers to the steam condensers where it cools and condenses the steam and is itself thereby heated, and finally returns to the ocean which is used as a heat sink for the power plant. The model shown is intended for analysis of the dynamic response of fluid flows and pressures only; the thermal aspects of the processes are essentially "decoupled" from the fluid mechanics in such a system. A thermal analysis (which would require its own model) might also be quite important in such a system, not only to ensure proper operation of the steam plant, but also to ensure that "thermal pollution" of the ocean is not excessive.

Large aircraft, such as the DC–10[3] of Fig. 4–3, use hydraulic power for many functions such as control surface positioning, wheel steering and braking (including anti-skid systems), and landing gear actuation. A modern jet transport may have up to 2000 hp of hydraulics on board. Since system reliability is so important to passenger safety, the use of redundant or backup systems, as in Fig. 4–3, is common. If engine and/or hydraulic failures occur in one of the systems, a backup system takes over to allow continued safe flight. Hydraulic power is generated by engine-driven pumps and distributed to motors, cylinders, and valves by metal tubing. The motors and cylinders convert the fluid power back into mechanical power used to move ailerons, rudder, etc. Some of these hydraulic systems are part of the autopilot system, which is a feedback system, and thus dynamic

---

[1] D. J. Wood and R. G. Dorsch, *Effect of Propellant-Feed-System Coupling and Hydraulic Parameters on Analysis of Chugging*, NASA TN D-3896, 1967.

[2] A. Reisman, *On a Systematic Approach to the Analysis and Synthesis of Complex Systems Involving the Unsteady Flow of Fluids*, ASME Paper 64-WA/FE-36, 1964.

[3] *The DC-10 and Its Hydraulic System*, Vickers Aerospace Fluid Power Conference, 1968.

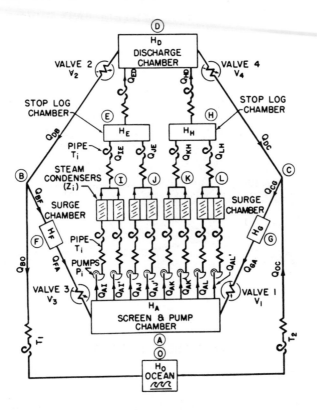

FIGURE 4-2

Power Plant Cooling Water System

behavior is very important to stability and control of the entire aircraft. Modeling of such systems requires mechanical elements, fluid elements, and fluid/mechanical transducers (energy-converters) such as pumps and motors.

The human body utilizes many flow processes in its operation. Perhaps the most obvious of these are the respiratory system, involving the flow of air, and the circulatory system, providing a vital flow of blood to all parts of the organism. Engineers are becoming more involved in the modeling of such systems, both because they have the fluid mechanics expertise which many medical people lack, and also because they are designing machines to aid doctors in their work, including artificial organs such as heart valves and even complete hearts. Figure 4-4[4] shows a rather compre-

---

[4] M. F. Snyder, V. C. Rideout, and R. J. Hillestad, "Computer Modeling of the Human Systemic Arterial Tree," *J. of Biomechanics*, Vol. 1, 1968, pp. 341–353.

HYDRAULIC POWER SYSTEM AND PIPING

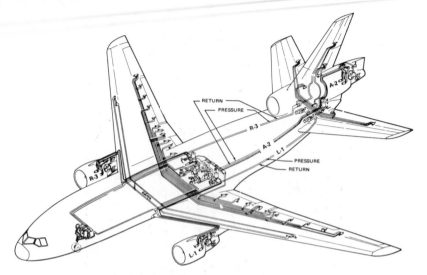

HYDRAULIC SYSTEM

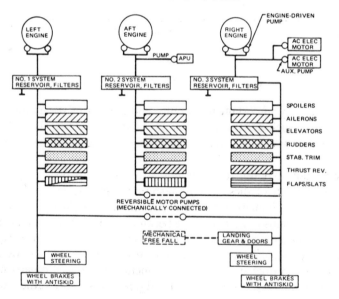

FIGURE 4–3

Aircraft Hydraulic System

hensive model of the human systemic arterial system based on lumped-parameter modeling methods. The diagram of Fig. 4–4b looks like an electric circuit, but only because the symbols for electrical resistance, capacitance, and inductance were used to represent fluid resistance, compliance, and inertance. Actually, the fluid mechanics equations were directly simulated on an electronic analog computer; an intermediate electrical model was not employed. The ultimate test of such a model is to see whether it can duplicate behavior measured in the real system. If it does this faithfully, it can be a very useful tool, since "experiments" can be run on the model much more easily, quickly, and safely than on a human being. Figure 4–4c shows some pressure-time records indicating this model to be quite realistic.

Polymerization processes in the chemical industry require careful temperature control, and are often carried out in jacketed kettles[5] as in Fig. 4–5. The kettle is loaded with monomer, catalyst, and water, and is initially heated to start the reaction. A constant desired temperature is maintained by proportioning the flow rates of steam and cold water to the kettle jacket. This temperature control becomes particularly difficult when the reaction becomes exothermic (producing heat rather than requiring heat addition), because this situation is inherently unstable; more heat produces a higher temperature, which causes a faster reaction, which produces more heat, etc. The control system stabilizes this process by providing just the right amount of cooling to maintain an essentially constant desired temperature. A careful dynamic analysis of this thermal system is necessary to achieve proper operation. The control loop on jacket temperature T_J can usually be made "tight" enough so that T_J follows its command very rapidly, and thus, these dynamics are neglected in analyzing the rather slow overall system. In *designing* the T_J loop for this behavior, however, one would have to consider its thermal dynamics. These involve heat flows due to the mass flows of water and steam, heat transfer through the walls of the jacket, and heat storage in the jacket fluid. Similarly, for the process vessel itself, there is heat transfer from the jacket, the endothermic or exothermic reaction heat, frictional heat due to the stirring device, and heat storage in the process fluid. Additional important thermal dynamics are found in the temperature measuring device and its protective well. Heat flows from the process fluid through the thermal resistance of the protective well, and is stored in the metal wall when its temperature rises. This heat flow continues through the oil film separating well and thermometer bulb, and into the thermometer bulb wall, where part of it is stored again when the bulb wall temperature rises. Since the thermometer

---

[5] G. L. Rock and Lee White, "Dynamic Analysis of Jacketed Kettles," *ISA Journal* (March–April, 1961).

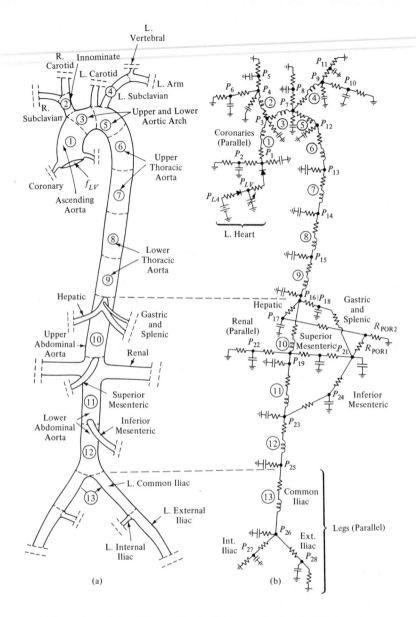

(a) Human Systemic Arterial System Showing Division into Segments for Modeling

(b) Lumped Fluid-Circuit Representation

FIGURE 4–4

Human Arterial Flow System

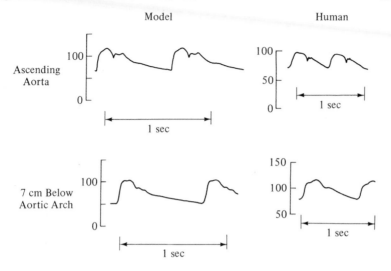

(c) Pressures in mmHg versus time

FIGURE 4–4 (cont.)

is one using the expansion of the gas in the bulb as its operating principle, heat must finally flow into this volume of gas, raise its temperature, and cause expansion.

Our final thermal system example is from the field of satellite temperature control. Equipment carried aboard satellites must often be kept within fairly narrow temperature ranges to ensure adequate performance and long life. Both "passive" and "active" control schemes are in use. In a passive temperature control scheme, no separate heating or cooling components are used; the vehicle's thermal properties are cleverly adjusted so that temperature stays within the desired range "naturally" for the anticipated thermal environment. When passive methods cannot meet requirements, active systems using feedback controlled heating and/or cooling may be employed. To design either active or passive systems, a thermal model of the vehicle is needed. Figure 4–6 shows a thermal model of a cylindrical spacecraft with four solar-cell paddles, such as the Advanced Orbiting Solar Observatory (AOSO).[6] The cylindrical body is divided into four sections, each of which is assumed at a uniform but time-varying temperature. Each section receives and transmits various heat fluxes from other sections, the sun, the earth, internal heat generated by instruments, etc. The difference between incoming and outgoing heat flux must be stored in the section, causing its temperature to rise and fall. Since many of the

---

[6] F. J. Cepollina, *Use of Analog Computation in Predicting Dynamic Temperature Excursions of Orbiting Spacecraft*, NASA TM-X-55432, 1966.

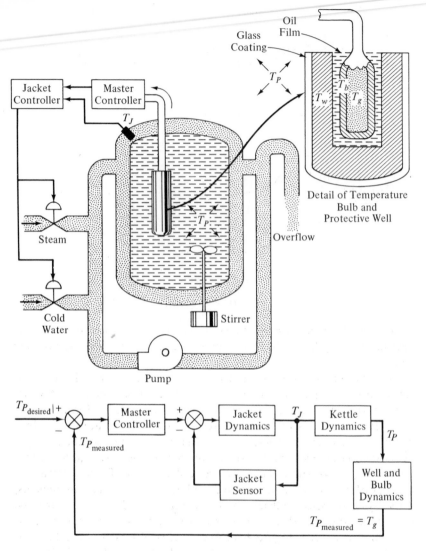

FIGURE 4–5

Chemical Process Temperature Control System

heat fluxes are by means of radiation (which depends on the fourth power of temperature) the problem has significant nonlinearity, which requires use of computer methods. The analog computer circuit for this model is shown in Fig. 4–7. Figures 4–8a and 4–8b show how adjustment of thermal properties can significantly reduce the excursions of the average temperature T_{AVE} during an orbit.

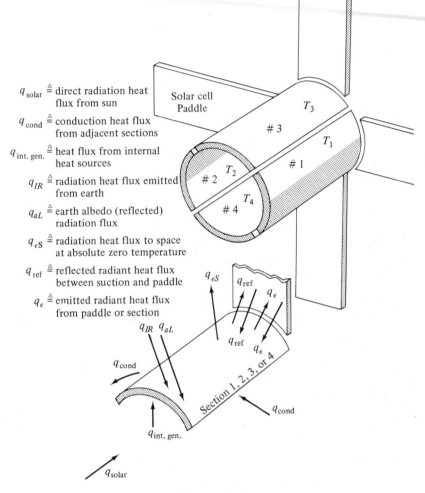

$q_{solar} \triangleq$ direct radiation heat flux from sun

$q_{cond} \triangleq$ conduction heat flux from adjacent sections

$q_{int. gen.} \triangleq$ heat flux from internal heat sources

$q_{IR} \triangleq$ radiation heat flux emitted from earth

$q_{aL} \triangleq$ earth albedo (reflected) radiation flux

$q_{\epsilon S} \triangleq$ radiation heat flux to space at absolute zero temperature

$q_{ref} \triangleq$ reflected radiant heat flux between suction and paddle

$q_\epsilon \triangleq$ emitted radiant heat flux from paddle or section

FIGURE 4-6

Satellite Thermal System Model

4-2. FLUID FLOW RESISTANCE AND THE FLUID RESISTANCE ELEMENT.

Consider the flow of a fluid in a constant area rigid-walled conduit as in Fig. 4-9. The variables of primary interest are the average fluid pressure p (lb$_f$/inch^2) and the volume flow rate q (inch^3/sec). The average flow velocity v (inch/sec) is defined as q/A where A is the conduit cross-sectional area (inch^2). While the actual fluid pressure and velocity vary from point

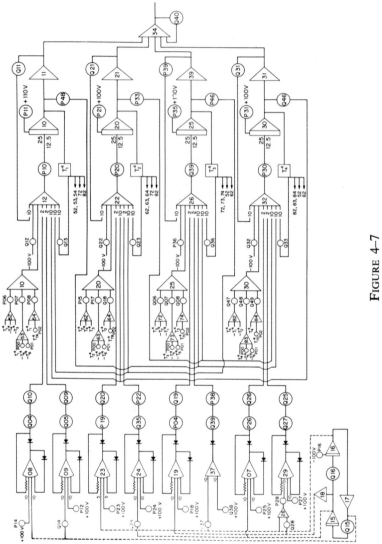

FIGURE 4-7

Analog Computer Setup for Satellite Thermal Analysis

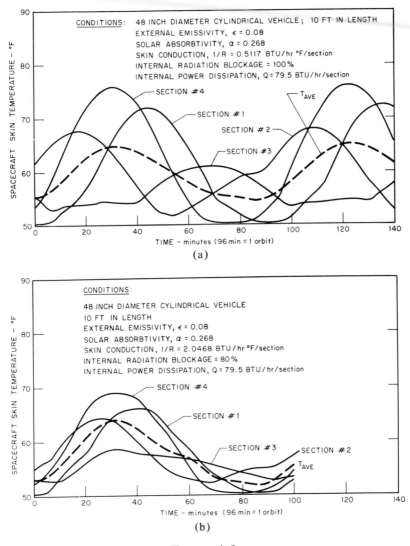

FIGURE 4-8

Orbiting Satellite Thermal Response Curves

to point over the flow cross section in a real fluid, we assume a so-called one-dimensional flow model in which the velocity and pressure are uniform over the area; thus, average velocity and pressure correspond to the values at any point in the cross section. This one-dimensional model has been found to give good results in many (but not all) applications. In a lumped-parameter analysis, the pipeline is broken into segments as shown

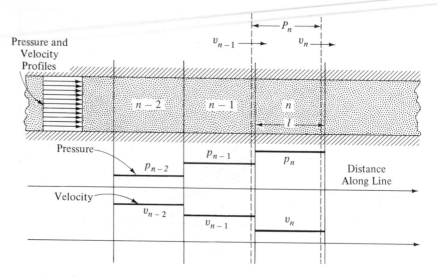

FIGURE 4–9

Lumped Model of Fluid Pipeline

in Fig. 4–9. Within each segment or lump, the pressure and velocity may vary arbitrarily with time, but are assumed uniform over the volume of the lump. By considering the behavior of one typical lump (the nth) we are led to definitions of the basic fluid elements.

If we assume thermal effects negligible, and only small changes in density, the two basic physical laws needed in our analysis are the conservation of mass and Newton's law. In addition we require data on fluid compressibility and friction. Fluid compressibility is described by the fluid property called the *bulk modulus*. Bulk modulus is measured experimentally by compressing a fluid sample of volume V and measuring the volume change ΔV caused by a pressure change ΔP:

$$B \triangleq \text{Bulk Modulus} \triangleq -\frac{\Delta P}{\Delta V/V}, \text{ psi} \qquad (4\text{–}1)$$

For hydraulic oil, for example, the bulk modulus is essentially constant with pressure and is approximately 250,000 psi, while for gases at constant temperature and for small pressure changes around an operating pressure P, the bulk modulus is equal to P. This variation of bulk modulus with pressure for gases is one of the nonlinearities which makes gas systems more difficult to analyze than liquid systems.

The topic of fluid friction is a complex one, which is treated at length in courses on fluid mechanics. We here take a simple phenomenological viewpoint, and state that experiments show that when fluid is forced through a pipe at a constant flow rate, a pressure drop related to that flow rate must be exerted to maintain the flow (see Fig. 4–10). One would intuitively expect that it would take larger pressure drops to cause larger flow rates, and this behavior is what is observed. In general the relation

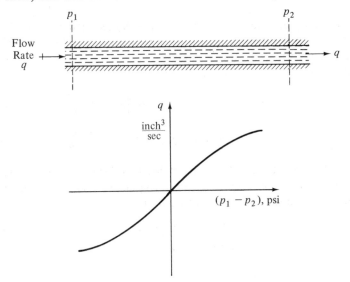

FIGURE 4–10

Experimental Determination of Fluid Flow Resistance

between pressure drop and flow rate is nonlinear; however, some situations give a nearly linear effect, and we take the ideal fluid friction or fluid resistance element to be linear, that is, $R_f \triangleq (p_1 - p_2)/q$, psi/(inch³/sec).

Returning now to Fig. 4–9, consider the conservation of mass as applied to the nth lump over an infinitesimal time interval dt. Mass enters the lump from the left at a rate $Av_{n-1}\rho$ and leaves at the right at a rate $Av_n\rho$ where ρ is the fluid mass density which we assume to be nominally constant. If the fluid were incompressible, these two mass flow rates would have to be identical, and thus v_n would equal v_{n-1}. However, real fluids (even liquids) are compressible, thus in general the difference between mass in and mass out shows up as a mass storage. Since mass and volume are related through the density $\rho \triangleq M/V$, and volume is related to pressure through the bulk modulus, we may write

$$(Av_{n-1}\rho - Av_n\rho)\, dt = dM = \rho\, dV = \rho\frac{V}{B}dp_n = \frac{\rho A\ell}{B}dp_n \qquad (4\text{--}2)$$

$$(q_{n-1} - q_n)\, dt = \frac{A\ell}{B}dp_n \qquad (4\text{--}3)$$

$$p_n = \frac{1}{C_f} \int (q_{n-1} - q_n) \, dt \qquad \text{(4-4)}$$

$$C_f \triangleq \frac{A\ell}{B} \triangleq \text{fluid compliance (fluid capacitance)}, \frac{\text{inch}^3}{\text{psi}} \qquad \text{(4-5)}$$

If we take the net volume flow rate $(q_{n-1} - q_n)$ as analogous to electric current, and the pressure p_n as analogous to voltage drop in Eq. (4–4), we see the analogy to electrical capacitance.

Newton's law for the nth lump gives us

$$Ap_{n-1} - Ap_n - R_f A q_n = \rho A \ell \frac{dv_n}{dt} = \rho \ell \frac{dq_n}{dt} \qquad \text{(4-6)}$$

$$(p_{n-1} - p_n) - R_f q_n = \frac{\rho \ell}{A} \frac{dq_n}{dt} \qquad \text{(4-7)}$$

Since Eq. (4–7) contains both the resistance (friction) and inertance (inertia) effects we consider each (in turn) negligible, to separate them. If the fluid had zero density (no mass) we would get

$$(p_{n-1} - p_n) = R_f q_n \qquad \text{(4-8)}$$

While if the resistance (friction) were zero

$$(p_{n-1} - p_n) = \frac{\rho \ell}{A} \frac{dq_n}{dt} = I_f \frac{dq_n}{dt} \qquad \text{(4-9)}$$

$$I_f \triangleq \frac{\rho \ell}{A} \triangleq \text{fluid inertance} \qquad \text{(4-10)}$$

Taking $(p_{n-1} - p_n)$ as analogous to voltage drop, and q_n as current, we again see the analogy to electrical resistance and inductance. Since we have earlier established electrical/mechanical analogies the fluid elements clearly have mechanical analogs also.

We will return to the fluid compliance and inertance elements in more detail in later sections. They were briefly introduced in this section on fluid resistance to allow a clearer discussion of resistance as a separate element. When a (one-dimensional) fluid flow is steady (velocity and pressure at any given point not changing with time), the inertance and compliance cannot manifest themselves, and only the resistance effect remains. We can thus experimentally determine fluid resistances by steady-flow measurements of volume flow rate and pressure drop, as in Fig. 4–11, or, if we attempt to calculate them theoretically, we must analyze a steady flow situation and find the relation between pressure drop and volume flow rate. If a nonlinear resistance operates near a steady flow q_0, we can define a linearized resistance good for small flow and pressure excursions from q_0 by the slope of the curve as in Fig. 4–11. Note that equation (4–8), which defines fluid resistance, is an algebraic (not differential) equation; thus,

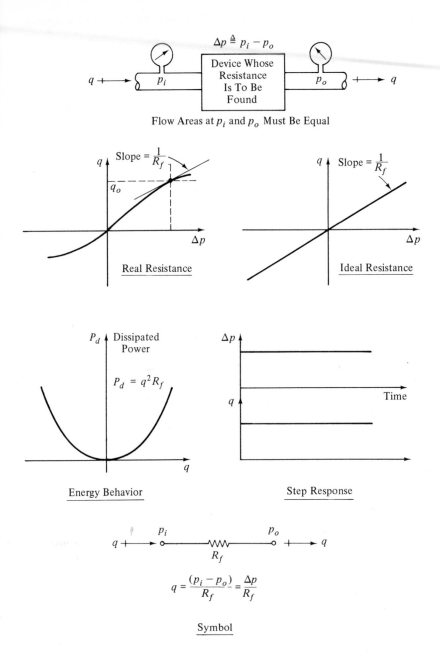

$$\Delta p \triangleq p_i - p_o$$

Device Whose Resistance Is To Be Found

Flow Areas at p_i and p_o Must Be Equal

Slope $= \dfrac{1}{R_f}$

q_o

Δp

Real Resistance

Slope $= \dfrac{1}{R_f}$

Δp

Ideal Resistance

P_d Dissipated Power

$P_d = q^2 R_f$

Energy Behavior

Δp

Time

q

Step Response

p_i p_o

R_f

$$q = \frac{(p_i - p_o)}{R_f} = \frac{\Delta p}{R_f}$$

Symbol

FIGURE 4–11

The Fluid Resistance Element

175

the pure fluid resistance element exhibits an instantaneous response of q to an applied Δp, or Δp to an applied q. The frequency response is thus, of course, flat from zero to infinite frequency with zero phase angle. Just as in electrical resistors, a fluid resistor dissipates into heat all the fluid power supplied to it. (This heat warms the fluid and any surrounding piping or machinery.) The fluid power associated with a flow is calculated by finding the rate at which energy crosses the boundaries of the element at the points where p_i and p_o exist. At the entrance, let the flow cross-sectional area be A_i. Then there is a force $A_i p_i$ acting at this boundary. In pushing the fluid a distance dx across the boundary, this force (which is provided by agencies external to the resistance element) does work $A_i p_i \, dx$. If the fluid at this point has velocity $v_i = dx/dt$, the rate at which work is being done on the element is $A_i p_i v_i \, dt/dt = A_i p_i v_i = q_i p_i$ inch-lb$_f$/sec. The power input to the element is thus $q_i p_i$. At the exit section, the element is pushing fluid out into the external system and thus doing work *on* this system at a rate $q_o p_o$. For a resistance element with incompressible fluid $q_1 = q_0 = q$, and the net power absorbed by the element is $q(p_i - p_o) = q\Delta p = q^2 R_f = \Delta p^2/R_f$.

While we can determine flow resistances by experimental steady-flow calibration, as in Fig. 4–11, it is, of course, also desirable to be able to calculate from theory, before a device has been built, what its resistance will be. For certain simple configurations and flow conditions, this can be done with fairly good accuracy. Flow conditions may be categorized in several useful ways, one of which is whether the flow is laminar or turbulent. *Laminar flow* occurs at relatively low flow velocities, and is characterized by an orderly and mathematically tractable motion of the fluid governed by viscosity effects, rather than inertia. *Turbulent flow* occurs at higher velocities, where inertia effects outweigh viscosity. While one can still speak of an average velocity at a cross section, the individual fluid "particles" have random transverse velocity components superimposed on their gross forward motion, making a detailed mathematical analysis which documents the motion of each particle effectively impossible. While certain aspects of turbulent flow are subject to analysis yielding useful results, frictional resistance effects generally require experimental study, the results of which may, however, often be generalized.

It has been found for steady flows that one can predict whether laminar or turbulent flow will occur by calculating a dimensionless parameter called the Reynold's number N_R, which is effectively a ratio of inertial to viscous forces. If N_R is low enough, laminar flow will occur. As N_R increases, a transition region is encountered in which a clear-cut distinction between laminar and turbulent operation is not possible. Above this transition region, turbulent flow definitely occurs. For flow inside a smooth, straight pipe of circular cross section, N_R is defined by

$$\text{Reynold's number} \triangleq N_R \triangleq \frac{\rho D v}{\mu} \qquad (4\text{--}11)$$

where

$\rho \triangleq$ fluid mass density, $lb_f\text{-sec}^2/inch^4$

$D \triangleq$ pipe inside diameter, inch

$v \triangleq$ fluid average velocity, inch/sec

$\mu \triangleq$ fluid viscosity, $lb_f\text{-sec}/inch^2$

This relation holds for both liquids and gases. If $N_R < 2000$ the flow will be laminar; if $N_R > 4000$ the flow will be turbulent, unless extreme care is taken to prevent disturbances which initiate turbulence. In most applications, $N_R > 4000$ essentially guarantees turbulent flow. In the transition region $2000 < N_R < 4000$, accurate resistance calculations are difficult and about all one can do is to "bracket" a numerical value between two calculated values, one based on laminar and the other on turbulent flow.

Laminar flow conditions produce the most nearly linear flow resistances, and these can also be calculated from theory for passages of simple geometrical shape. The most common case is a long, thin flow passage called a *capillary tube*. For a circular cross section, theoretical analysis gives the result (ideally for incompressible fluids, but usable for gases also, as long as density changes are small)

$$\text{volume flow rate} = q = \frac{\pi D^4}{128\mu L} \Delta p, \text{ inch}^3/\text{sec} \qquad (4\text{--}12)$$

where

$D \triangleq$ pipe diameter, inch

$\mu \triangleq$ fluid viscosity, $lb_f\text{-sec}/inch^2$

$L \triangleq$ length of pipe, inch

$\Delta p \triangleq$ pressure drop, psi

The fluid resistance is thus

$$R_f \triangleq \frac{\Delta p}{q} = \frac{128\mu L}{\pi D^4}, \frac{\text{psi}}{\text{inch}^3/\text{sec}} \qquad (4\text{--}13)$$

Note that increasing the length increases the resistance in direct proportion, while decreasing the diameter by, say, 2 to 1 causes a $2^4 = 16$ to 1 increase in resistance. Also, while Eq. (4–12) allows *any* numerical value of Δp to be inserted, if this Δp causes a flow rate with $N_R > 2000$, we do not have laminar flow and the formula is *not* applicable.

The theory leading to Eq. (4–12) actually assumes the length L is a section of an infinitely long pipe, so that end effects can be neglected. In real capillary tubes, these end effects cause nonlinearity and may not be negligible. A formula for estimating these effects is[7]

$$R_f = \frac{128\mu L}{\pi D^4} \left[1 + 0.0434 \frac{D}{L} N_R \right] \tag{4-14}$$

If $0.0434 D N_R/L$ is negligible relative to 1, then the capillary will be nearly linear. If not, the nonlinearity is made obvious by noting that $N_R = 4\rho q/\pi\mu D$, and thus (4–14) becomes

$$R_f = \frac{128\mu L}{\pi D^4} \left[1 + \frac{0.1736}{\pi} \frac{\rho}{L\mu} q \right] \tag{4-15}$$

which shows R_f to depend on flow rate q, and is thus nonlinear. Figure 4–12 shows theoretical laminar flow resistances for some other shapes of flow passages. In estimating N_R for such shapes, to check for laminar flow, use the so-called hydraulic diameter D_h for D in the N_R formula.

$$D_h \triangleq \frac{4(\text{cross-section area})}{\text{perimeter}} \tag{4-16}$$

For turbulent flow of liquids or of gases with small density change in smooth circular pipes, the pressure/flow relation has been found by experiment to be well fitted by the relation[8]

$$\Delta p = \frac{0.242 L \mu^{0.25} \rho^{0.75}}{D^{4.75}} q^{1.75} \tag{4-17}$$

Note that the fluid mass density ρ is now present, and also q is raised to the 1.75 rather than the 1.0 power; thus the $\Delta p/q$ relation is nonlinear. It is still, however, an algebraic relation, so that the time response is still instantaneous. When computers are used for fluid system dynamics studies, the nonlinear relation of Eq. (4–17) is easily handled. If a linear model is desired for studies of small flow excursions about a steady-flow operating point q_0 we may linearize in the usual way and define an incremental linearized resistance as $d(\Delta p)/dq$.

$$\Delta p = \frac{0.242 L \mu^{0.25} \rho^{0.75}}{D^{4.75}} q^{1.75} \triangleq K_q q^{1.75} \tag{4-18}$$

$$\frac{d(\Delta p)}{dq}\bigg|_{q=q_0} = 1.75 K_q q_0^{0.75} = R_f \tag{4-19}$$

$$K_q \triangleq \frac{0.242 L \mu^{0.25} \rho^{0.75}}{D^{4.75}} \tag{4-20}$$

[7] H. E. Merritt, *Hydraulic Control Systems* (New York: John Wiley & Sons, Inc., 1967), p. 33.
[8] *Ibid.*, p. 39.

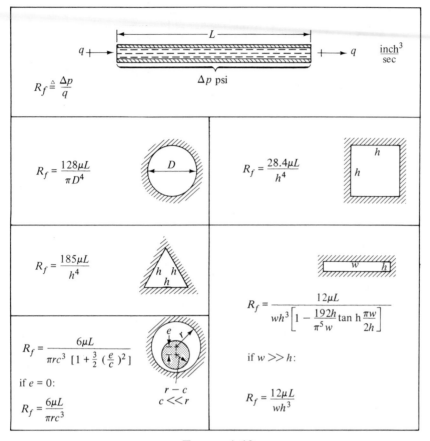

FIGURE 4–12

Laminar Flow Resistances

For non-circular pipes use the hydraulic diameter (Eq. 4–16) in place of D for estimating pressure drops and flow resistances. As a practical matter it should be mentioned that in fluid systems turbulent flow is considerably more common than laminar.

When we say that the response of q to an applied Δp is instantaneous for the laminar and turbulent pipe-flow resistance elements discussed above, it is important to remember that the bodies of fluid within the pipe length L also have compliance and inertance. Thus if one applies a sudden Δp to the fluid at rest, it does *not* suddenly achieve a flow rate q, since friction is *not* the only effect present. A simple frequency response study may be helpful in illustrating this point. In Fig. 4–13, an air pressure controller allows us to apply a sinusoidal pressure difference $p_1 - p_2 = \Delta p = A_p \sin \omega t$ to the ends of a pipe of length L connected between two

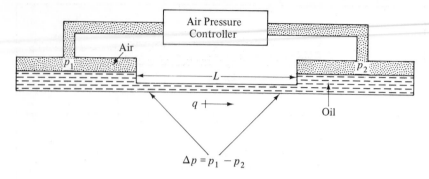

FIGURE 4–13

Oscillating Flow System

large shallow tanks. The fluid in the pipe will be considered incompressible (compliance = 0), but will exhibit friction and inertance. The difference between the applied Δp and the frictional pressure drop $R_f q$ is available to accelerate the inertance. In established laminar flow, the effective mass is 4/3 the "physical" mass ρAL; thus the inertance is $4\rho L/3A$.

$$\Delta p - R_f q = \frac{4\rho L}{3A} \frac{dq}{dt} \tag{4-21}$$

$$\left(\frac{4\rho L}{3AR_f} D + 1\right) q = \frac{1}{R_f} \Delta p \tag{4-22}$$

$$\frac{q}{\Delta p}(D) = \frac{1/R_f}{\dfrac{4\rho L}{3AR_f} D + 1} \tag{4-23}$$

$$\frac{q}{\Delta p}(i\omega) = \frac{1/R_f}{\dfrac{4\rho L}{3AR_f} i\omega + 1} \tag{4-24}$$

In Eq. (4–24) note that, if 1.33 $\omega\rho L/AR_f$ is small compared to 1, then

$$\frac{q}{\Delta p}(i\omega) \approx \frac{1}{R_f} \tag{4-25}$$

that is, the pipe is essentially a resistance element for sufficiently small 1.33 $\omega\rho L/AR_f$. To get some numbers, let us assume laminar flow in a pipe with $D = 0.05$ inch and $L = 20$ inch. For a typical oil at room temperature, $\rho = 7.95 \times 10^{-5}$ lb$_f$-sec^2/in^4 and $\mu = 4 \times 10^{-6}$ lb$_f$-sec/in^2; thus

$$R_f = \frac{128\mu L}{\pi D^4} = \frac{128 \times 4 \times 10^{-6} \times 20}{3.14 \times 625 \times 10^{-8}} = 522 \frac{\text{psi}}{\text{inch}^3/\text{sec}} \tag{4-26}$$

and

$$\frac{1.33\rho L}{AR_f} = \frac{7.95 \times 10^{-5} \times 20 \times 1.33}{522 \times 1.96 \times 10^{-3}} = 2.06 \times 10^{-3} \text{ sec} \tag{4-27}$$

If we want $1.33 \, \omega\rho L/AR_f$ to be, say, 0.1 or less, then frequency ω must be 64.4 rad/sec or less. Thus the given tube acts essentially like a flow resistance if frequencies are less than about 10 Hz. For higher frequencies, its inertance becomes significant.

In all the above calculations of flow resistances, the approach has been to use formulas relating flow and pressure drop derived theoretically or measured experimentally for *steady* flows, as if they held for general (unsteady) flows. This approach is widely used, and usually of sufficient accuracy; however, it should be recognized as an approximation. A more rigorous approach would be to set up partial differential equations using a distributed-parameter model to find the sinusoidal transfer function relating Δp to q. This function will in general have the form

$$\frac{\Delta p}{q}(i\omega) = R_f + i\omega I_f \qquad (4\text{--}28)$$

where the real part R_f is the fluid resistance, and the imaginary part (divided by ω) is the fluid inertance I_f. For our simple analysis above, for example,

$$\frac{\Delta p}{q}(i\omega) = \frac{128\mu L}{\pi D^4} + i\omega \frac{4\rho L}{3A} \qquad (4\text{--}29)$$

For laminar incompressible flow in circular tubes, the distributed-parameter model can be solved, giving an involved solution in terms of Bessel functions.[9] For low frequencies $(\omega < 32\mu/\rho D^2)$ the exact expression becomes our simple result in Eq. (4–29). For high frequencies $(\omega > 7200\mu/\rho D^2)$ the result is

$$\frac{\Delta p}{q}(i\omega) = \frac{8L}{\pi D^3}(2\rho\mu\omega)^{1/2} + i\omega \frac{\rho L}{A} \qquad (4\text{--}30)$$

Note that at high frequencies the inertance corresponds to the "physical" mass ρAL. The reference gives plots showing a smooth transition between the low and high frequency regions; thus, the inertance is always bracketed between $4\rho L/3A$ and $\rho L/A$. Perhaps the most significant feature is that the resistance is frequency dependent, increasing with $\sqrt{(\omega)}$.

We now leave pipes (where the resistance is distributed over considerable length L), and consider *orifices* (where the resistance is concentrated in a short distance). In Fig. 4–14a, a flowing liquid discharges through a sharp-edge orifice into atmospheric air. Note that the liquid pressure p drops from the upstream value p_u to p_{at} over a very short distance. Figure 4–14b shows the more common situation, where an orifice is inserted in a line. Downstream of the orifice the liquid flow spreads out so that it again fills the pipe. The pressure drop in an orifice is basically due to a conversion

[9] C. K. Stedman, *op. cit.*

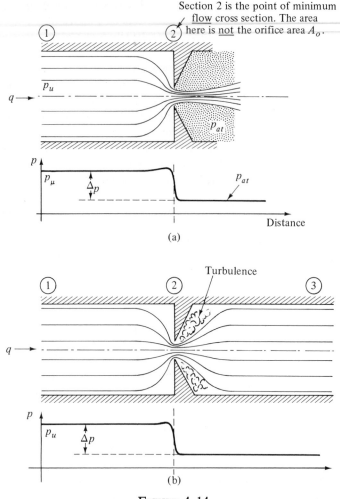

FIGURE 4–14

Orifice Flow Characteristics

of energy from the form of "flow power" (pressure) (volume flow rate) to the power of kinetic energy. Conservation of energy for level flow of a frictionless incompressible fluid gives

flow power + kinetic energy power = a constant   **(4–31)**

Considering any two locations 1 and 2, we may write

p_1q + (kinetic energy per unit time)₁

$$= p_2q + \text{(kinetic energy per unit time)}_2 \quad \textbf{(4–32)}$$

For a section of liquid of length L, the kinetic energy is $\rho A L v^2/2$. This kinetic energy passes a cross section in time interval AL/q, so the kinetic energy crossing the boundary per unit time is $\rho v^2 q/2$. Our power equation is then

$$p_1 q + \frac{\rho v_1^2 q}{2} = p_2 q + \frac{\rho v_2^2 q}{2} \tag{4-33}$$

$$p_1 - p_2 = \Delta p = \frac{\rho}{2}(v_2^2 - v_1^2) \tag{4-34}$$

showing that if v_2 is to be larger than v_1, we must expect a pressure drop Δp. Since $q = A_1 v_1 = A_2 v_2$,

$$\frac{2\Delta p}{\rho} = \left(\frac{q}{A_2}\right)^2 - \left(\frac{q}{A_1}\right)^2 = q^2 \left[\frac{1}{A_2^2} - \frac{1}{A_1^2}\right] = \frac{1 - \left(\frac{A_2}{A_1}\right)^2}{A_2^2} q^2 \tag{4-35}$$

and thus

$$q = \frac{A_2}{\sqrt{1 - \left(\frac{A_2}{A_1}\right)^2}} \sqrt{\frac{2\Delta p}{\rho}} \tag{4-36}$$

This is the basic pressure/flow relation for an orifice. Note that fluid viscosity is not present, but density is. Also the relation is nonlinear. It is conventional to write this relation in the form

$$q = C_d A_o \sqrt{\frac{2\Delta p}{\rho}} \tag{4-37}$$

where

$$A_o \triangleq \text{orifice cross-sectional area}$$

$$C_d \triangleq \text{discharge coefficient}$$

The discharge coefficient is partly theoretical and partly experimental, taking into account the difference between A_2 and A_o, and also any other deviations of the theory from actuality. Numerical values of C_d depend mainly on the area ratio A_1/A_o and the Reynold's number. In many situations $A_1/A_o > 25$ and $N_R > 10000$ and C_d can be taken as 0.61. In Fig. 4–14b, the velocity at section 3 must be the same as at section 1, since the area is the same. The theory of Eq. (4–34) applied to sections 1 and 3 would indicate that $\Delta p_{1-3} = 0$ and thus $p_3 = p_1$. Experiments show, however, that turbulence downstream of the orifice prevents the kinetic energy at section 2 from being fully reconverted into pressure at section 3. In fact, most of this energy is converted into heat by the turbulence, and thus p_3 is very nearly equal to p_2, as shown in the graph of Fig. 4–14b.

While sharp-edge orifices have more predictable characteristics and are less viscosity (and thus temperature) sensitive, manufacturing costs are lower for simple drilled holes, giving an orifice in the form of a short length of pipe (see Fig. 4–15). Discharge coefficients may be estimated from the formulas[10]

$$C_d = \frac{1}{\sqrt{1.5 + 13.74 \dfrac{L}{DN_R}}} \qquad \frac{DN_R}{L} > 50 \qquad \text{(4–38)}$$

$$C_d = \frac{1}{\sqrt{2.28 + 64 \dfrac{L}{DN_R}}} \qquad \frac{DN_R}{L} < 50 \qquad \text{(4–39)}$$

where the Reynold's number is calculated from

$$N_R = \frac{\rho Dq}{A_o \mu} \qquad \text{(4–40)}$$

Sharp edge orifices are the most nearly pure fluid resistance elements, since the space over which the pressure drop takes place is very small. Thus only

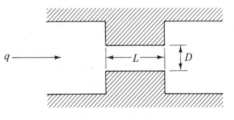

FIGURE 4–15

Short-Tube Orifice

a small volume of fluid is involved, and the compliance and inertance will be correspondingly small. For small flow changes around a steady-flow operating point, the orifice resistance can be linearized in the usual way.

$$q = C_d A_o \sqrt{\frac{2\Delta p}{\rho}}$$

$$\left. \frac{dq}{d\Delta p} \right|_{\Delta p = \Delta p_0} = \frac{1}{R_f} = \frac{C_d A_o}{\Delta p_0} \sqrt{\frac{2}{\rho}} \qquad \text{(4–41)}$$

$$R_f = \frac{\Delta p_0}{C_d A_o} \sqrt{\frac{\rho}{2}} \qquad \text{(4–42)}$$

If an R_f is wanted for operation about a zero flow condition, $\Delta p_0 = 0$ and Eq. (4–42) gives $R_f = 0$. Actually, as Δp is decreased, the orifice flow becomes laminar and the flow resistance becomes linear and *not* equal to zero. A short-tube orifice (Fig. 4–15) can be treated like a capillary tube

---

[10] H. E. Merritt, *Hydraulic Control Systems* (New York: John Wiley & Sons, Inc., 1967), p. 42.

if DN_R/L is less than about 2; Eq. (4–39) actually becomes $C_d = \sqrt{DN_R/L}/8$ for small DN_R/L, and if this is substituted into (4–37), we get (4–12), the capillary tube formula. For sharp-edge orifices, a theoretical laminar flow result is available[11]

$$q = \frac{\pi D^3}{50.4\mu} \Delta p \qquad \qquad \text{(4–43)}$$

for estimating flow resistance near zero flow. The transition from laminar to turbulent flow occurs at about $N_R = 9.3$. Above 9.3, Eq. (4–37) with $C_d = 0.61$ can be used.

Figure 4–16 shows some pressure flow curves actually measured for water flowing in capillary tubes and an orifice. Note that the capillary tube of length 5.5 inch is really quite nonlinear even at $N_R = 1000$. To get the same flow resistance with better linearity, three tubes of the same diameter but three times as long were connected in parallel. This design change has two good effects as predicted by Eq. (4–14). First, D/L will now be $\frac{1}{3}$ of what it was before. Also, for a given total flow rate, each tube now carries only $\frac{1}{3}$ the flow, and thus has $\frac{1}{3}$ the Reynold's number. The calibration data show that the predictions of Eq. (4–14) are indeed borne out, and linearity is now much better. The orifice shown is in the form of a short tube, with $D/L \approx 1$; thus, Eqs. (4–38) and (4–39) are applicable for C_d calculations. A curve fit of the data shows good correlation with the square root relation predicted by Eq. (4–37).

4–3. FLUID COMPLIANCE AND THE FLUID COMPLIANCE ELEMENT.

We already have seen that a fluid *itself* exhibits compliance due to its compressibility, whether a liquid or a gas. Certain *devices* may also introduce compliance into a fluid system, even if the fluid were absolutely incompressible. Metal tubing and (in particular) rubber hoses will expand when fluid pressure increases, allowing an increase in the volume of liquid stored. Accumulators use spring-loaded cylinders or rubber air bags to provide intentionally large amounts of compliance. A simple open tank exhibits compliance, since an increase in volume of a contained liquid results in a pressure increase due to gravity. In general, the compliance of a device is found by forcing into it a quantity of fluid and noting the corresponding rise in pressure. For liquids, the input quantity is a volume of fluid V, and the ideal compliance is defined by

---

[11] H. E. Merritt, *op. cit.*, p. 44.

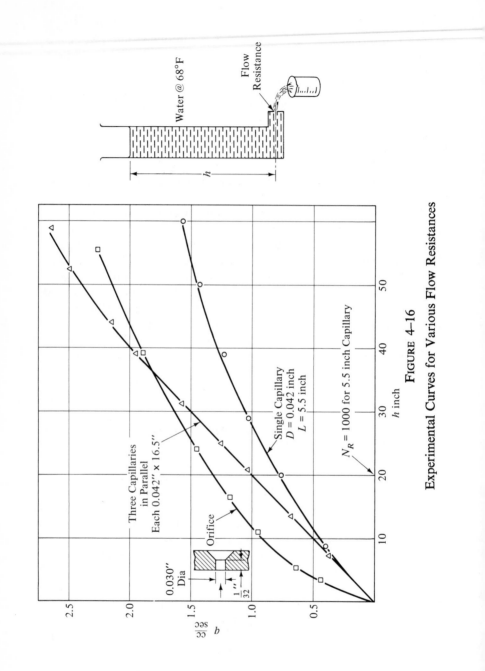

FIGURE 4-16

Experimental Curves for Various Flow Resistances

186

$$C_f \triangleq \frac{V}{p}, \quad \text{inch}^3/\text{psi} \tag{4-44}$$

or, in terms of flow rate q,

$$C_f \triangleq \frac{\int q \, dt}{p} \tag{4-45}$$

For nonlinear compliances, the actual p/V curve can be implemented on a computer, or, if linearized analysis is desired, the local slope can be used to define an incremental compliance. A standardized symbol for fluid compliance has not been agreed upon, although some writers use the electrical capacitance symbol with one end of the "capacitor" always connected to ground. However, since it is preferable to work from the fluid equations directly, rather than to use analogies, the symbol of Fig. 4–17 is suggested as consistent with fluid circuit diagrams.

In Eq. (4–5) we calculated the compliance of a section of hydraulic line due to the bulk modulus of the liquid itself as $A\ell/B$. Additional contributions to compliance which may be significant here are due to entrained air bubbles and the flexibility of the tubing. Suppose air bubbles take up x percent of a total volume V. A pressure increase of Δp will cause a change of liquid volume of $\Delta p(1 - x)V/B_\ell$ and of gas volume $\Delta pxV/B_g$. The fluid volume change is thus

$$\Delta V_{\text{fluid}} = \left[\frac{(1 - x)V}{B_\ell} + \frac{xV}{B_g} \right] \Delta p \tag{4-46}$$

For thin-walled tubing the tubing volume change is given by

$$\Delta V_{\text{tubing}} = \frac{VD}{tE} \Delta p \tag{4-47}$$

where

$$t \triangleq \text{wall thickness}$$

$$D \triangleq \text{tube diameter}$$

$$E \triangleq \text{tube modulus of elasticity}$$

The total volume change is

$$\Delta V_{\text{total}} = \left[\frac{D}{tE} + \frac{(1 - x)}{B_\ell} + \frac{x}{B_g} \right] V\Delta p \tag{4-48}$$

and the total compliance is

$$C_f = \left[\frac{D}{tE} + \frac{1 - x}{B_\ell} + \frac{x}{B_g} \right] V, \frac{\text{inch}^3}{\text{psi}} \tag{4-49}$$

As a numerical example, let us assume steel pipe with $D/t = 10$, 1% air bubbles, $B_\ell = 250{,}000$ psi, and $B_g = 500$ psi. The B_g of 500 psi assumes the system pressure is 500 psi and slow (isothermal) pressure changes.

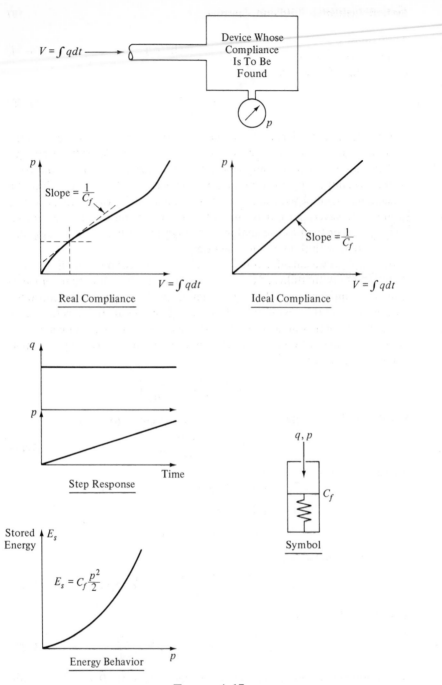

FIGURE 4–17

The Fluid Compliance Element

(For rapid changes [adiabatic] B_g would be $1.4p = 700$ psi.) The relative importance of the three terms in Eq. (4–49) is:

$$\text{steel tube} \qquad \frac{D}{tE} = \frac{10}{3 \times 10^7} = 0.33 \times 10^{-6}$$

$$\text{oil} \qquad \frac{1 - x}{B_t} = \frac{0.99}{2.5 \times 10^5} = 0.40 \times 10^{-5}$$

$$\text{air bubbles} \qquad \frac{x}{B_g} = \frac{0.01}{5 \times 10^2} = 0.2 \times 10^{-4}$$

It is clear from this that even a small amount of entrained air can severely reduce the "stiffness" of hydraulic fluid. Another way of looking at this problem is to define an "effective" or equivalent bulk modulus which includes all three effects. From Eq. (4–48)

$$\frac{1}{\text{effective bulk modulus}} \triangleq \frac{1}{B_e} \triangleq \frac{\Delta V/V}{\Delta p} = \left[\frac{D}{tE} + \frac{(1 - x)}{B_t} + \frac{x}{B_g} \right] \quad \text{(4–50)}$$

For the numerical example $B_e = 41,200$ psi, much lower than the 250,000 of the liquid itself.

Figure 4–18 shows some devices, called accumulators, which are designed intentionally to exhibit fluid compliance. In 4–18a, we have a simple spring-loaded piston and cylinder; 4–18b shows a flexible metal bellows; while in 4–18c, the compliant element is a nitrogen-filled rubber bag. The devices of Fig. 4–18a and c can be designed to store large amounts of fluid energy and are widely used in hydraulic power systems for short-term power supplies, pulsation smoothing, and to reduce pump size in systems with intermittent flow requirements. Since metal bellows are somewhat limited in volume change, they are more often found used as dynamic elements in low-power devices such as instruments. Due to their complex shape, the compliance of metal bellows is difficult to calculate but it can be measured easily once a bellows has been constructed. For the spring-loaded piston

$$C_f = \frac{\Delta V}{\Delta p} = \frac{A \Delta x}{K_s \Delta x / A} = \frac{A^2}{K_s}, \frac{\text{inch}^3}{\text{psi}} \quad \text{(4–51)}$$

where

$$A \triangleq \text{piston area, in}^2$$

$$K_s \triangleq \text{spring constant, lb}_f/\text{in}^2$$

In Fig. 4–18c, the rubber bag is initially pressurized (through a "tire-valve" not shown) with nitrogen, so that the bag completely fills the steel pressure vessel. Then hydraulic oil is forced in, compressing the gas more and partially filling the vessel with liquid. If the accumulator is now connected to a hydraulic system so that liquid can flow in and out, the following relations hold (see Fig. 4–19).

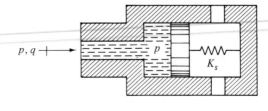

(a)

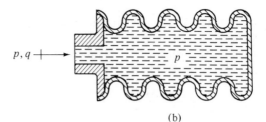

(b)

(c)

FIGURE 4-18

Accumulators (Liquid Compliances)

perfect gas law $pV = MRT$ **(4-52)**

When the rubber bag initially is charged, a definite mass M of gas is put into it; since the bag is then sealed, this M is constant. For slow pressure changes the temperature T is assumed to stay constant (isothermal process); the gas constant R is very nearly constant for the conditions under which accumulators are used, and thus MRT is constant. At any instant t

$$V_{gas} = \frac{MRT}{p_{gas}} = V_{gas,op} - \int_0^t q_{liq}\, dt \qquad \textbf{(4-53)}$$

where $V_{gas,op}$ is the gas volume at time $t = 0$ corresponding to an operating point. As long as there is any liquid in the vessel, $p_{liq} \equiv p_{gas}$ and thus

$$p_{liq} = \frac{MRT}{V_{gas,op} - \int_0^t q_{liq}\, dt} = \frac{MRT}{V_{gas,0} - V_{liq}} \qquad \textbf{(4-54)}$$

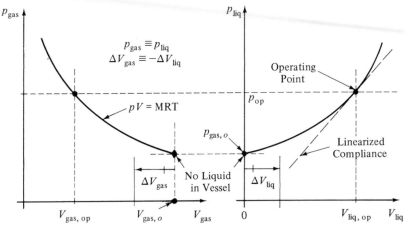

FIGURE 4-19

Characteristics of Gas-Bag Accumulator

This is the pressure/flow relation defining the compliance effect of the accumulator, and is seen to be nonlinear in the p_{liq} versus V_{liq} relation. This nonlinearity generally prevents analytical study but is easily implemented on a computer. For small changes about an operating point, a linearized compliance may be defined by taking the slope of the curve.

$$C_f \triangleq \frac{dV_{\text{liq}}}{dp_{\text{liq}}} = \frac{(V_{\text{gas},0} - V_{\text{liq},\text{op}})^2}{MRT} = \frac{V_{\text{gas},\text{op}}^2}{MRT} = \frac{MRT}{p_{\text{op}}^2} \qquad \textbf{(4-55)}$$

In Fig. 4-20a, a vertical cylindrical tank of area A is supplied with a flow q; pressure at the tank inlet is p, liquid height is h. The vertical motion of the liquid in such tanks is usually slow enough that velocity and acceleration have negligible effects on the pressure p and it is simply given by $p = \gamma h$, where γ is the specific weight of the liquid in $\text{lb}_f/\text{inch}^3$. If we add a volume V of liquid to the tank, the level h goes up an amount V/A and pressure p rises an amount $V\gamma/A$. The compliance is thus

$$C_f = \frac{\text{volume change}}{\text{pressure change}} = \frac{V}{V\gamma/A} = \frac{A}{\gamma}, \frac{\text{inch}^3}{\text{psi}} \qquad \textbf{(4-56)}$$

For non-cylindrical tanks such as in Fig. 4-20b, the compliance effect is nonlinear. In Fig. 4-20c, a rigid tank of volume V contains a gas at pressure p. For slow (isothermal) pressure changes in which fluid density is nearly constant, we may write

$$pV = MRT$$

If we force a mass $dM = \rho\, dV$ of fluid into the tank we cause a pressure change dp given by

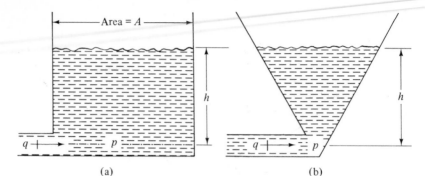

FIGURE 4–20

Liquid and Gas Tanks as Fluid Compliances

$$dp = \frac{RT}{V} dM = \frac{RT}{V} \rho \, dV = \frac{RT}{V} dV \frac{p}{RT} \qquad \text{(4-57)}$$

and

$$C_f \triangleq \frac{dV}{dp}\bigg|_{p=p_0} = \frac{V}{p_0}, \frac{\text{inch}^3}{\text{psi}} \qquad \text{(4-58)}$$

This is a linearized compliance useful for small changes near an operating pressure p_0. For rapid (adiabatic) but still small pressure changes, analysis shows the compliance to be given by $C_f = V/kp_0$ where k is the ratio of specific heats for the gas (1.4 for air, for example).

4–4. FLUID INERTANCE.

While devices for introducing fluid resistance (capillaries, orifices) and fluid compliance (tanks, accumulators) are often intentionally designed into fluid systems, the inertia effect is more often than not a parasitic one; thus inertance "devices" are relatively unknown as commercial components. In terms of analytical treatment, the inertance associated with flow in *pipes* is perhaps most commonly encountered, and we shall limit ourselves to this subject. Any flowing fluid will have stored kinetic energy because of its density (mass) and velocity. The inertance of a finite-size lump of fluid represents a summing up of this kinetic energy over the volume of the lump. The simplest assumption possible here is that of one-dimensional flow (Fig. 4–21a), where all fluid particles have identical velocities at any instant of time. Since every fluid particle has the same velocity, a lump of fluid can be treated as if it were a rigid body of mass $M = \rho AL$. A pressure drop Δp across a pure inertance element will cause a fluid acceleration according to Newton's law:

$$A\Delta p = \rho AL \frac{dv}{dt} = \rho AL \left(\frac{1}{A} \frac{dq}{dt} \right) \qquad (4\text{–}59)$$

$$\Delta p = \frac{\rho L}{A} \frac{dq}{dt} \triangleq I_f \frac{dq}{dt} \qquad (4\text{–}60)$$

where

$$I_f \triangleq \text{fluid inertance} \triangleq \frac{\rho L}{A} \qquad (4\text{–}61)$$

Equation (4–60) is of course analogous to $e = L(di/dt)$ for inductance in electrical systems and $f = M(dv/dt)$ for mass in mechanical systems. The electrical inductance symbol (Fig. 4–22) is widely used for fluid inertance, and, since it causes no conceptual difficulties in setting up fluid equations, we will adopt it also.

Theoretical analysis of steady laminar flow shows the velocity profile to be parabolic, as in Fig. 4–21b. By computing the kinetic energy of a unit length of fluid with this velocity profile, and equating it to that of a uniform-velocity fluid mass with the same *average* velocity, we find that the uniform-velocity mass must be 4/3 the actual mass. This result also is a close approximation for unsteady flows of sufficiently low frequency content. As frequency increases (Fig. 4–21c) the velocity profile becomes more square and the correct mass approaches the "physical" mass ρAL. The inertance for laminar flow is thus always between $(4/3)\rho L/A$ and $\rho L/A$, the midpoint $(7/6)\rho L/A$ occurring at about $\omega = 50\mu/R^2\rho$.[12] For a typical hydraulic oil ($\mu = 2 \times 10^{-6}$ lb$_f$-sec/in^2, $\rho = 0.8 \times 10^{-4}$ lb$_f$-sec^2/in^4) flowing in a tube of radius $R = 0.05$ in., the frequency ω is 500 rad/sec, for example.

In steady turbulent flow, velocity profiles cannot be calculated from theory but experiments have shown that, for rough or smooth pipes, the measured profiles are quite accurately fitted by equations of form

$$\frac{v_r}{v_c} = \left(\frac{r}{R} \right) \frac{1}{n} \qquad (4\text{–}62)$$

---

[12] C. K. Stedman, *op. cit.*

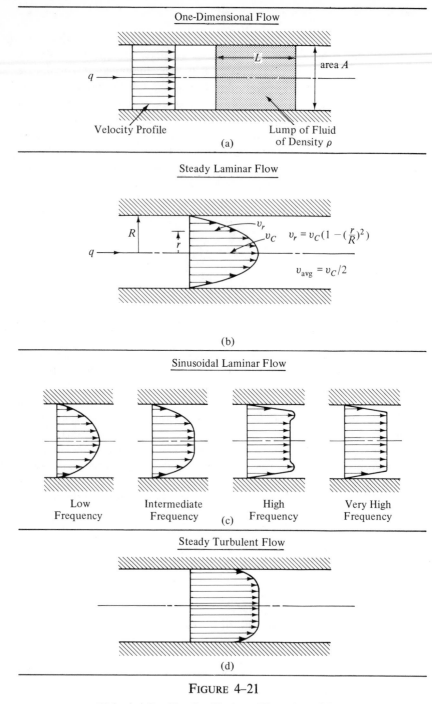

One-Dimensional Flow

(a)

Steady Laminar Flow

$v_r = v_C(1 - (\frac{r}{R})^2)$

$v_{avg} = v_C/2$

(b)

Sinusoidal Laminar Flow

| Low Frequency | Intermediate Frequency | High Frequency | Very High Frequency |

(c)

Steady Turbulent Flow

(d)

FIGURE 4–21

Velocity Profiles for Various Flow Conditions

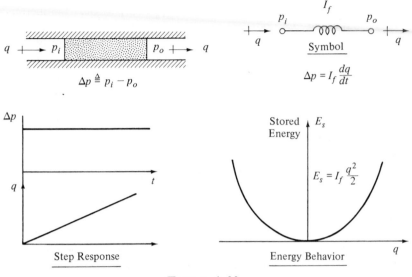

FIGURE 4–22

The Fluid Inertance Element

The parameter n varies over the range 4 to 5 for rough pipes and from 6 (Reynold's number 4000) to 10 (Reynold's number 3×10^6) for smooth pipes. In all cases the profile is quite square and calculations of kinetic energy show that the effective mass is nearly equal to the "physical" mass ρAL. For turbulent flow it is thus quite reasonable to use the value of inertance corresponding to one-dimensional flow, that is, $\rho L/A$.

While measurements of fluid resistance can be made without having compliance and inertance effects present (in *steady* flow dp/dt and dq/dt are zero), and fluid compliance measurements are similarly separable (for *no* flow, q and dq/dt are zero), the separate "steady-state" measurement of fluid inertance would require a non-zero dq/dt with a zero q and a zero dp/dt. These conditions are generally not realizable in practice, and thus steady-state measurements of fluid inertance are not commonly made. For this reason a "characteristic curve" relating Δp and dq/dt for inertance (corresponding to the $q/\Delta p$ curve for resistance [Fig. 4–11] and the $p/\int q\,dt$ curve [Fig. 4–17] for compliance) is not given in Fig. 4–22. For the pure and ideal inertance element, the Δp versus dq/dt curve would, of course, be a straight line. Since real resistance and compliance generally exhibit some (sometimes considerable) nonlinearity in their characteristic curves, one wonders whether real inertance behaves similarly. Unfortunately, the above-mentioned lack of an experimental method for getting the required characteristic curve for inertance prevents a straightforward comparison.

If we appeal to theory we see that, for a fluid "particle," the inertance effect is nothing more nor less than the force/acceleration characteristic of Newton's law, which experiments have proven to be extremely linear, except when the velocities approach the velocity of light. Since fluid systems operate with velocities which are entirely negligible relative to the speed of light, the inertance effect for a particle should be essentially linear. In practical problems, however, we are concerned with the inertance effect of the total flow in a pipe or machine, which clearly involves a complicated summing-up of the effects of myriad "particles," the motions of which are influenced not just by inertia but also by resistance, for example. The basic concept of linearity requires that a change in the level of the input quantity produces a strictly *proportional* change in the level of the output quantity. For fluid inertance, the input would be an applied Δp and the output would be the rate of change of flow rate, dq/dt. Suppose we apply to a "lump" of fluid in a pipe a sinusoidal pressure difference $\Delta p = \Delta p_0 \sin \omega t$ of fixed frequency ω. If the amplitude Δp_0 is sufficiently small, we would expect the peak flow rate to also be small enough to allow laminar flow at all times, and if ω is small enough, the inertance will be very nearly that of steady laminar flow, $4\rho L/3A$. If input Δp_0 is now increased (keeping ω fixed) it is intuitively clear that this larger accelerating force will cause larger peak flow rates, and at some point the flow will become turbulent rather than laminar. While little is known about the velocity profiles of unsteady turbulent flow, it would be unreasonable to expect them to be identical with those of laminar flow; thus, the inertance must change as Δp_0 is increased, showing a nonlinear inertance behavior. Thus, while the fundamental inertia law for a particle is linear, the gross inertance effect in a real fluid system may be nonlinear. Fortunately, this nonlinearity is not excessive, since it is very likely bounded by the inertance values $4\rho L/3A$ and $\rho L/A$ in all practical situations, a 33% range at worst.

4–5. FLUID IMPEDANCE.

Most fluid system problems do not really *require* the separation of pressure/flow relations into their resistive, compliant, and inertial components; this separation is mainly one of analytical convenience. For complex fluid systems where experimental measurements may be a necessity, the measurement of *overall* pressure/flow characteristics has become a useful tool. The term fluid impedance is directly analogous to mechanical and electrical impedance discussed earlier, and is defined as the transfer function relating pressure drop (or pressure) as output to flow rate as input, that is,

$$\text{fluid impedance} \triangleq \frac{\Delta p}{q}(D), \frac{\text{psi}}{\text{inch}^3/\text{sec}} \qquad (4\text{--}63)$$

For the individual fluid elements we have

fluid resistance $\quad \dfrac{\Delta p}{q}(D) = R_f \qquad \dfrac{\Delta p}{q}(i\omega) = R_f \qquad$ **(4-64)**

fluid compliance $\quad \dfrac{p}{q}(D) = \dfrac{1}{C_f D} \qquad \dfrac{p}{q}(i\omega) = \dfrac{1}{\omega C_f}\,\underline{/-90°} \quad$ **(4-65)**

fluid inertance $\quad \dfrac{\Delta p}{q}(D) = I_f D \qquad \dfrac{\Delta p}{q}(i\omega) = \omega I_f\,\underline{/+90°} \quad$ **(4-66)**

and Fig. 4–23 shows the frequency response curves.

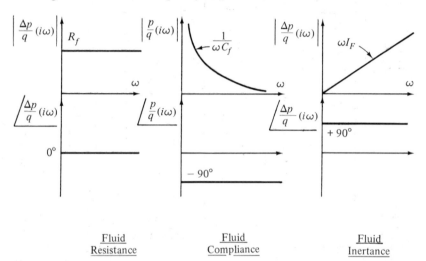

Fluid
Resistance

Fluid
Compliance

Fluid
Inertance

FIGURE 4–23

Fluid Impedance of Basic Elements

A good example of a complex fluid system[13] in which impedance measurements have been found useful is shown in Fig. 4–24. The system under
study is a boiler using Freon (a class of fluids used in refrigeration systems)
as the working fluid. Since in a boiler the working fluid enters as a liquid
and leaves as a gas (vapor), the flow situation (called two-phase flow) is
quite complex and difficult to analyze. Accurate dynamic performance
data on the boiler is needed to properly design the larger thermal power
system of which it will be a part, so an experimental fluid impedance study
was run. To minimize the effect of nonlinearities, the impedance is taken
for small perturbations around an equilibrium operating point. By running
several such tests at different operating points, one can explore the degree
of nonlinearity. (If all operating points gave exactly the same impedance

---

[13] E. A. Krejsa, J. H. Goodykoontz, and G. H. Stevens, "Frequency Response of
Forced-Flow Single Tube Boiler," NASA TN-D-4039 (June, 1967).

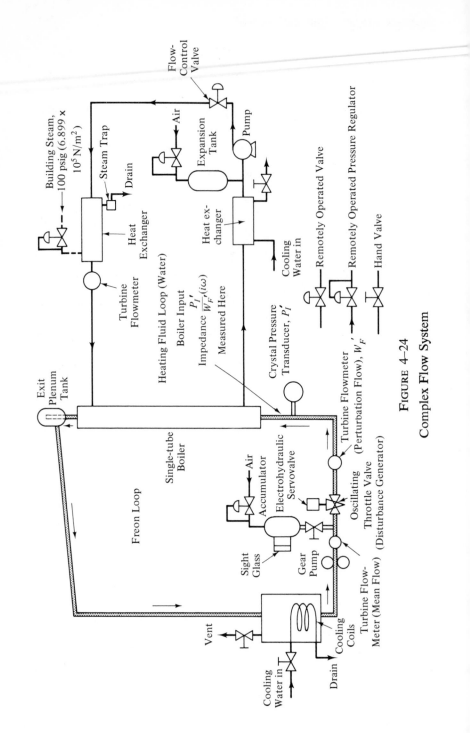

FIGURE 4-24

Complex Flow System

curves, the system would be perfectly linear.) The test procedure is to set the flow-control servovalve at some fixed position and establish steady flow. Then the valve is oscillated sinusoidally about the original position, causing a small oscillation of flow rate and, thereby, pressure. (For example, the data of Fig. 4–25 had an equilibrium flow rate of 445 lb_m/hr and a

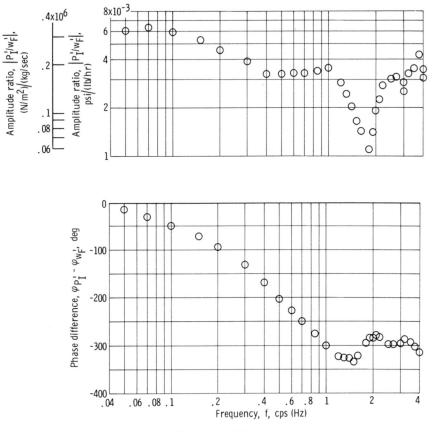

FIGURE 4–25

Impedance of a Freon Boiler

pressure of 25.5 psia. The flow rate oscillation was set at an amplitude of about 40 lb_m/hr; the resulting pressure oscillation amplitude ranged from 0.04 to 0.28 psi.) The flow rate oscillation was measured with a turbine flowmeter, and the pressure with a crystal pressure transducer.[14] Measure-

[14] E. O. Doebelin, *Measurement Systems* (New York: McGraw-Hill Book Company, 1966).

ments were made at 30 frequencies between 0.05 and 4.0 Hertz, giving the
sinusoidal transfer function of the impedance (P_l'/W_F') $(i\omega)$ as shown in
Fig. 4–25. (Note that the flow rate W_F' is given as a mass flow rate $[\text{lb}_m/\text{hr}]$
rather than a volume flow rate $[\text{ft}^3/\text{hr}]$, but one can easily convert if the
fluid density is known.) A theoretical model of this system has also been
derived and comparison with measured behavior made.[15] Good agreement
was achieved under some, but not all, operating conditions.

If a fluid impedance is known as an operational transfer function
$(\Delta p/q)$ (D), it should be clear that one can then calculate the response to
any given input by solving the appropriate differential equation. For example, if

$$\frac{\Delta p}{q}(D) = \frac{10}{5D + 1}, \frac{\text{psi}}{\text{in}^3/\text{sec}} \tag{4-67}$$

and an input flow rate $q = 1 + 2t$ is applied, the differential equation is

$$(5D + 1)\Delta p = 10q = 10 + 20t \tag{4-68}$$

$$5\frac{d\Delta p}{dt} + \Delta p = 10 + 20t \tag{4-69}$$

The solution of this equation (with initial condition $\Delta p = 0$ at $t = 0$) is

$$\Delta p = 90(e^{-0.2t} - 1) + 20t \tag{4-70}$$

When a fluid impedance is measured by the frequency response technique,
we do not have a transfer function in equation form; we have only the
curves such as in Fig. 4–25. The response to sinusoidal inputs is, of course,
easily calculated from such curves. What is not obvious, but nevertheless
true, is that the response to *any* form of input can be calculated when these
frequency response curves are known, using the method of Fourier transforms. Thus the measured impedance curves give a complete and general
description of the pressure/flow dynamics of the fluid system, and are not
limited to just sinusoidal inputs. Of course this same statement applies
also to mechanical and electrical impedances.

4–6. FLUID SOURCES, PRESSURE AND FLOW.

An ideal pressure source produces a specified pressure at some point in a
fluid system, no matter what flow might be required to maintain this pressure. Similarly, an ideal flow source produces a specified flow, irrespective
of the pressure required to produce this flow. In man-made fluid systems
the most common source of fluid power is a pump or compressor of some

---

[15] E. A. Krejsa, "Model for Frequency Response of a Forced Flow, Hollow, Single
Tube Boiler," NASA TM-X-1528 (March, 1968).

sort. A positive-displacement liquid pump draws in, and then expels, a fixed amount of liquid for each revolution of the pump shaft. When driven at a constant speed, such a pump closely approximates an ideal constant-flow source over a considerable pressure range. Its main departure from ideal behavior is a decrease in flow rate as load pressure increases due to leakage through clearance spaces. This leakage flow is proportional to pressure; thus one can represent a real pump as a parallel combination of an ideal flow source and a linear (and large) flow resistance R_{fl} as in Fig. 4–26. If the inlet flow impedance of the load is low relative to R_{fl}, most of

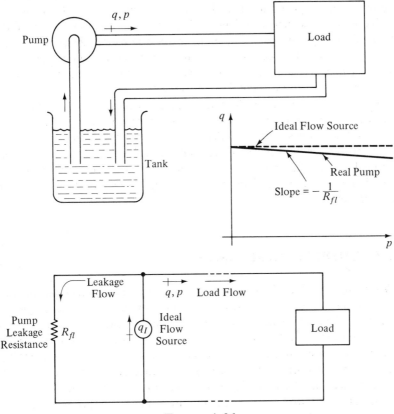

FIGURE 4–26

Positive Displacement Pump as a Flow Source

the flow q goes into the load rather than the pump leakage path, and the pump acts nearly as an ideal flow source. When time-varying flows are required, a number of approaches are possible. A fixed displacement pump may be driven at a time-varying speed, or a variable-displacement pump may be employed. In a variable-displacement pump, the shaft speed is constant but pump output per revolution can be varied, while the pump is running, by moving a stroke mechanism. This mechanism allows pump flow rate to be varied smoothly and quickly from full flow in one direction through zero flow to full flow in the reverse direction.

By combining a positive displacement pump with a relief valve, one can achieve a practical constant-pressure source. This real source will not have the perfect characteristic of an ideal pressure source, but can be modeled as a combination of an ideal source with a flow resistance. A relief valve is a spring-loaded valve which remains shut until the set pressure is reached. At this point it opens partially, adjusting its opening so that the pump flow splits between the demand of the load and the necessary return flow to the tank. To achieve this partial opening against the spring, the pressure must change slightly; thus we do not get an exactly constant pressure (see Fig. 4–27). This real source can be modeled as a series combination of an ideal pressure source with a small flow resistance.

The above two examples do not, of course, exhaust the possibilities with regard to power sources in fluid systems, but they should give some idea of how real sources may be modeled in terms of ideal sources and passive elements. Other fluid power sources encountered in practice include centrifugal pumps, accumulators (used for short-term power supplies), elevated tanks or reservoirs (gravity is the energy "source"), the human heart (a complex pump), etc.

4–7. THERMAL RESISTANCE.

Whenever two objects (or two portions of the same object) have different temperatures, there is a tendency for heat to be transferred from the hot region to the cold region in an attempt to equalize the temperatures. For a given temperature difference, the rate of heat transfer varies depending on the thermal resistance of the path between the hot and cold region. The nature and magnitude of the thermal resistance depend on the modes of heat transfer involved; conduction, convection, or radiation.

In Fig. 4–28, two bodies at temperature T_1 and T_2, respectively, are connected by a solid rod of constant cross-sectional area A and length L. The rod is made of a material with thermal conductivity k. Fourier's law of heat conduction may be written in the following form as a means of defining thermal resistance.

$$\text{heat transfer rate} \triangleq q = \frac{kA}{L}(T_1 - T_2) = \frac{kA}{L}\,\Delta T, \frac{\text{Btu}}{\text{sec}} \quad \textbf{(4–71)}$$

where

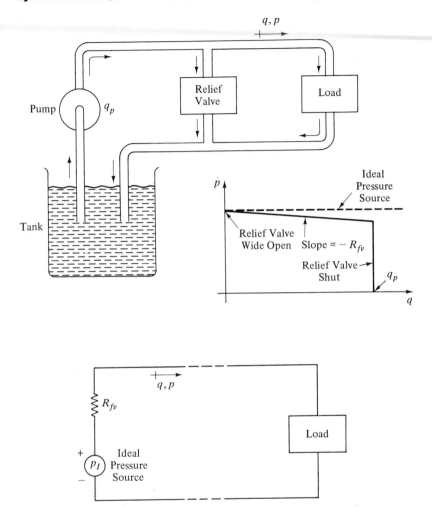

FIGURE 4–27

Positive Displacement Pump and Relief Valve as Pressure Source

$$A \triangleq \text{area, inch}^2$$

$$L \triangleq \text{length, inch}$$

$$k \triangleq \text{thermal conductivity, } \frac{\text{Btu}}{\text{in}^2 - {}^\circ\text{F/in}}$$

$$T_1, T_2 \triangleq \text{temperature, } {}^\circ\text{F}$$

Thermal conductivity is a material property which is found by experiments based on Eq. (4–71). That is, q, A, L, and ΔT are all measured for a steady-state situation and k is calculated from Eq. (4–71). Ideally, k is a constant,

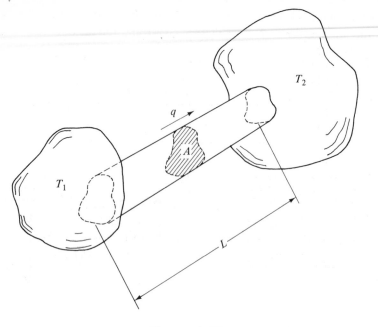

FIGURE 4–28

Heat Transfer by Conduction

but in reality it may vary with temperature, position in the body, and direction of the heat flow. The assumption of constant k is often adequate in practical problems, and is necessary for a linear model. For solutions of specific numerical problems, computer methods to handle non-constant k are available. Equation (4–71) also indicates an instantaneous relation between ΔT and q. This would be true if the rod had no thermal capacitance (heat storage capability). Since a real rod will have thermal capacitance, we must think of Eq. (4–71) as defining only the resistive component of the rod dynamics; its thermal capacitance will be taken into account separately. We can now define the pure and ideal thermal resistance for conduction heat transfer as follows.

$$q = \frac{\Delta T}{R_t}$$

$$R_t \triangleq \frac{\Delta T}{q} = \frac{L}{kA}, \ °F/(Btu/sec) \qquad (4\text{–}72)$$

The analogy to electrical resistance is clear if we think of ΔT as the driving force (voltage) and heat flux q as the current.

Many practical situations involve heat flow through fluid/solid interfaces by convection. Here the heat flows by conduction through a thin layer of fluid (called the boundary layer) which adheres to the solid wall. At the interface between the boundary layer and the main body of fluid, the heat is carried away by the constantly moving fluid particles into the main stream. This overall process is called convection heat transfer, and is illustrated in Fig. 4–29. Experiments have shown that this process may be

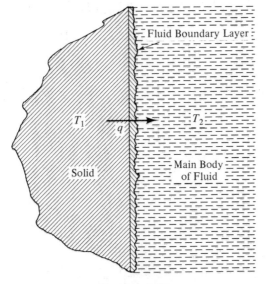

FIGURE 4-29

Heat Transfer by Convection

described by the equation

$$q = hA(T_1 - T_2) = hA\Delta T \tag{4-73}$$

where

$h \triangleq$ film coefficient of heat transfer, $\dfrac{\text{Btu}}{\text{sec-in}^2\text{-}°F}$

The film coefficient h must be found by experiment; however, the experiments have been generalized so that h may often be predicted with fair accuracy from calculations using basic fluid properties such as conductivity, density, viscosity, etc. Also, h varies somewhat, with temperature for example, so Eq. (4–73) is really a linearized version of reality, but the accuracy is often quite adequate. The thermal resistance associated with convection is seen to be

$$R_t \triangleq \frac{\Delta T}{q} = \frac{1}{hA}, \quad °F/(\text{Btu/sec}) \tag{4-74}$$

Often conduction and convection are combined, and we can define an overall heat transfer coefficient and thereby an overall resistance. Figure

4-30 shows a cross section of an automobile radiator in which heat flows from the hot internal liquid through the metal radiator wall and into the air forced over the radiator by the fan. Since the same heat flux q goes through all three of the resistances we may write

$$q\left(\frac{1}{h_W A}\right) + q\left(\frac{L}{kA}\right) + q\left(\frac{1}{h_A A}\right) = T_W - T_A \qquad (4\text{-}75)$$

$$q = \frac{T_W - T_A}{\dfrac{1}{h_W A} + \dfrac{L}{kA} + \dfrac{1}{h_A A}} = \frac{\Delta T}{R_t} \qquad (4\text{-}76)$$

$$\text{overall resistance} \triangleq R_t = \frac{1}{h_W A} + \frac{L}{kA} + \frac{1}{h_A A} \triangleq \frac{1}{U} \qquad (4\text{-}77)$$

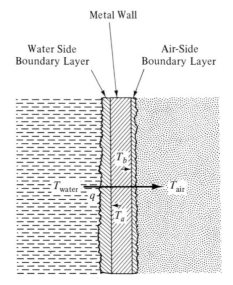

FIGURE 4-30

Overall Heat Transfer

where U is called the overall coefficient of heat transfer. We see that the overall resistance is simply the sum of the individual resistances, as we might have expected from the electrical analogy.

Two bodies can exchange thermal energy with no physical contact whatever by the process of radiation (see Fig. 4-31). The rate of heat transfer depends on a surface property of each body called the emissivity, geometrical factors involving the portion of emitted radiation from one body that actually strikes the other body, the surface areas involved, and the temperatures of the two bodies. For a given configuration and materials, the defining equation takes the form

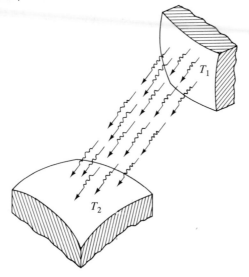

FIGURE 4-31

Radiation Heat Transfer

$$q = C(T_1^4 - T_2^4) \tag{4-78}$$

where C includes all the effects other than the temperatures, and the temperatures are absolute (°Kelvin or °Rankine). This mode of heat transfer is clearly nonlinear, but can be linearized for approximate analyses as long as the temperatures do not vary greatly from defined operating points, T_{10} and T_{20}.

$$q \approx C(T_{10}^4 - T_{20}^4) + \left.\frac{\partial q}{\partial T_1}\right|_{\substack{T_1 = T_{10} \\ T_2 = T_{20}}} (T_1 - T_{10}) + \left.\frac{\partial q}{\partial T_2}\right|_{\substack{T_1 = T_{10} \\ T_2 = T_{20}}} (T_2 - T_{20}) \tag{4-79}$$

$$q \approx -3CT_{10}^4 + 3CT_{20}^4 + (4CT_{10}^3)T_1 - (4CT_{20}^3)T_2 \tag{4-80}$$

While Eq. (4-80) is linear in T_1 and T_2, it does not allow definition of a thermal resistance unless $T_{10} = T_{20} = T$, that is, the operating point must be one of zero heat transfer. For this case,

$$q \approx 4CT^3(T_1 - T_2) \tag{4-81}$$

and

$$R_t \approx \frac{\Delta T}{q} = \frac{1}{4CT^3} \tag{4-82}$$

While the Taylor series linearization of Eq. (4-79) does not lead to a general resistance expression, another approach provides a useful result. Equation (4-78) can be factored exactly as

$$q = C(T_1 + T_2)(T_1^2 + T_2^2)(T_1 - T_2) \qquad (4\text{-}83)$$

which gives

$$R_t = \frac{\Delta T}{q} = \frac{1}{C(T_1 + T_2)(T_1^2 + T_2^2)} \qquad (4\text{-}84)$$

This resistance varies with T_1 and T_2, and is thus nonlinear; however, a linear approximation near a given operating point is

$$R_t \approx \frac{1}{C(T_{10} + T_{20})(T_{10}^2 + T_{20}^2)} \qquad (4\text{-}85)$$

The number C generally is quite small, so R_t for radiation is large (compared to typical conduction and convection values), unless operating temperatures are very high. Thus at low and medium temperatures, the radiation mode of heat transfer often contributes little to the overall heat flow, and is neglected. At high temperatures radiation may be very significant, and, in some applications, such as a satellite in orbit, it is even important at low temperatures, because the other modes of heat transfer cannot take place in the absence of an atmosphere and physical contact with other objects.

Figure 4-32 shows the symbol used for thermal resistance in system diagrams; it is identical with that for electrical resistance. The temperature difference $(T_1 - T_2)$ is a driving potential for heat flux q, just as a voltage difference $(e_1 - e_2)$ is a driving potential for current i. Since the equation relating q to ΔT is an algebraic one, the response of q to ΔT is instantaneous.

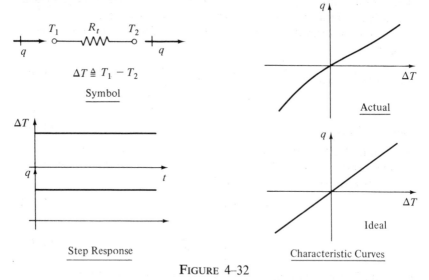

FIGURE 4-32

Thermal Resistance Characteristics (See Table of Properties in Appendix D.)

When energy behavior is considered, the thermal/electrical analogy breaks down, since the heat flux q (Btu/sec) is *already* power, whereas the analogous current i is not. Also, all the heat flux entering the thermal resistance at one end leaves it at the other, and none of it is lost or dissipated; whereas the electrical energy applied to a resistor is all converted to heat, and is thus lost to the electrical system. Appendix D also lists some numerical values relating to conduction, convection, and radiation resistance, to give some idea of orders of magnitude, ranges, and relative sizes.[16]

4-8. THERMAL CAPACITANCE AND INDUCTANCE.

When heat flows into a body of solid, liquid, or gas, this thermal energy may show up in various forms such as mechanical work or changes in kinetic energy of a flowing fluid. If we restrict ourselves to bodies of material for which the addition of thermal energy does not cause significant mechanical work or kinetic energy changes, the added energy shows up as stored internal energy and manifests itself as a rise in temperature of the body. For an ideal thermal capacitance, the rise in temperature is directly proportional to the total quantity of heat transferred into the body.

$$T - T_0 = \frac{1}{C_t} \int_0^t q \, dt \qquad (4\text{-}86)$$

where

$T \triangleq$ temperature of body at time t

$T_0 \triangleq$ temperature of body at time $t = 0$

$C_t \triangleq$ thermal capacitance, Btu/°F

Since we refer to *the* temperature of the body, we are assuming that, at any instant, the body's temperature is uniform throughout its volume. For fluid bodies, this ideal situation is closely approached if the fluid is thoroughly and continuously mixed. For solid bodies, uniform temperature requires a material with infinite thermal conductivity k, since then, for any heat flow rate q through the body, the temperature difference $\Delta T = -q\Delta x/kA$ would be zero. No real material has infinite k; thus, there is always some nonuniformity of temperature in a body during transient temperature changes. Many practical problems involve solid bodies immersed in fluids, and for this situation a useful criterion for judging the validity of the uniform-temperature assumption is found in the Biot number, N_B (see Fig. 4-33).

---

[16] A. I. Brown and S. M. Marco, *Introduction to Heat Transfer* (New York: McGraw-Hill Book Company, 1951).

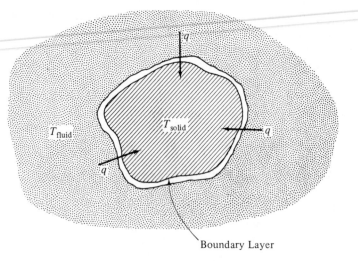

Boundary Layer

FIGURE 4–33

Configuration for Biot Number Calculation

$$N_B \triangleq \frac{hL}{k} \qquad (4\text{–}87)$$

$h \triangleq$ film coefficient at surface

$L \triangleq$ volume/surface area

$k \triangleq$ thermal conductivity of solid body

For bodies whose shape approximates a plate, sphere, or cylinder, if $N_B < 0.1$, the error in assuming the solid to have uniform temperature is less than about 5%. For example, a one-inch diameter spherical steel ball being heated in stagnant air has $h \approx 2.$, $k \approx 35.$, and $L = 1/6$, making $N_B = 0.0095$; thus the ball's internal temperature may safely be assumed uniform. The thermal capacitance of the boundary layer fluid film is generally negligible, since the film is very thin; thus convective resistances are very nearly pure resistances.

The calculation of numerical values of thermal capacitance is relatively straightforward, since the temperature rise of a body when heat is added is given by:

$$\text{heat added} = \int q \, dt = (\text{mass})(\text{specific heat})(\text{temperature rise}) \qquad (4\text{–}88)$$

Thus

$$C_t \triangleq \frac{\text{heat added}}{\text{temperature rise}} = (\text{mass})(\text{sp. heat}) = Mc \qquad (4\text{–}89)$$

The specific heat c of real materials varies somewhat with temperature,

but a constant mean value may often be assumed. If the temperature variation of c is known and considered not negligible, computer approaches
allow simulation of Eq. (4–88) directly. For fluids (particularly gases) the
specific heat is often measured for two conditions: constant volume and
constant pressure. Since these values are quite different (for air at 32°F,
$c_p = 0.240$ and $c_v = 0.171$), one must be careful to use the one which
corresponds to the particular application. When heat is added to or taken
away from a material which is changing phase (melting or freezing, vapor-
izing or condensing) the thermal capacitance is essentially infinite, since
one can add heat without causing *any* temperature rise.

In Eq. (4–86) we may define $T_0 \equiv 0$ if we wish, giving the transfer function

$$\frac{T}{q}(D) = \frac{1}{C_t D} \tag{4–90}$$

and the step response of Fig. 4–34. A standard symbol for thermal capacitance has not been defined; that given in Fig. 4–34 is suggested as a simple
and reasonable one.

By analogy to electrical systems, a thermal inductance would have a
q/T characteristic given by $T_1 - T_2 = L_t(dq/dt)$. No physical effect dis-

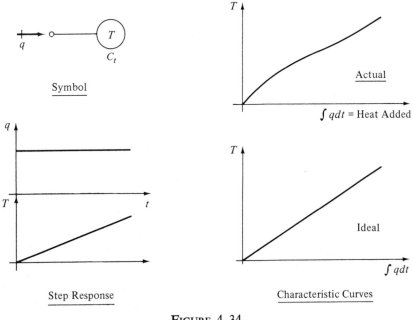

FIGURE 4–34

Thermal Capacitance Characteristics (See Table of Properties in
Appendix D.)

playing this relation has yet been discovered; thus, thermal inductance is not necessary for the description of thermal system behavior and is not defined or used.

4-9. THERMAL SOURCES, TEMPERATURE AND HEAT FLOW.

The ideal temperature source maintains a prescribed temperature irrespective of how much heat flow it must provide, while an ideal heat flow source produces a prescribed heat flow irrespective of the temperature required. Constant-temperature sources may often be quite well approximated by utilizing materials undergoing phase change. A well-stirred bath of ice and water remains very nearly at 32°F, even if heat flows are entering or leaving it; similarly for water boiling at atmospheric pressure and 212°F. The melting points of various metals and salts are similarly used to establish desired constant temperatures over a wide range. A large vessel of liquid, even if not changing phase, will maintain a nearly constant temperature for short time intervals as long as the heat flows in or out are not too large. When a specific time-varying temperature is required, a liquid bath with a feedback temperature control system may be necessary (Fig. 4–35).[17] Here the bath temperature is measured and converted to a proportional voltage, which is then compared to a command voltage representing the desired temperature. If desired and actual temperatures are not equal, the controller modulates the power to the electric heater so as to provide more or less heat, as needed.

Perhaps the most convenient heat flow source for many applications is electrical resistance heating. A constant or time-varying voltage $e(t)$ applied to a resistance heating coil produces an electrical heat generation rate of $e^2(t)/R$ if inductance is negligible. Suppose such a coil is immersed in a fluid bath to act as a heat-flow source. The electrically generated heat goes partly into heating the metal of the coil itself, and the rest flows away to the fluid as intended. If the thermal capacitance of the coil metal is sufficiently small, practically all the electrically generated heat flows into the fluid, and we have a good approximation to a heat-flow source. Radiant heat flux may also be usable as a heat-flow source. The sun, for example, provides about 400 Btu/hr-ft² at high altitudes, and as much as 300 Btu/hr-ft² at the earth's surface at noon on a clear day. These heat flows are of course quite unaffected by the presence or absence of an object to receive them, and thus represent nearly ideal heat-flow sources. Radiant heat lamps are available when a controllable source of this type is needed.

---

[17] E. O. Doebelin, *Dynamic Analysis and Feedback Control* (New York: McGraw-Hill Book Company, 1962).

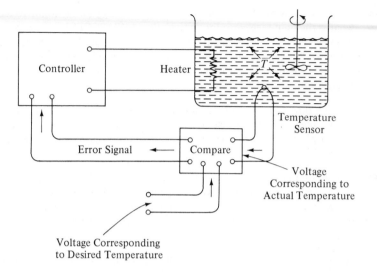

FIGURE 4–35

Feedback System for Temperature Source

Mechanical shutters can "turn on" radiant beams very quickly if step changes in heat flow are needed.

BIBLIOGRAPHY

1. Andersen, B. W., *The Analysis and Design of Pneumatic Systems*. New York: John Wiley & Sons, Inc., 1967.

2. Kreith, F., *Principles of Heat Transfer* (2nd ed.). Scranton: International Textbook Company, 1965.

3. Merritt, H. E., *Hydraulic Control Systems*. New York: John Wiley & Sons, Inc., 1967.

4. Vennard, J. K., *Elementary Fluid Mechanics* (4th ed.). New York: John Wiley & Sons, Inc., 1961.

PROBLEMS

4–1. In Fig. 4–12, compare the fluid resistance R_f for circular, square, and triangular "pipes" of the same area and length.

4–2. For water flowing in a 0.25 inch diameter smooth pipe of length 10 feet, what is the fluid resistance for laminar flow? What is the largest pressure drop which will give laminar flow and what is the flow rate for this condition? If the pressure drop is made ten times this value, will the flow rate also increase by 10 times? What *will* the flow rate now be? Use viscosity of 3.8×10^{-7} lb_f-sec/ft^2 and density 2.3×10^{-3} lb_f-sec^2/ft^4.

4–3. For the pipe of prob. 4–2, calculate and plot versus flow rate the incremental linearized flow resistance for turbulent flow. For a pressure drop of 300 psi, how much does R_f change for a $\pm 10\%$ change in pressure drop?

4–4. For a laminar pipe flow, if viscosity changes by 100%, how much does R_f change? Compare this with the change in R_f caused by a similar viscosity change, but in turbulent flow. If one desires to minimize the effects of viscosity changes, should he design for laminar or turbulent flow?

4–5. In Eq. (4–14) if N_R is 500, what is the maximum allowable D/L to give 5% nonlinearity? Using this D/L value and $\mu = 5 \times 10^{-6}$ lb_f-sec/in^2, design a capillary tube to give $R_f = 100$ psi/(in^3/sec). What is the maximum flow rate this tube can pass with laminar flow? Use density of 6×10^{-5} lb_f-sec^2/in^4.

4–6. In Eq. (4–24) reduce the term $4\rho L/3AR_f$ to its simplest form in terms of basic constants. Now state the requirement for negligible inertance effect in terms of ρ, D, μ, and ω. Why does length L have no effect on this? If ω_{max} is the frequency at which the imaginary term in Eq. (4–24) is 0.1, plot a curve of $D^2\rho/\mu$ versus ω_{max}. What is the usefulness of such a curve in designing flow resistances?

4–7. Using Eq. (4–37), plot q versus Δp for water flowing through an orifice of 0.1 inch diameter, taking $C_d = 0.6$. What is the linearized resistance in the neighborhood of $\Delta p = 100$ psi? How much does it change for a ± 10 psi change in Δp?

4–8. For the single capillary in Fig. 4–16, compute a theoretical pressure/flow curve and compare with the actual measured behavior, using
   a. Equation (4–12).
   b. Equation (4–15).

4–9. Repeat prob. 4–8 for the three capillaries in parallel in Fig. 4–16.

4–10. Using the appropriate formulas from the text, plot a theoretical pressure/flow curve for the orifice of Fig. 4–16, and compare with the measured result.

4–11. Using Eq. (4–37) fit an empirical curve to the orifice data of Fig. 4–16. This is, find K in $q = K\sqrt{\Delta p}$. How well does this curve fit the data? For the K you found, what is the corresponding value of C_d?

4–12. What D/t ratio is required in steel tubing for the tubing compliance to equal the compliance of oil with bulk modulus of 200,000 psi? What would D/t be for aluminum tubing?

4–13. Rubber hose is often a complex composite material with layers of rubber, fabric, and woven metal reinforcement. This makes theoretical calculation of compliance a practical impossibility. Explain how you would set up experiments to find the compliance, remembering that it will probably be nonlinear.

4–14. Design a piston-type accumulator as in Fig. 4–18a to supply a hydraulic load which consumes 0.2 horsepower for a 1-minute period. Assume the load can use all the stored energy and design for a maximum pressure of 3000 psi. Find all combinations of A and K_s which meet these requirements. If space limits A to 5 in², find K_s and the total stroke.

4–15. Repeat prob. 4–14 for the case where the load can only use the energy supplied when the pressure is between 2000 and 3000 psi.

4–16. Find an expression for the linearized compliance of the conical tank in Fig. 4–20b.

4–17. Repeat prob. 4–16 for a spherical tank.

4–18. Using Eq. (4–62), compute the kinetic energy, effective mass, and inertance for any value of n. Evaluate the effective mass for $n = 4$ and $n = 10$.

4–19. For the system of Fig. 4–25, find the flow rate if:
a. Pressure $= 1 \sin 0.1\pi t$, psi, t in seconds.
b. Pressure $= 1 \sin \pi t$.
c. Pressure $= 1 \sin 3.6\pi t$.
d. Pressure $= 1 \sin 0.1\pi t + 1 \sin \pi t + 1 \sin 3.6\pi t$.

4–20. In the system of Fig. 4–26, $q_I = 30$ inch³/sec, $R_{fl} = 1000$ psi/(in³/sec), while the load is a flow resistance of 10 psi/(in³/sec). What pressure will this system run at? What flow is supplied to the load? Plot a q versus p curve for this pump. What is the power?

4–21. In Fig. 4–27, $p_I = 1000$ psi, $R_{fv} = 1$ psi/(in³/sec), and the pump flow is 30 inch³/sec. Plot a p versus q curve for this system. What is the percent pressure change over the full range of this power supply? What is the efficiency at $q = 0$? At $q = 30$ in³/sec? At $q = 15$ in³/sec? What happens to the wasted portion of the energy?

4–22. The brass rod of Fig. P4–1 carries a steady heat flux of 50 Btu/hr. Compute the thermal resistance of each section of the rod and its total resistance. If the left end is at 400°F, calculate and plot the variation of temperature from left to right. If we arbitrarily call the bar's energy content zero when it is all at 0°F, compute the stored energy when it is heated as above. Compute the thermal capacitance of each section of the rod, and also the total capacitance. Can you think of a way to use these

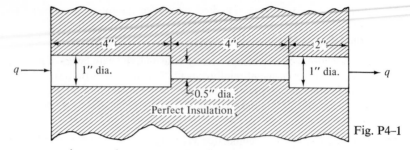

Fig. P4–1

capacitance values to compute the stored energy when the rod is at a nonuniform temperature as in this example?

4–23. In Fig. 4–30, let $h_w = 500$ Btu/(hr-ft²-°F), $h_A = 20$ Btu/(hr-ft²-°F), and let the metal wall be 0.03-inch-thick brass. Compute the total thermal resistance per square foot of area. Which component of resistance dominates the total? How much error is caused by entirely neglecting the others?

4–24. If $h = 2$ Btu/(hr-ft²-°F), what is the size of the largest steel cube for which the internal temperature may be assumed uniform? Repeat for a silver cube. How do these results change if $h = 200$?

4–25. Based on the data of App. D , what material should one use for thermal energy storage if he wishes to use up the minimum space?

4–26. A 500-ohm electric resistance heater is embedded in a brass sphere of 5 inch diameter. The surface of the sphere is perfectly insulated. If the temperature is 70°F at time zero when 100 volts is applied to the heater, calculate and plot the temperature rise of the sphere versus time. If the 100-volt source is turned on and off in a repetitive cycle (2 minutes on, 3 minutes off), again calculate and plot temperature versus time.

5

BASIC ENERGY
CONVERTERS

5-1. INTRODUCTION.

Chapters two, three, and four have introduced basic elements of mechanical, electrical, fluid, and thermal systems. Practical machines and processes often include hardware which involves several of these fields, since the different functions in a process may each be best accomplished in a particular way. If different parts of a process operate with different forms of energy but must all work together, it means that devices called *energy converters* must be available to couple the diverse parts. Also, systems often exhibit coupling between different forms of energy which was not intentionally designed into the system, but must nevertheless be accounted for in analysis. This chapter will give a brief introduction to some of these devices and effects which accept an energy input in one form and produce an output in another form. In most cases just a word description or diagram will be given, but in a few instances the analytical models are sufficiently simple and useful that they will be stated and explained. Our main purpose here is not to develop a broad theory of energy conversion, but merely to introduce the reader to some basic devices which will be useful in our study of complete systems in later chapters. Some of these devices will already have been mentioned earlier in the portions of chapters devoted to sources. For example, the DC generator mentioned as a voltage source in chapter three is also a basic mechanical-to-electrical energy converter.

5–2. CONVERTING MECHANICAL ENERGY TO OTHER FORMS.

We define mechanical power as the product of torque and angular velocity for a rotating shaft, or force and translational velocity for a translating shaft. Figure 5–1 summarizes some of the methods of converting power available in mechanical form to other forms.

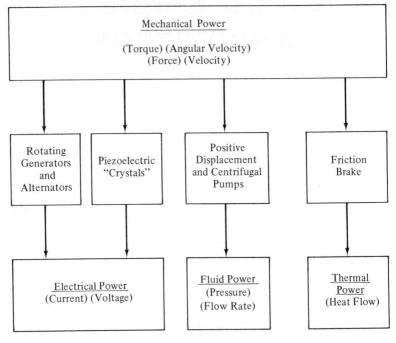

FIGURE 5–1

Converting Mechanical Power to Other Forms

The most important mechanical-to-electrical energy conversion process undoubtedly is that associated with the rotating electrical machines called generators (DC) or alternators (AC). In these machines the mechanical power is applied to the rotating member (rotor) and the electrical power is taken off from windings on either the stationary member (stator) or the rotor. For the DC machine the relationships are relatively simple, and we now state them without detailed derivation. In Fig. 5–2 application of

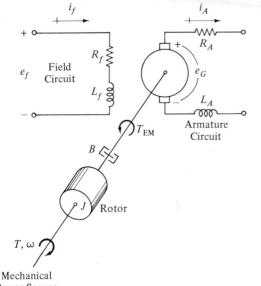
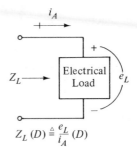

$$Z_L (D) \triangleq \frac{e_L}{i_A} (D)$$

FIGURE 5–2

Mechanical-to-Electrical Energy Conversion: The DC Generator

Newton's law to the rotor gives us

$$\sum \text{Torques} = J\alpha \qquad (5\text{–}1)$$

$$T - B\omega - T_{EM} = J\dot{\omega} \qquad (5\text{–}2)$$

where

$T \triangleq$ applied torque, inch-lb_f

$B \triangleq$ viscous friction of rotor bearings and windage, inch-lb_f/(rad/sec)

$\omega \triangleq$ rotor speed, rad/sec

$J \triangleq$ rotor moment of inertia, inch-lb_f-sec^2

$T_{EM} \triangleq$ electromagnetic reaction torque due to interaction of magnetic field and current in the armature.

The torque T_{EM} is related to machine dimensions, material properties, magnetic field strength and armature current in a complicated way which can, however, be estimated by application of basic physical laws. An engineer who *designs* generators would have to become familiar with the details of such an analysis. A system engineer, however, is not generally a designer of generators, but rather a *user* of generators which have already been designed and built. As such, his knowledge of generator characteristics may be limited to those overall performance indices which are necessary to an intelligent *selection* of a particular generator from a family of machines available from the manufacturer. Furthermore, the accurate numerical values needed for system analysis rarely can be obtained from theory alone; thus the system engineer often relies on experimental testing as a means of obtaining needed values of system parameters. For the torque T_{EM}, theory indicates that it should be of the form

$$T_{EM} = K_T i_f i_A \qquad (5\text{-}3)$$

where

$i_f \triangleq$ instantaneous field current

$i_A \triangleq$ instantaneous armature current

$K_T \triangleq$ a composite machine constant including several basic constants.

If we have a generator in hand, an experiment to verify the form of Eq. (5-3) and find a numerical value for K_T can be run. Since Eq. (5-3) is independent of speed, we can run our experiment at any speed, and the easiest speed is zero. The experiment thus consists of holding the rotor fast with a torque-measuring device (like a spring scale) and applying and measuring a range of field currents and armature currents. The data can then be plotted either as in Fig. 5-3a or 5-3b. If the data plot essentially as equally spaced straight lines (as in Fig. 5-3), the form of Eq. (5-3) is validated and the numerical value of K_T can be calculated from any set of values of T_{EM}, i_f, and i_A. If the data are somewhat scattered, *all* the sets of T_{EM}, i_f, and i_A can be used to compute an average value of K_T. Note in Fig. 5-3a that as field current is increased a point is reached beyond which further increase does not result in a torque increase. This corresponds to the saturation of the iron in the field; here an increase in field current does not actually cause an increase in the magnetic field, and Eq. (5-3) no longer holds. In Fig. 5-3b, for any field current (including values which cause saturation), an increase in i_A always causes a proportional increase in torque. This does not, however, mean that arbitrarily large values of i_A may be permitted, since these would cause overheating of the machine.

Assuming the field to be supplied by a voltage source, it would be modeled as in Fig. 5-2. Both the resistance and inductance shown there would normally be significant. In the armature circuit the inductance L_A might often be negligible. The generated voltage e_G can again be estimated from basic physics, and theory gives the form

$$e_G = K_E \omega i_f \qquad (5\text{-}4)$$

where

$K_E \triangleq$ a composite machine constant including several basic constants.

We again appeal to experiments to validate Eq. (5-4) and find an accurate value for K_E. The generator is run at various speeds and field currents, and

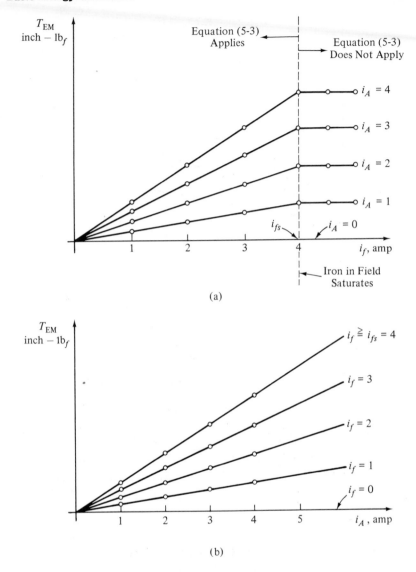

FIGURE 5–3

Generator Characteristic Curves

the voltage e_G is measured with a high-resistance voltmeter so as to keep i_A essentially zero, since the resistance R_A is *internal* to the generator and would cause an incorrect reading of voltage if current flowed. The data may be plotted as in Fig. 5–4 and interpreted just as was Fig. 5–3. Equations (5–2), (5–3), and (5–4), together with Fig. 5–2, are sufficient to

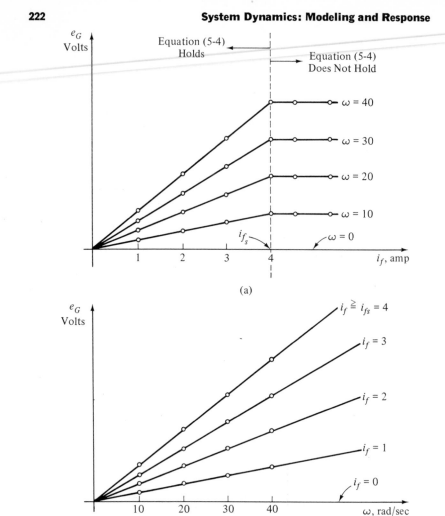

(a)

(b)

FIGURE 5–4

Generator Characteristic Curves

describe the behavior of a DC generator, and will be used in later chapters
in studies of complete systems.

In Fig. 5–1, the other mechanical-to-electrical energy converter listed is
the piezoelectric "crystal." Let us first emphasize that this physical effect
(the piezoelectric effect) is not at all widely used as a means of generating
electrical power, because of certain inherent limitations when compared
with rotating generators. It *is*, however, usefully exploited in a number of
important applications, mainly in the instrumentation field, where large

amounts of continuous power are not required. Many force, pressure, and acceleration measuring instruments use piezoelectric "crystals" as basic sensing elements for converting mechanical force and motion into a proportional electrical signal. The principle *has* been used to produce electrical power, but only in very specialized applications particularly suited to it. One of these is a power supply for a portable pulse x-ray unit.[1] The requirement here is for a short power pulse of very high voltage but low current in a compact unit—conditions which closely match the capabilities of piezoelectric generators.

Figure 5–5 shows perhaps the simplest configuration of piezoelectric

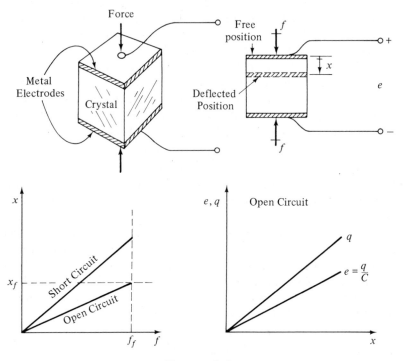

FIGURE 5–5

Piezoelectric Energy Converter

generator; a "crystal" sandwiched between metal electrodes and subject to direct tension or compression. Materials which exhibit the piezoelectric effect include natural crystals (quartz and Rochelle salt are examples), and man-made polycrystalline ceramics (barium titanate, lead zirconate titanate, etc.) which can be made piezoelectric by suitable processing. We use

---

[1] Ohio State University, Welding Engineering Department.

the word "crystal" somewhat loosely in our discussions to cover all piezoelectric materials. Since all these materials are good electrical insulators (dielectrics), when one applies the electrode plates as in Fig. 5–5, he creates a capacitor. The capacitor is now quite unusual, however, since it is also a *generator* of electrical charge whenever the crystal is deflected by the application of mechanical force. While the detailed analysis of piezoelectric devices can become very complicated, some of the observed overall behavior can be described in simple terms. If we gradually apply a force to the crystal of Fig. 5–5 with its electrical terminals open-circuit, the crystal deflects in proportion to the applied force (like a spring) and simultaneously a charge q and voltage $e = q/C$ appear on the "capacitor," the amount of charge being directly proportional to the deflection. The mechanical work done by the applied force is $f_f x_f/2$, and a portion of this work has been converted into electrical energy, since the charged capacitor has stored energy $q^2/2C$. If the deflection is reversed from compression to tension, the polarity of the charge and voltage also reverses. The fraction of the mechanical work which can be converted into electrical energy varies from material to material. A material constant k, called the *electromechanical coupling coefficient*, is defined by

$$k^2 \triangleq \frac{\text{electrical energy}}{\text{input mechanical energy}} \tag{5–5}$$

and varies from $k \approx 0.1$ (1% energy conversion) for quartz, through 0.5 to 0.7 for synthetic ceramics, to 0.9 (81% energy conversion) for Rochelle salt.

Due to the coupling between mechanical and electrical effects, some unusual behavior is observed. If we apply a force with the electrical terminals short-circuited (so that the capacitor cannot be charged), the crystal will be observed to be a "softer" spring (more compliant) than if the terminals are open-circuit. The relation between the spring constants K_{oc} and K_{sc} (lb$_f$/inch) for the two conditions is given by

$$\frac{K_{oc}}{K_{sc}} \triangleq \frac{K_{\text{open circuit}}}{K_{\text{short circuit}}} = \frac{1}{1 - k^2} \tag{5–6}$$

This relation can be derived from Eq. (5–5) by the following reasoning. With the terminals open-circuit, apply a force f which causes a deflection x. The total energy put into the crystal by the mechanical power source is $x^2 K_{oc}/2$, the electrical energy produced is $k^2(x^2 K_{oc}/2)$ and the stored mechanical energy is $x^2 K_{oc}(1 - k^2)/2$. If we now hold the crystal fast so x cannot change (and thus no mechanical work can be done), and then short-circuit the terminals, the electrical energy will be dissipated, and the force will relax to the value associated with the deflection x and the short-circuit spring constant K_{sc}. We may then equate the stored mechanical energy to that associated with the short-circuit spring constant

$$\frac{x_f^2 K_{oc}(1 - k^2)}{2} = \frac{x_f^2 K_{sc}}{2} \qquad (5\text{--}7)$$

and thus

$$\frac{K_{oc}}{K_{sc}} = \frac{1}{1 - k^2} \qquad (5\text{--}8)$$

Figure 5–6 illustrates this process. We should mention here (even though the topic of electrical-to-mechanical energy will have its own section) that

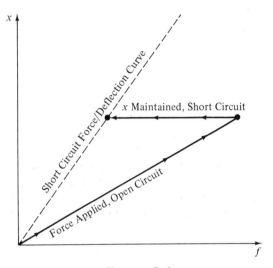

FIGURE 5–6

Relation Between Spring Constants
for Open-Circuit and Short-Circuit

the piezoelectric effect is *reversible*, and one can apply an external voltage to the crystal and thereby cause it to mechanically deflect.

If a piezoelectric generator is connected to an electrical load with input impedance Z_L, as in Fig. 5–7, and then a driving force $f(t)$ is applied, the following equations hold.

$$\sum \text{Forces} = (\text{Mass})(\text{Acceleration}) \qquad (5\text{--}9)$$

$$-B\dot{x} - K_s x - C_1 e + f(t) = M\ddot{x} \qquad (5\text{--}10)$$

where

$$B \triangleq \text{effective damping coefficient of crystal}$$

$$M \triangleq \text{effective inertia (mass) of crystal}$$

The spring constant K_s is defined by applying a static force f with the terminals short-circuited so that $e = 0$. Since $\dot{x} = \ddot{x} = 0$ for a static force, Eq. (5–10) gives $K_s = f/x$, which can be experimentally measured

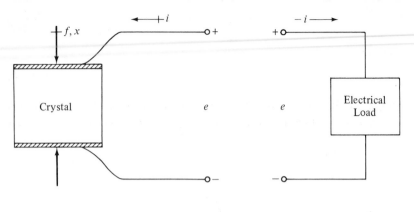

$$Z_L(D) = \frac{-i}{e}(D)$$

FIGURE 5-7

Piezoelectric Crystal with Electrical Load

by applying dead weights and measuring the deflection. The constant C_1 is found by clamping the crystal rigidly so that $x = \dot{x} = \ddot{x} = 0$, applying a known voltage e, and measuring the force produced by the crystal pushing against the clamp. Equation (5-10) then gives $C_1 = f/e$. The electric circuit equations are

$$C\dot{e} - C_q\dot{x} = i \qquad (5\text{-}11)$$

for the crystal and

$$\frac{e}{i}(D) = -Z_L \qquad (5\text{-}12)$$

for the electrical load. In Eq. (5-11), if we clamp the crystal so that $\dot{x} = 0$, then we find that $C = i/\dot{e} = \left(\int i\, dt\right)/e$, showing that C is just the capacitance of the crystal, which can be measured with an ordinary capacitance measuring device, but *must* be done with the motion constrained. (If we measure C with the crystal free to deflect, we will get a *different* number for C.) The constant C_q is defined by leaving the terminals open-circuit, so that $i \equiv 0$, and then applying a known deflection x which will produce a voltage e which can be measured, giving $C_q = C\dot{e}/\dot{x} = Ce/x$. Finally, note that in Eq. (5-12) the minus sign is necessary since the definition of positive current in the crystal is opposite to that which would be used to define the load impedance Z_L.

We have defined the constants C_1, K_s, C and C_q in terms of measurements performed on a device already built. All these constants can actually be estimated, before a device is constructed, from dimensions and fundamental material properties. These material properties must themselves be measured experimentally, but this need be done only *once* for each material, not for each new device. Before leaving this topic let us cast Eq. (5–10) into a more usable form. The effective damping B and inertia M are not readily interpreted physically, so let us divide through by the spring constant K_s, giving

$$M\ddot{x}/k_s + B\dot{x}/k_s + x + C_1e/k_s = f(t)/k_s \qquad \textbf{(5–13)}$$

which we rewrite as

$$\ddot{x}/\omega_n^2 + 2\zeta\dot{x}/\omega_n + x + C_1e/k_s = f(t)/k_s \qquad \textbf{(5–14)}$$

where

$\omega_n \triangleq$ undamped natural frequency of crystal vibration (short circuit), rad/sec

$\zeta \triangleq$ crystal damping ratio, dimensionless

Methods for calculating and/or measuring ω_n and ζ are available and relatively convenient.

Returning to Fig. 5–1, let us now consider methods of converting mechanical power to fluid power, with emphasis on equipment designed for use with liquids, since the relations for compressible fluids are more complex and beyond our intended scope. Two main types of pumps account for the majority of such energy conversion; positive displacement (Fig. 5–8) and centrifugal (Fig. 5–9). While positive displacement pumps take a variety of forms (piston, vane, gear, etc.), their overall characteristics are basically similar, and a general model adequate for dynamic analysis purposes can be formulated. Figure 5–8 shows a multiple-piston pump with a rotary mechanical input which we shall use as a concrete example in developing the general model. As the input shaft is rotated, the individual pistons are sequentially forced in and out, drawing fluid from the input port and expelling it at the output. Valves (not shown in Fig. 5–8) are properly sequenced with the rotation so that each cylinder is alternately exposed to the inlet port and then the discharge. The outflows from each cylinder are summed at the discharge port, so that, while each individual cylinder flow rate is pulsating, the total pump flow rate is relatively smooth. Intuitively one would guess that smoothness would increase with the number of cylinders; commercial pumps with as many as 12 cylinders are not unusual.

A fundamental parameter of a positive displacement pump is its displacement of fluid per radian of shaft rotation, D_p, inch³/radian. It is easily found for an existing pump by measuring its flow rate at constant speed with no back pressure (so that leakage is negligible). The torque T_p required to drive the pump is directly related to D_p and the pressure drop Δp, as revealed by the following analysis. Assuming perfect energy conversion (no losses due to friction, etc.), all the mechanical energy put into

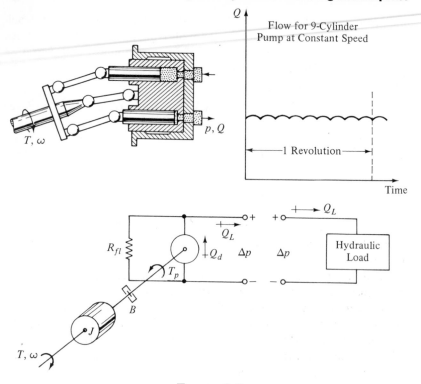

<figure>FIGURE 5–8</figure>

Positive Displacement Pumps

the pump shows up as fluid energy. In turning the shaft through a small angle $d\theta$, a torque T_p does mechanical work $T_p \, d\theta$. At the same time, a volume of fluid $D_p \, d\theta$ has been forced through the pump against the pressure drop Δp. The mechanical power would be $T_p \, d\theta/dt$, and the fluid power (flow rate) (pressure drop) would be $D_p(d\theta/dt) \Delta p$, thus, for no losses

$$T_p \frac{d\theta}{dt} = D_p \, \Delta p \, \frac{d\theta}{dt} \tag{5–15}$$

$$T_p = D_p \, \Delta p \tag{5–16}$$

The instantaneous torque felt by the drive shaft is thus given by Eq. (5–16) in terms of the instantaneous pressure drop.

We can now state the equations describing system behavior as

$$T - B\omega - D_p \, \Delta p = J\dot\omega \tag{5–17}$$

where

$$B \triangleq \text{viscous damping of pump moving parts}$$

$$J \triangleq \text{moment of inertia of pump moving parts}$$

For the fluid circuit

$$\omega D_p - \Delta p / R_{fl} = Q_L \tag{5-18}$$

$$\text{fluid impedance of load} \triangleq Z_L(D) \triangleq \frac{\Delta p}{Q_L}(D) \tag{5-19}$$

The pump leakage resistance R_{fl} can be measured experimentally for a given pump. As soon as the hydraulic load is specified in detail, Eq. (5-19) can be made specific. For example, if the load were simply a flow resistance R_L, we would have $R_L = \Delta p / Q_L$.

Turning now to centrifugal pumps (Fig. 5-9), we first note that while a positive displacement pump is basically a flow source, delivers a flow rate proportional to speed and independent (except for the small leakage) of

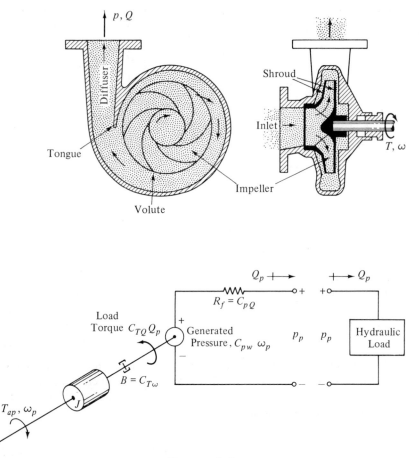

FIGURE 5-9

Centrifugal Pumps, Small-Signal Model

pressure, and will stall if the flow is shut off, a centrifugal pump is more like a pressure source. If the flow is shut off, it continues to run and develop pressure. In applications, positive displacement pumps often are used in fluid-power systems for actuating machinery. The same fluid is circulated back to the pump when discharged from the hydraulic load, such as a hydraulic motor, and used over and over again. Centrifugal pumps are not generally used in this way; rather they are employed to *move fluids* from one place to another in chemical processing plants, refineries, power plants, sewage treatement plants, municipal water supplies, hydroelectric pumped-storage systems, etc. In these applications, the fluid is not usually recirculated through the pump; the processes are "once-through" processes. While turbomachinery theory may be applied to the analysis of pumps to predict characteristics, system analyses often rely on use of measured characteristic curves. Figure 5–10 shows a test setup used to

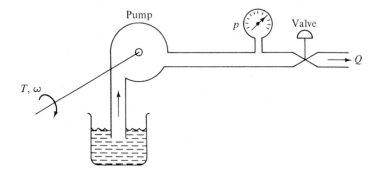

FIGURE 5–10

Centrifugal Pump Test Setup

determine the two families of curves needed to describe a pump. These curves are a graphical presentation of the functional relations

$$p = p(Q, \omega) \qquad \qquad (5\text{--}20)$$

and

$$T_p = T_p(Q, \omega) \qquad \qquad (5\text{--}21)$$

While mathematical formulas for these functions are not available, computer studies can work directly from the experimental curves using "table lookup" or curve fitting techniques. Using such an approach, the system equations would be

$$T_a - T_p(Q, \omega) = J\dot{\omega} \qquad \qquad (5\text{--}22)$$

$$p = p(Q, \omega) \qquad \qquad (5\text{--}23)$$

$$Z_L(D) = \frac{p}{Q}(D) \qquad \qquad (5\text{--}24)$$

An analysis suitable for studies of small perturbations from an operating point can be carried out by linearizing the nonlinear functions $p(Q, \omega)$ and $T_p(Q, \omega)$. The resulting linear equations allow a more general evaluation of the effect of system parameters on response than do numerical computer studies, and are particularly useful in the early stages of system analysis and design. We assume steady-state operation with constant values T_{po}, Q_o, ω_o, p_o of torque, flow rate, speed, and pressure when at $t = 0$ a small change T_{ap} is made in the driving torque T_a. Using a Taylor series expansion, the nonlinear functions may be approximated as

$$T_p(Q, \omega) \approx T_p(Q_o, \omega_o) + \frac{\partial T_p}{\partial Q}\bigg|_{\substack{Q_o \\ \omega_o}}(Q - Q_o) + \frac{\partial T_p}{\partial \omega}\bigg|_{\substack{Q_o \\ \omega_o}}(\omega - \omega_o) \quad \text{(5-25)}$$

$$T_p(Q, \omega) \approx T_p(Q_o, \omega_o) + \frac{\partial T_p}{\partial Q}\bigg|_{\substack{Q_o \\ \omega_o}}Q_p + \frac{\partial T_p}{\partial \omega}\bigg|_{\substack{Q_o \\ \omega_o}}\omega_p \quad \text{(5-26)}$$

where

$$Q_p \triangleq Q - Q_o \triangleq \text{perturbation in } Q$$
$$\omega_p \triangleq \omega - \omega_o \triangleq \text{perturbation in } \omega$$

Similarly,

$$p(Q, \omega) \approx p(Q_o, \omega_o) + \frac{\partial p}{\partial Q}\bigg|_{\substack{Q_o \\ \omega_o}}Q_p + \frac{\partial p}{\partial \omega}\bigg|_{\substack{Q_o \\ \omega_o}}\omega_p \quad \text{(5-27)}$$

The numerical values of the partial derivatives in Eqs. (5-26) and (5-27) are found from the experimental curves as in Fig. 5-11. We can then write the system equations as

$$(T_{po} + T_{ap}) - (T_{po} + C_{TQ}Q_p + C_{T\omega}\omega_p) = J\dot{\omega}_p \quad \text{(5-28)}$$

$$T_{ap} - C_{TQ}Q_p - C_{T\omega}\omega_p = J\dot{\omega}_p \quad \text{(5-29)}$$

$$p - p_o \triangleq p_p \approx -C_{pQ}Q_p + C_{p\omega}\omega_p \quad \text{(5-30)}$$

$$Z_L(D) = \frac{p_p}{Q_p}(D) \quad \text{(5-31)}$$

These equations may be interpreted so as to yield the model of Fig. 5-9. The term $-C_{TQ}Q_p$ represents a load torque presented by the pump, while $-C_{T\omega}\omega_p$ has the form of a viscous damping torque. In Eq. (5-30) $C_{p\omega}\omega_p$ is the generated pressure, while $-C_{pQ}Q_p$ represents a pressure drop due to flow resistance.

The final conversion process considered in Fig. 5-1 is that of mechanical power into thermal power. Perhaps the most common instance of this is found in various frictional processes. Generally, friction is considered an undesirable parasitic effect; however, certain practical devices, notably brakes and clutches, rely on it for their principle of operation. The heat generated by the friction is usually an undesirable, but unavoidable, byproduct, and must be taken into account in the design of frictional

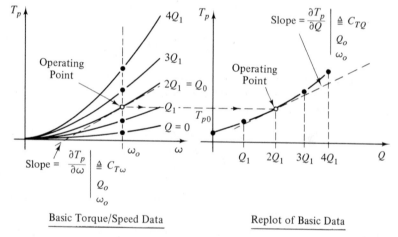

$$\textsc{Figure } 5\text{--}11$$

Definition of Pump Parameters

devices. A good example is found in disk brakes for aircraft wheels. These brakes must absorb and dissipate into heat a portion of the large amount of mechanical kinetic energy possessed by the moving aircraft to bring it to a controlled stop. In doing this, the temperature of the brake lining and other parts must be kept low enough to prevent damage or excessive fading of the brakes. We have thus an example of a mechanical/thermal system in which the heat flow into the brakes is provided by the conversion of mechanical energy into heat by the friction process.

The simplest model of solid (rather than fluid) friction assumes the friction force directly proportional to normal force, and independent of rubbing speed, temperature, or any other influences. Thus, in Fig. 5–12

$$f_f = \mu f_n \qquad (5\text{–}32)$$

where

$f_f \triangleq$ friction force, lb_f

$f_n \triangleq$ normal force, lb_f

$\mu \triangleq$ friction coefficient, assumed constant

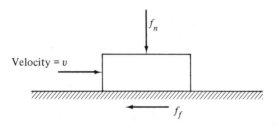

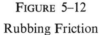

FIGURE 5–12

Rubbing Friction

At any instant when the rubbing velocity is v inch/sec, the mechanical friction power would be $\mu f_n v$, inch-lb_f/sec. Since 1 inch-lb_f/sec is equal to 0.000107 Btu/sec, the frictional dissipation represents a heat flow source for which

$$\text{heat flow rate} = q = 0.000107 \ \mu f_n v, \text{ Btu/sec} \qquad (5\text{–}33)$$

In more complex models the friction coefficient may be taken as some function of rubbing speed, temperature, and other pertinent factors as revealed by experimental testing. When these complex friction models are embedded in an overall system model, computer analysis techniques may be necessary.

5–3. CONVERTING ELECTRICAL ENERGY TO OTHER FORMS.

Figure 5–13 lists some devices for converting electrical power to mechanical, fluid, and thermal form. Beginning with electro-mechanical conversion, the electric motor in its various forms is obviously of overriding importance. One need only begin to count the motors present in any building he happens to be in, to appreciate the widespread usefulness of this device.

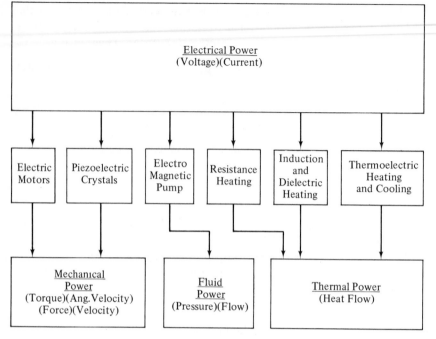

FIGURE 5–13

Converting Electrical Power to Other Forms

Motors take many different detail forms; DC varieties include permanent-magnet field, wound-field separately excited, shunt, series and compound, while induction and synchronous are common AC types. A more recent development is the stepper motor, which operates from electrical pulses and produces mechanical motion in discrete steps. It is particularly suited for systems using digital computers, since the normal output of a computer is in the form of pulses. While rotary motors most often come to mind, the operating principles of most types can be adapted to a translational configuration; the linear induction motors for rapid-transit vehicles of the tracked air cushion type being currently investigated are an example.

We will discuss in detail only the separately excited DC motor, since its simplicity is in keeping with the scope and level of this text. Since it is, in fact, a direct inversion of the DC generator of Fig. 5–2, very little new will need to be said. In Fig. 5–14, the motor could be controlled by maintaining a constant armature current and manipulating the field, holding the field constant (permanent-magnet field is one way to do this) and manipulating the armature voltage, or varying both field and armature. The first two schemes lead to linear equations, the last to nonlinear which may require computer analysis.

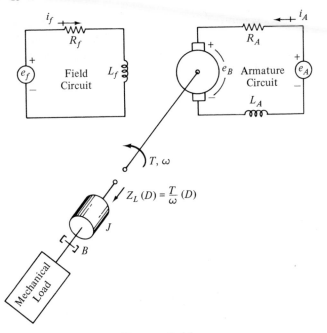

FIGURE 5–14

Model for DC Motor and Load

At any instant, the torque developed by the motor is

$$T = K_T i_f i_A \tag{5-34}$$

while the back emf e_B is

$$e_B = K_E \omega i_f \tag{5-35}$$

where K_T and K_E are exactly the same constants defined in Eqs. (5–3) and (5–4) for the machine when used as a generator. A Newton's equation may be written for the motor shaft as soon as the mechanical load is defined. The total mechanical impedance Z_L is that of the motor inertia and damping (J and B) in combination with the load. Figure 5–14 together with Eqs. (5–34) and (5–35) allow analysis of the motor/load system.

Just as in the DC motor/generator, the piezoelectric energy converter is a reversible device which obeys the same equations when the input is electrical energy and output is mechanical, as when the roles are reversed. If the crystal is unrestrained by any attached mechanical load, its motion in response to an applied voltage e is given by Eq. (5–14) with $f(t) \equiv 0$. If a current source is applied, both Eqs. (5–11) and (5–14) must be used. If one face of the crystal is fixed and the other drives a mechanical load of impedance $Z_L(D) = (f/\dot{x})(D)$, the crystal will feel a force $f = -Z_L\dot{x}$ which should be inserted for $f(t)$ in Eq. (5–14).

The direct conversion of electrical power to fluid power can be accomplished by the electromagnetic pump (see Fig. 5–15), but this mode of energy conversion is of limited practical importance. Only fluids of high electrical conductivity, such as certain liquid metals, can be pumped

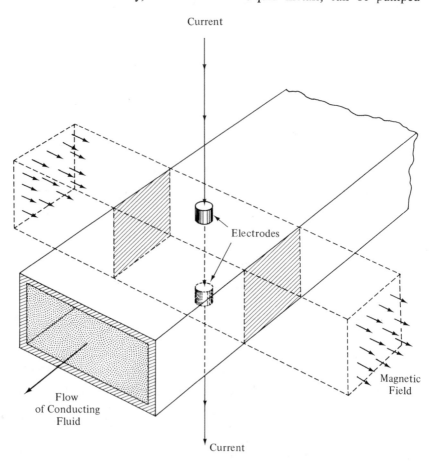

FIGURE 5–15

Principle of the Electromagnetic Pump

efficiently in this way. Without going into detail, the principle is the same as that of the electric motor, in that a current-carrying conductor in a magnetic field feels a force, except here the "conductor" is the fluid being pumped. If the flow is blocked, the force is still felt; thus, the pump can produce pressure at zero flow. When flow is allowed, a "back emf," which opposes the applied voltage, develops just as in a motor.

Electrical heating processes of various types are of considerable commercial importance. For the simplest form, resistance heating, the conversion of electrical to thermal power follows the simple relation

$$\text{rate of heat generation} = 0.000949 \, i^2R, \text{Btu/sec} \qquad (5\text{--}36)$$

where i^2R is in watts. While resistance heating can be accomplished using either DC or AC, induction and dielectric heating basically require AC. In induction heating, a coil carrying AC power induces eddy currents into the piece to be heated and these currents cause i^2R heat generation within the piece, frequencies in the range 480 Hz to 450 KHz being employed. The workpiece must be a reasonably good electrical conductor. At high frequencies the "skin effect" crowds the current near the surface, allowing concentration of the heating for surface heating processes such as case-hardening of metal parts. In dielectric heating (2 to 40 MHz) the workpiece is a fairly good electrical insulator, and the heating effect is uniformly distributed over its volume. The heating effect is produced by the dielectric loss coefficient of the material being heated. Thermoelectric heaters and coolers employ circuits of two properly chosen dissimilar materials to convert electric power directly to heat flow.

5-4. CONVERTING FLUID ENERGY TO OTHER FORMS.

Positive displacement machines (cylinders for translation and motors for rotation) and turbines are widely used to convert fluid power to mechanical form. While there are differences in details of construction, these devices are essentially the same as the corresponding pumps (positive displacement and centrifugal), except that fluid energy is now the input, and mechanical energy the output. Our discussion can thus be fairly brief, since the models are so similar. The torque developed by a positive displacement motor is given by

$$T_M = D_M \Delta p, \text{inch-lb}_f \qquad (5\text{--}37)$$

where

$D_M \triangleq$ motor displacement, $\text{inch}^3/\text{radian}$

$\Delta p \triangleq$ instantaneous pressure drop across motor, psi

Newton's law for the motor and attached mechanical load is thus easily written; the motor contributes its own inertia J and damping B to the total load. The motor flow rate in terms of motor speed ω and pressure drop is given by $D_M\omega + \Delta p/R_{fl}$ where R_{fl} is the motor leakage resistance. As soon as the pressure/flow characteristics of the fluid power source driving the motor are given, they may be combined with those just given for the motor to get an overall fluid system equation.

Turbines and centrifugal pumps have much in common; both fall in the category called turbomachines, the pump converting mechanical power to fluid power, the turbine accomplishing the reverse conversion. Water turbines, such as are used to drive generators in hydroelectric plants, may be modeled in a fashion quite similar to that used for centrifugal pumps, using experimental characteristic curves analogous to those of Fig. 5–11. Gas and steam turbines are somewhat more complex, since both thermal and fluid aspects must be considered; however, the linearization of families of characteristic curves is still a useful technique.

The electromagnetic flow meter of Fig. 5–16 is really a measuring instrument, rather than a power generating device; however, it does convert

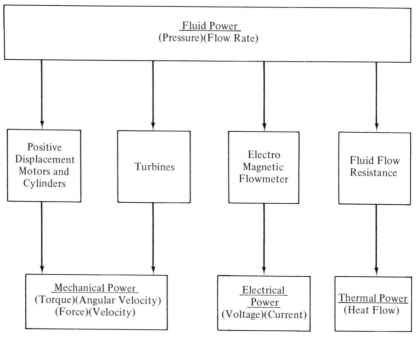

FIGURE 5–16

Converting Fluid Power to Other Forms

fluid power into an electrical signal. The principle is identical to the pump of Fig. 5–15, except that now the flow is the input, which produces a voltage output at the electrodes in direct proportion to the flow rate. Both the voltage produced and the resulting current flow into the external voltage-measuring circuitry are very small, and thus the power output is negligible; however, this type of flowmeter has considerable practical importance, because of certain advantages it has over other methods of measuring flow rates.

Fluid flow resistance (friction) as a means of generating heat is usually an undesirable parasitic effect. If frictional pressure drops are not kept small enough in fluid power systems the working fluid may heat up to the point where it deteriorates, or critical components such as seals fail. To prevent this, heat exchangers with cooling water may be added to maintain fluid temperature low enough to ensure long equipment life. Since frictional pressure drop times volume flow rate has the dimensions of mechanical power (inch-lb_f/sec), one can convert to heat flow using familiar conversion factors.

5-5. CONVERTING THERMAL ENERGY TO OTHER FORMS.

The direct conversion of thermal energy to mechanical energy may be accomplished through the phenomenon of thermal expansion. Any solid body subjected to heat addition will experience a temperature rise, and the accompanying expansion may be caused to do mechanical work by letting it push a load. This process is not widely used to generate mechanical power, but does serve a useful function in many temperature-measuring and control devices such as thermostats. Since electrical power is of such great importance, the direct generation from thermal energy has received considerable attention and several classes of useful devices are in various stages of practical development (see Fig. 5-17). While some of these are now in actual service, it is only fair to say that their total contribution to the world's electrical generating capacity is exceedingly small. They do, however, find application in certain specialized situations where other methods are at a disadvantage. The thermoelectric effect, in which the application of heat to an electric circuit made up of two properly chosen dissimilar materials results in a current flow, was known for many years and is still usefully employed as a temperature-measuring instrument, the thermocouple. More recently the effect has been practically employed to generate small quantities of electric power in specialized applications. Thermionic generating devices employ the same sort of principle as used at the cathode of an electronic vacuum tube—the "boiling off" of electrons from a suitable material by the application of heat. Again, only small amounts of power are presently produced in this way and applications are very limited. Magnetohydrodynamic (MHD) methods of power generation have the potential for large-scale power production and are being actively researched to bring this potential to practical realization. At present,

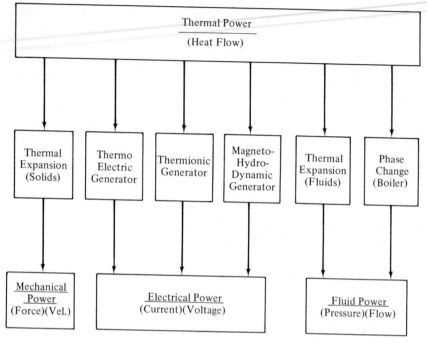

FIGURE 5-17

Converting Thermal Power to Other Forms

however, no practical plants of this sort are in existence. The reader interested in learning more details about these processes of direct conversion of heat to electricity will find a number of texts devoted entirely to this subject.[2]

Thermal expansion of fluids allows conversion of thermal power to fluid power. Such energy conversion is limited to low power levels, such as in measuring instruments, when liquids are involved. In the case of gases and also when phase change from liquid to gas occurs, large amounts of power can be converted. Actually, the addition of heat to a gas does not result in a straightforward conversion to fluid energy; several forms of energy are present and must be properly taken into account, using principles of thermodynamics and fluid mechanics. Internal combustion engines (gasoline, diesel, natural gas, etc.) and gas turbines are good examples of such energy conversion. The combustion process converts the chemical energy of the fuel into heat, which in turn is converted to fluid pressure and flow, and ultimately into mechanical power available at the shaft. In a steam turbine, the combustion occurs external to the turbine in a furnace

---

[2] S. Angrist, *Direct Energy Conversion* (Boston: Allyn and Bacon, Inc., 1965).

where the heat is applied to a boiler. The boiler accepts a heat input and uses it to vaporize the water, increasing its temperature and pressure. We thus have a conversion of thermal energy into fluid energy which the turbine can in turn convert to mechanical shaft power. Again, several forms of energy (not just fluid energy in the form of [pressure] [flow]) are present and must be properly accounted for.

5-6. OTHER SIGNIFICANT ENERGY CONVERSIONS.

Since earlier chapters concentrated on mechanical, electrical, fluid, and thermal forms of energy, the present chapter emphasized the interactions and couplings present among these. While we do not intend this brief treatment to be comprehensive in the field of energy conversion, certain processes not falling in the above categories are of sufficient interest that we wish to at least mention them (see Fig. 5–18). The importance of

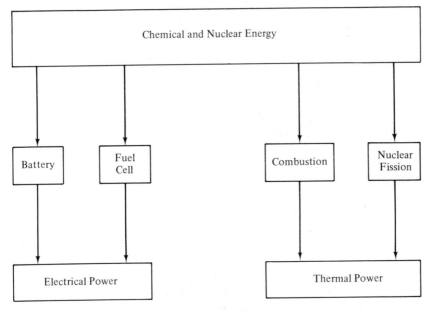

FIGURE 5–18

Conversion of Chemical and Nuclear Energy

chemical and nuclear fuels can hardly be overemphasized, since they are the fundamental source of practically all of our useful power. Conversion of this stored energy into directly usable form involves combustion of chemical fuels (coal, oil, gas, etc.) and fission of nuclear fuels to produce thermal energy. System studies of overall power plants must thus take into account the dynamic behavior of these combustion or nuclear processes. Direct conversion of chemical energy to electricity is accomplished by various forms of batteries and the more-recently-developed fuel cell. Dynamic behavior of batteries has not received (or apparently needed) much consideration; however, fuel cell dynamics are of considerable importance, since control systems are needed to obtain proper cell operation.

BIBLIOGRAPHY

1. Angrist, S., *Direct Energy Conversion*. Boston: Allyn and Bacon, Inc., 1965.

2. Blackburn, J. F., G. Reethof, and J. L. Shearer, *Fluid Power Control*. New York: John Wiley & Sons, Inc., 1960.

3. Gehmlich, D. K. and S. B. Hammond, *Electromechanical Systems*. New York: McGraw-Hill Book Company, 1967.

4. Shepherd, D. G., *Principles of Turbomachinery*. New York: The Macmillan Company, 1961.

5. White, D. C. and H. H. Woodson, *Electromechanical Energy Conversion*. New York: John Wiley & Sons, Inc., 1959.

PROBLEMS

5–1. Explain what happens in Eqs. (5–1) and (5–2) if a *motion* source (rather than a torque source) is used to drive the generator. Does this make the analysis simpler or more complex? Which model do you think is more realistic? Why?

5–2. When the field is fixed (as in a permanent-magnet field, for example), Eqs. (5–3) and (5–4) become $T_{EM} = K_{T1} i_A$ and $e_G = K_{E1}\omega$. If T_{EM}, i_A, e_G, and ω are all constant, what mechanical power is being supplied to overcome T_{EM}? What electrical power does e_G supply? For an ideal (lossless) machine, all the mechanical power input is converted into electrical power output. Explain how this shows that K_{T1} and K_{E1} are not independent parameters.

5–3. A positive-displacement pump with $D_p = 0.2$ in^3/rad runs at 1000 rpm against a pressure drop of 500 psi. How much fluid horsepower is being produced? What is the flow rate? If the overall efficiency of energy conversion is 80%, how much mechanical power must be supplied? If the pump is driven by an electric motor of 70% efficiency, how many watts will the motor draw?

5-4. How do Eqs. (5–17) and (5–18) change if a motion source is substituted for the torque source in the pump drive? Is the analysis made simpler or more complex?

5-5. A 3000 lb_f car traveling at 60 mph makes an emergency stop with brakes locked and all 4 tires sliding on the concrete pavement with friction coefficient of 0.5.
   a. How long does it take to stop?
   b. How far does it travel in stopping?
   c. What is the rate of heat generation at the tire-road interface?
   d. What is the total heat generated? If this heat went entirely into raising the temperature of the tires (assume a total of 80 lb of pure rubber), how hot would the tires get? Discuss the complicating factors which would have to be included in an *accurate* analysis to find tire temperature.

5-6. The car of prob. 5–5 is stopped by applying the brakes (but not locking them) and allowing the wheels to roll, not slide. If the deceleration is constant and requires 8 seconds to stop the car, find the rate at which heat is generated in the brakes.

5-7. In Fig. 5–14, discuss the effects of replacing the voltage source e_f with a current source.

5-8. In Fig. 5–14, discuss the effects of replacing the voltage source e_A with a current source.

5-9. A hydraulic motor has $D_M = 0.1$ in^3/rad and $R_{fl} = 10,000$ psi/(in^3/sec). If the pressure drop across it is 1000 psi, and it runs at 500 rpm, how much fluid power is being taken from the fluid source? What portion of this is wasted in motor leakage? What happens to this wasted energy?

6

SOLUTION METHODS FOR DIFFERENTIAL EQUATIONS

6-1. INTRODUCTION.

Application of lumped-parameter models to dynamic analysis of physical systems leads to a system description in terms of ordinary differential equations. These equations may be solved to find the system behavior by three general methods: analytical, analog computer, and digital computer. Analytical methods are essentially limited to linear equations with constant coefficients, while computer methods handle both linear and nonlinear problems. Analytical methods generally are preferred, since they allow general solutions which show the effects of system parameters directly. Both analog and digital computers can "solve" only specific numerical problems; thus the effects of parameters are revealed only by running many special cases. We will, in this chapter, show how all three approaches are used, in the case of digital methods concentrating on the popular digital simulation approach.

6-2. ANALYTICAL SOLUTION OF LINEAR, CONSTANT-COEFFICIENT EQUATIONS.

While certain nonlinear and variable-coefficient linear differential equations have closed-form analytical solutions, the majority of these equations which arise in engineering practice have no such analytical solution and yield only to computer methods. Only for linear equations with constant coefficients do general solution techniques exist which "always work." We will here briefly review the classical operator method of solution; a condensed treatement of an alternative technique, the Laplace transform, is given in the appendix. The general form of equation which we treat is

$$a_n \frac{d^n x}{dt^n} + a_{n-1} \frac{d^{n-1} x}{dt^{n-1}} + \ldots + a_1 \frac{dx}{dt} + a_0 x = f(t) \tag{6-1}$$

where the a's are constants and $f(t)$ is a known function of time. The solution proceeds in three steps:

1. Find the complementary function part of the solution, x_c.
2. Find the particular solution, x_p.
3. Add x_c to x_p to get the total solution x and apply initial conditions to evaluate the constants of integration.

A method for finding x_c which always works is available. Using the operator $D \triangleq d/dt$ we first write Eq. (6-1) as

$$(a_n D^n + a_{n-1} D^{n-1} + \ldots + a_1 D + a_0) x = f(t) \tag{6-2}$$

The *characteristic equation*

$$a_n D^n + a_{n-1} D^{n-1} + \ldots + a_1 D + a_0 = 0 \tag{6-3}$$

is treated as an algebraic equation in the unknown D and we must solve it for its n roots $s_1, s_2, \ldots, s_n$. For $n > 4$, numerical approximate root-finding methods must be used and these require that the coefficients a_n to a_0 be known as *numbers* rather than letters. This need to work with specific numbers is undesirable but unavoidable. Fortunately, root-finding methods can be implemented on digital computers to reduce the time and effort required at this stage of the solution. Once the roots are known, we *immediately* write down the solution x_c using a set of rules whose validity is here taken for granted but which is proven in most first courses on differential equations. For any real root s_1 which is not repeated, the solution is $C_1 e^{s_1 t}$ where C_1 is a constant of integration as yet unknown, and e is the base of natural logarithms. For a double root s_1, s_1 the solution is $C_1 e^{s_1 t} + C_2 t e^{s_1 t}$, for a triple root, $C_1 e^{s_1 t} + C_2 t e^{s_1 t} + C_3 t^2 e^{s_1 t}$ and so forth. If complex roots arise, they always come in pairs of the form $a \pm ib$ and the solution for such a pair is $C e^{at} \sin(bt + \phi)$ where C and ϕ are constants of integration. If a complex root pair is repeated $a \pm ib$, $a \pm ib$ the solution is $C_1 e^{at} \sin(bt + \phi_1) + C_2 t e^{at} \sin(bt + \phi_2)$; however, this occurs very rarely in practical problems. We thus see that once the roots are known, the solution x_c follows at once. For example, if the roots are $-1, +3, +3$, $-2, \pm i4, +3 \pm i7$ corresponding to a seventh order equation, the solution is

$$x_c = C_1e^{-t} + C_2e^{3t} + C_4te^{3t} + C_5e^{-2t}\sin(4t + \phi_1) + C_6e^{3t}\sin(7t + \phi_2) \tag{6-4}$$

While the above method for finding x_c always works, no such universal method exists for finding x_p, since it depends on the form of the forcing function $f(t)$ in Eq. (6-1). No matter what method one might propose, a mathematician can always concoct a sufficiently "pathological" $f(t)$ to thwart it. Thus, one must be satisfied with methods which handle a certain class of function. Fortunately, a simple method (the method of undetermined coefficients) suffices for most $f(t)$'s of engineering interest. This method will work if successive derivatives of $f(t)$ ultimately become zero or repeat themselves. For example, if $f(t) = 2t^3$, all derivatives beyond the third will be identically zero and the method works. For $f(t) = 2\sin 3t$ successive derivatives give rise only to $\sin 3t$ and $\cos 3t$ functions and the method works. If $f(t) = e^{t^2}$, successive differentiation, no matter how far it is carried, continues to produce new functional forms and the method will not work. For those cases where the method works, the particular solution x_p is written down as a sum of terms made up of every different kind of function found in $f(t)$ and its derivatives, each multiplied by an undetermined coefficient. These coefficients can be found immediately by substituting x_p into the differential equation. An example illustrates the procedure.

$$\frac{d^2x}{dt^2} + 3\frac{dx}{dt} + 2x = 4e^{-5t} \tag{6-5}$$

when $t = 0^+$, $\dfrac{dx}{dt} = 0$, $x = 2.0$ (The symbol $t = 0^+$ refers to a time an infinitesimal amount after $t = 0$ and is the time instant at which "initial" conditions must be evaluated when using the classical solution method.) The characteristic equation is

$$D^2 + 3D + 2 = 0 \tag{6-6}$$

with roots $s_1 = -2$, $s_2 = -1$. The complementary function solution is $x_c = C_1e^{-2t} + C_2e^{-t}$. Repeated differentiations of the forcing function $4e^{-5t}$ clearly give only terms of the form Ae^{-5t}, so the method of undetermined coefficients will work. The solution $x_p = Ae^{-5t}$ is substituted into Eq. (6-5) to give

$$25Ae^{-5t} - 15Ae^{-5t} + 2Ae^{-5t} \equiv 4e^{-5t} \tag{6-7}$$

$$12Ae^{-5t} \equiv 4e^{-5t} \tag{6-8}$$

$$A = 1/3 \tag{6-9}$$

The complete solution is thus

$$x = x_c + x_p = C_1e^{-2t} + C_2e^{-t} + (1/3)e^{-5t} \tag{6-10}$$

To find C_1 and C_2 we apply the initial conditions.

$$x(0) = 2 = C_1 + C_2 + 1/3 \qquad (6\text{–}11)$$
$$\dot{x}(0) = 0 = -2C_1 - C_2 - 5/3 \qquad (6\text{–}12)$$
$$C_1 = -10/3 \qquad C_2 = 5$$

The complete specific solution for the given initial conditions is thus

$$x = -(10/3)e^{-2t} + 5e^{-t} + (1/3)e^{-5t} \qquad (6\text{–}13)$$

One nice feature of the topic of differential equations is that there never is any need to accept an incorrect answer. If the solution, such as Eq. (6–13), is substituted into the original equation (6–5) and makes it an identity, and if it satisfies the initial conditions, then it *must* be the one and only correct solution.

The above simple routines will enable one to solve any ordinary linear differential equation with constant coefficients irrespective of its order (the order of the highest derivative), as long as $f(t)$ can be handled by the method of undetermined coefficients. Two special cases which occur rarely, but should be mentioned, require a slightly modified procedure. If a term in x_p has the same functional form as one in x_c, the term in x_p should be multiplied by the lowest power of t which will make it different from all the x_c terms associated with the root which produced the x_c term. For example, if the right-hand side of Eq. (6–5) had been $4e^{-t}$, then x_p would have had the form Ae^{-t}, the same as C_2e^{-t} in x_c. We should thus modify x_p to be Ate^{-t} before finding A. If in addition the left-hand side of Eq. (6–5) had been $D^2 + 2D + 1$, with roots $s_1 = s_2 = -1$ and $x_c = C_1e^{-t} + C_2te^{-t}$, then x_p would have to be modified to At^2e^{-t}. The second special case arises if the characteristic equation has the form

$$D^m(a_n D^{n-m} + a_{n-1}D^{n-m-1} + \ldots + a_{m+1}D + a_m) = 0, \qquad a_m \neq 0$$
$$(6\text{–}14)$$

When writing x_p for such a situation we must include, in addition to the usual terms, terms in the first, second, ..., mth *integral* of $f(t)$.

We should at this point mention the important *principle of superposition*, which applies only to linear differential equations. If the driving function $f(t)$ in Eq. (6–1) is composed of a sum of terms, $f_1(t), f_2(t)$, etc., this principle allows us to find the particular solution x_p for each term of the driving function separately, and then get the total x_p by simply adding all the individual solutions. In addition to its direct mathematical utility in getting equation solutions, this principle also has two important general consequences relative to the behavior of linear systems. The first might be called the "amplitude insensitivity" of linear systems. By this we mean that if we have found the response of a system to a driving function, say, $4e^{-5t}$, if we scale this driving function up to $8e^{-5t}$ or down to $2e^{-5t}$, the response will similarly scale up or down. Nothing "new" is thus found out about the response of linear systems by changing the *size* of the driving inputs; as long as the *form* of the driving input remains the same, the responses all are directly proportional. This follows from the superposition principle by noting that, for example, $8e^{-5t}$ can be written as $(4e^{-5t} + 4e^{-5t})$; thus the x_p for $8e^{-5t}$ is just twice that for $4e^{-5t}$. Such statements *cannot* be made for nonlinear systems; the response to an input of doubled size may be *entirely different in form* from the response to the original input. The other general consequence of superposition is that if we know how a system responds to each of two different inputs when they are separately applied, then there will be no "surprises" when they are *simultaneously* applied. That is, the behavior for the combined inputs is just the sum of the responses to the individual inputs. Again, nonlinear systems do not behave so simply; the response to a combination of inputs may show features found in *none* of the individual responses. In nonlinear systems which can go unstable, for instance, the system may be stable for each input applied separately, but unstable when they are applied together.

6–3. SIMULTANEOUS EQUATIONS.

A physical system need not be very complex for its description to require several simultaneous equations (rather than a single equation). In Fig. 6–1a, application of Newton's law to each mass in turn leads to

$$f - K_{s1}(x_1 - x_2) - B(\dot{x}_1 - \dot{x}_2) = M_1\ddot{x}_1 \qquad \textbf{(6–15)}$$

$$K_{s1}(x_1 - x_2) + B(\dot{x}_1 - \dot{x}_2) - K_{s2}x_2 = M_2\ddot{x}_2 \qquad \textbf{(6–16)}$$

Neither of these equations can be solved separately since each contains *both* of the unknowns x_1 and x_2. The pair of equations *can*, however, be solved simultaneously. Similarly, in Fig. 6–1b we get

$$e - Ri_1 - L\left(\frac{di_1}{dt} - \frac{di_2}{dt}\right) = 0 \qquad \textbf{(6–17)}$$

$$-L\left(\frac{di_1}{dt} - \frac{di_2}{dt}\right) + \frac{1}{C}\int i_2\,dt = 0 \qquad \textbf{(6–18)}$$

and again the equations must be solved simultaneously. Whether the classical or Laplace transform method is used, the procedure in solving simultaneous equations basically involves reducing a set of n equations in n unknowns to a single equation in one unknown. When the classical method is used, the equations are written in operator form, whereupon they may be treated as a set of simultaneous *algebraic* equations and reduced to one equation in one unknown by any valid algebraic method, determinants being the most systematic.

An example will illustrate the procedure. Suppose we have two equations in two unknowns as below.

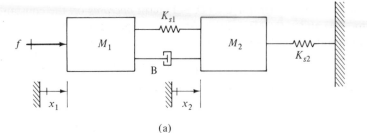

(a)

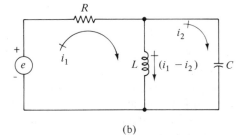

(b)

FIGURE 6-1

Systems Leading to Simultaneous Equations

$$\dot{x}_1 + 2x_1 - 2\dot{x}_2 + 3x_2 = 4 \qquad \text{(6-19)}$$

$$2\dot{x}_1 + x_1 + \dot{x}_2 - x_2 = 2t \qquad \text{(6-20)}$$

when $t = 0^+$, $x_1 = 1$, $x_2 = -2$.

In operator form these become

$$(D + 2)x_1 + (-2D + 3)x_2 = 4 \qquad \text{(6-21)}$$

$$(2D + 1)x_1 + (D - 1)x_2 = 2t \qquad \text{(6-22)}$$

We now treat these as algebraic equations in the unknowns x_1 and x_2 with the D operator carried along as if it were an ordinary parameter. We wish to reduce the set of equations to a single equation in x_1 and another single equation in x_2. Using determinants as in algebra we get

$$x_1 = \frac{\begin{vmatrix} 4 & -2D + 3 \\ 2t & D - 1 \end{vmatrix}}{\begin{vmatrix} D + 2 & -2D + 3 \\ 2D + 1 & D - 1 \end{vmatrix}} = \frac{(D - 1)4 - (-2D + 3)2t}{(D + 2)(D - 1) - (2D + 1)(-2D + 3)}$$

$$= \frac{-6t}{5D^2 - 3D + 5} \qquad \text{(6-23)}$$

Note that in the numerator the $(D - 1)$ term *operates* on the constant 4 (giving -4) and $-(-2D + 3)$ *operates* on $2t$ (giving $+4 - 6t$).
Cross-multiplying Eq. (6–23) gives

$$(5D^2 - 3D + 5)x_1 = -6t \qquad (6\text{–}24)$$

the desired single equation in x_1. Similarly, for x_2

$$x_2 = \frac{\begin{vmatrix} D + 2 & 4 \\ 2D + 1 & 2t \end{vmatrix}}{\begin{vmatrix} D + 2 & -2D + 3 \\ 2D + 1 & D - 1 \end{vmatrix}} = \frac{(D + 2)2t - (2D + 1)4}{5D^2 - 3D + 5}$$

$$= \frac{4t - 2}{5D^2 - 3D + 5} \qquad (6\text{–}25)$$

$$(5D^2 - 3D + 5)x_2 = 4t - 2 \qquad (6\text{–}26)$$

Note that the characteristic equation $5D^2 - 3D + 5 = 0$ is the same no matter which unknown, x_1 or x_2, is being solved for. We can see that this will be true for the general case of n equations in n unknowns since the denominator determinant is the same no matter which unknown is being considered. Since the physical system is described by the whole set of equations, this means a linear system, no matter how complex, has only one characteristic equation. To complete the solution we find the characteristic equation roots to be 1.34 and -0.74 and thus

$$x_{1c} = C_1 e^{1.34t} + C_2 e^{-0.74t} \qquad (6\text{–}27)$$

$$x_{2c} = C_3 e^{1.34t} + C_4 e^{-0.74t} \qquad (6\text{–}28)$$

We also find $x_{1p} = -6t + 18$ and $x_{2p} = 4t + 10$, giving

$$x_1 = -6t + 18 + C_1 e^{1.34t} + C_2 e^{-0.74t} \qquad (6\text{–}29)$$

$$x_2 = 4t + 10 + C_3 e^{1.34t} + C_4 e^{-0.74t} \qquad (6\text{–}30)$$

It appears we have four constants of integration to be found and only two initial conditions; however, the four constants of integration are not really all independent, so the problem is solvable. Perhaps the easiest way to find the needed constants is to generate additional initial conditions from those given and the system equations. That is, the basic equations (6–19) and (6–20) are true at every instant of time, including $t = 0$, and can be used to find the needed initial conditions. At $t = 0$, (6–19) and (6–20) give

$$\dot{x}_1 + 2 - 2\dot{x}_2 - 6 = 4 \qquad (6\text{–}31)$$

$$2\dot{x}_1 + 1 + \dot{x}_2 + 2 = 0 \qquad (6\text{–}32)$$

and these yield $\dot{x}_2(0) = -3.8$ and $\dot{x}_1(0) = 0.4$. We now have sufficient initial conditions to solve for C_1, C_2, C_3, and C_4 and get the complete specific solutions for x_1 and x_2.

6–4. ANALOG COMPUTER METHODS.

When we first embark on a description of computer methods of differential equation "solution" it is important to be clear on the fundamental difference between analytical solutions and the results of analog or digital computer approaches. While either type of computer would have no difficulty "solving" the equation $\dot{y} + y = 0$, $y(0) = 1.0$, there is no computer that will tell us the functional form $y = e^{-t}$ of the solution. An analog computer would give us a graph of y versus t and a digital computer a table of y and t at discrete values of t; however, neither of these tells us that the function is e^{-t}. If the equation were $a\dot{y} + 1 = 0$, $y(0) = b$, the analytical solution is $be^{-t/a}$, whereas the computer cannot be used at all until a and b are given as *numbers*. Thus computers can only solve specific numerical equations with numerical initial conditions; they *cannot* get a solution in letter form which one can then examine to evaluate the effect of system parameters. To evaluate the effect of parameters with computer methods, one must run a complete specific solution for each value of the parameter one is interested in.

Turning specifically, now, to analog methods, an analog computer is a physical device which obeys the same form of equation as does the unknown in the equation to be solved. Analog computers are used both as components of operating machines and processes (special purpose computers) and as equation-solving machines in computer laboratories (general purpose computers). Special purpose computers may employ mechanical, pneumatic, electronic, thermal, hydraulic, and optical techniques to perform the needed computations, whereas general purpose computers today are nearly all electronic. Two simple mechanical analog computing devices for performing multiplication by a constant and integration are shown in Fig. 6–2. The ball-and-disk integrator was actually used in the first general-purpose computers (which were all mechanical) in the 1930s, but today is used only in special-purpose applications.

We will now consider in some detail the general-purpose electronic analog computer. Its basic capability extends over all types of ordinary differential equations, linear or nonlinear, single equations or simultaneous sets of equations. For linear constant coefficient equations only three basic components are necessary: the summer, the coefficient potentiometer, and the integrator. To these must be added the variable multiplier and the arbitrary function generator to expand into linear equations with time-varying coefficients and nonlinear equations. The summer and integrator are constructed around operational amplifiers exactly as described in Sec. 3–8. Since the variable multiplier and arbitrary function generator are electronically somewhat complex, we describe them only functionally since this is adequate for their use. The coefficient potentiometer is a

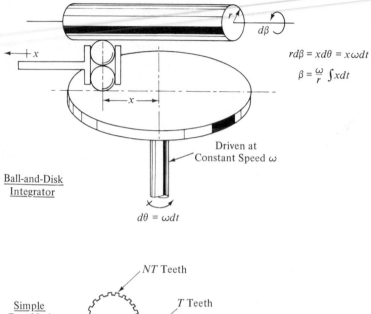

$$rd\beta = xd\theta = x\omega dt$$

$$\beta = \frac{\omega}{r} \int x \, dt$$

Driven at
Constant Speed ω

Ball-and-Disk
Integrator

$$d\theta = \omega dt$$

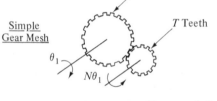

NT Teeth

Simple
Gear Mesh

T Teeth

θ_1

$N\theta_1$

FIGURE 6–2

Mechanical Analog Computing Elements

simple voltage divider, usually a ten-turn precision potentiometer. All
these devices are shown in Fig. 6–3 together with their describing equations. The voltages in most present-day computers are either 10 or 100
volts full scale. The variable multiplier and divider of Fig. 6–3 have scale
factors (1/100 and 100) so that when e_{i1} and e_{i2} are both at their full scale
values (100 volts), the output is also at full scale. The arbitrary function
generator allows synthesis of arbitrary single-valued functions by fitting
together 10 or 20 straight line segments of adjustable slope and intercept
to approximate a desired analytical or experimentally measured function.
For common functions such as sines, cosines, logs, and exponentials,
"committed" function generators usable only for these functions but
requiring a minimum of adjustment and setup time are available.

The basic setup of the analog computer is relatively simple; however,
time scaling and magnitude scaling may require some cutting and trying
before a satisfactory solution is obtained. For our first example, we will
neglect scaling entirely so as to concentrate on the basic setup procedure.

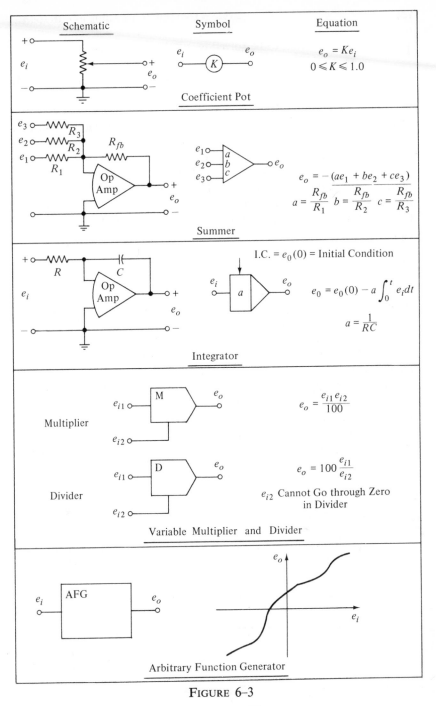

$$\text{Schematic} \qquad \text{Symbol} \qquad \text{Equation}$$

$e_o = Ke_i$

$0 \leqslant K \leqslant 1.0$

Coefficient Pot

$e_o = -(ae_1 + be_2 + ce_3)$

$a = \dfrac{R_{fb}}{R_1} \quad b = \dfrac{R_{fb}}{R_2} \quad c = \dfrac{R_{fb}}{R_3}$

Summer

I.C. $= e_0(0) = $ Initial Condition

$e_0 = e_0(0) - a \displaystyle\int_0^t e_i dt$

$a = \dfrac{1}{RC}$

Integrator

Multiplier

$e_o = \dfrac{e_{i1}e_{i2}}{100}$

Divider

$e_o = 100\dfrac{e_{i1}}{e_{i2}}$

e_{i2} Cannot Go through Zero in Divider

Variable Multiplier and Divider

Arbitrary Function Generator

FIGURE 6–3

Analog Computer Elements

253

Suppose the equation to be solved is

$$A \frac{d^2e}{dt^2} + B \frac{de}{dt} + Ce = F \sin \omega t \tag{6-33}$$

with initial conditions $\dot{e}(0) = +5$, $e(0) = -10$. We will assume that the unknown e is a voltage (thus Eq. (6–33) is already a voltage equation), and that the speed of response of e is within the range of the computer, so that no time scaling is necessary. (Analog computers will ordinarily complete a solution in a minute or less, but some fast models designed for *repetitive operation* take as little as 1/1000 second.) Irrespective of whether the equation is linear or nonlinear, the first step is always to arrange it so that the term with the highest derivative stands alone on the left side of the equation.

$$A \frac{d^2e}{dt^2} = F \sin \omega t - B \frac{de}{dt} - Ce \tag{6-34}$$

One now starts drawing the computer connection diagram with a summer whose output is to be $A(d^2e/dt^2)$. If this is to be so, its inputs *must* be $-F \sin \omega t$, $+B(de/dt)$, and $+Ce$. The voltage $-F \sin \omega t$ is a completely known driving input which can be obtained from a *signal generator* external to the computer; however $B(de/dt)$ and Ce are as yet unavailable. In fact, we will "manufacture" them on the computer by suitably processing the signal $A(d^2e/dt^2)$. Figure 6–4 shows how this is done. The inverter needed to change the sign of $-B\dot{e}$ was not shown in Fig. 6–3, since it is just a special case of the summer in which only e_1 is present and $R_{fb} = R_1$. The initial conditions are, in effect, applied by charging the capacitors in the integrators to the proper initial voltages and

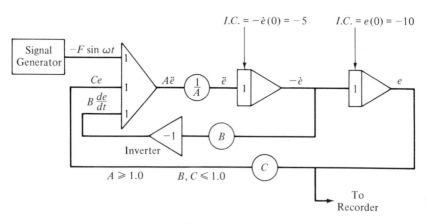

FIGURE 6–4

Computer Setup Diagram

then open-circuiting them. At $t = 0$, pushing a switch reconnects them into the integrators and the integrator output voltages thus start from their proper values. A recorder is shown connected to e, but of course any other voltage of interest, such as $A\ddot{e}$ or $B\dot{e}$, could also be recorded.

The procedure for nonlinear equations is not essentially different; one must merely apply the components so as to properly implement the nonlinearities present in the equation. To illustrate, Fig. 6–5 shows the equation

$$(1 - At)M\ddot{x} + B(\dot{x})^3 + K \sin x = F \qquad \textbf{(6–35)}$$

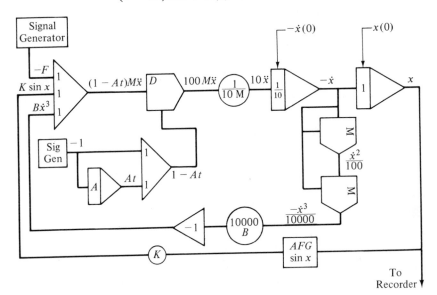

FIGURE 6–5

Nonlinear and Time-Varying-Coefficient Equation

which has both a time-varying coefficient and two nonlinear terms. Simultaneous equations are simulated directly; there is no need to reduce to one equation in one unknown. Figure 6–6 shows the computer diagram for Eqs. (6–19) and (6–20). Note that it is not necessary to calculate additional initial conditions (as was the case for the analytical solution); the two basic initial conditions are sufficient. Also, solutions for both x_1 and x_2 are obtained simultaneously.

Now that we have demonstrated the basic simplicity of the analog approach, it is necessary to turn to the details of time and magnitude scaling. The response speed of physical systems varies widely; an all-electronic system may complete its response in microseconds while the response of a system of rivers and dams to rainfall extends over days.

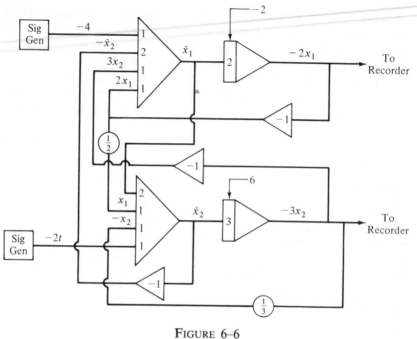

FIGURE 6–6

Simultaneous Equations

When we study such problems on an analog computer there is no need to run in "real time"; we can speed up or slow down the simulation if we so desire. For physical systems which react in hours or days, a computer speedup is desirable, since it saves time (money) and reduces the effects of drift in computer components. When the real system responds in microseconds or milliseconds, it usually is necessary to slow down the simulation, since the computer and/or associated recorders cannot accurately perform at such speeds. We find, thus, that most conventional analog systems are time scaled so that the problem runs to completion in a few seconds or minutes. Exceptions are found in the repetitive operation machines which can complete a solution in a few milliseconds. Such machines are quite useful in optimization studies where large ranges of many parameters must be surveyed to find the best combination.

The usual procedure in time scaling is to write a transformation equation relating the real time t of the physical system to the computer time variable τ. For example

$$\text{computer time} \triangleq \tau = 10 \text{ real system time} = 10t \qquad (6\text{–}36)$$

Note that when $t = 1$ sec in the real system, $\tau = 10$ sec, and thus an event which takes 1 sec in the real system occurs in 10 sec on the computer;

we have achieved a 10-to-1 slowdown. While Eq. (6–36) allows direct transformation of t to τ wherever t appears explicitly in an equation, we must also transform all our derivatives properly. We may write

$$\frac{dx}{dt} = \frac{dx}{d\tau}\frac{d\tau}{dt} = 10\frac{dx}{d\tau} \tag{6–37}$$

$$\frac{d^2x}{dt^2} = \frac{d}{dt}\left(10\frac{dx}{d\tau}\right) = 100\frac{d^2x}{d\tau^2}, \text{ etc. for higher derivatives} \tag{6–38}$$

We see then that as soon as the time scale transformation is decided upon, the system equations can be rewritten properly using the new time variable τ in place of t.

Magnitude scaling is needed for several reasons. First, we must establish a relation between the computer variables, which are volts, and the physical problem variables, which may be temperature, pressure, velocity, etc. That is, each volt in the computer represents so many °F, psi or inch/ sec. Furthermore, computers are designed for definite voltage ranges, usually ± 10 or ± 100 volts. If a problem is scaled so that one or more of the variables exceeds the design range, the machine will not compute properly. (Analog computers are not usually damaged by over-ranging. They also generally provide a visible [light] and/or audible [beep] alarm showing which signal is overloading to help in re-scaling.) On the other hand, if a problem is scaled so that the variables are only a small fraction of the design range (say ± 1 volt in a 100-volt computer) then the accuracy will suffer, since machine errors are given as a percentage of full scale. That is, a 0.1% error for a 100-volt machine would be 0.1 volt and if the problem variables were at a 1-volt level, this would be a 10% error.

To illustrate the above principles of scaling we shall now carry through a complete problem. Analysis of a building heating system has led to the following mathematical model relating interior temperature T, °F to furnace heating rate Q, Btu/hr when the outside temperature is assumed constant at 0°F:

$$0.04\frac{d^3T}{dt^3} + 0.53\frac{d^2T}{dt^2} + 1.4\frac{dT}{dt} + T = 0.0005Q, \text{ } t \text{ in hours} \tag{6–39}$$

We need to know how the temperature T rises if the house is initially at 0°F when the furnace is suddenly turned on and left on with $Q = 150,000$ Btu/hr. Since time scaling affects magnitude scaling, it is generally best to do the time scaling first. If one has no physical feeling or previous experience with a problem, both time and magnitude scaling will involve an element of guesswork; however, if initial estimates are wrong the computer quickly tells us whether things are happening too fast or too slow and if voltages are too big or small, so that we can improve the scaling. Usually, however, a problem is not entirely unfamiliar and we can base our initial scaling on this past experience. In the present case, our familiarity with heating systems indicates that the house may take several hours to heat up so we will wish to speed up our computer simulation. If we estimate about two hours (7200 seconds), we might choose a 1000-to-1 speedup to get a computer running time of about 7.2 seconds. Our derivatives must thus be rewritten according to the transformation

$$\tau = 0.001t \tag{6-40}$$

and thus

$$\frac{dT}{dt} = 0.001\frac{dT}{d\tau}, \qquad \frac{d^2T}{dt^2} = 10^{-6}\frac{d^2T}{d\tau^2}, \qquad \frac{d^3T}{dt^3} = 10^{-9}\frac{d^3T}{d\tau^3} \tag{6-41}$$

Since the computer problem will take a matter of seconds, it may be convenient to rewrite Eq. (6-39) in second units before we apply the time scale transformations of Eqs. (6-41). We have

$$(4 \times 10^{-2})(46.656 \times 10^9)\frac{d^3T}{dt^3} + (0.53)(12.96 \times 10^6)\frac{dT^2}{dt^2}$$

$$+ (1.4)(3600)\frac{dT}{dt} + T = 75, \; t \text{ in seconds} \tag{6-42}$$

which can then be rewritten in terms of computer time τ as

$$(0.04)(46.656)\frac{d^3T}{d\tau^3} + (0.53)(12.96)\frac{d^2T}{d\tau^2} + (1.4)(3.6)\frac{dT}{d\tau} + T = 75 \tag{6-43}$$

$$1.87\frac{d^3T}{d\tau^3} + 6.86\frac{d^2T}{d\tau^2} + 5.04\frac{dT}{d\tau} + T = 75 \tag{6-44}$$

We can now begin the magnitude scaling procedure. Just as in time scaling, several different methods are possible and we merely show one direct and simple approach. The first step is to change the temperature equation (6-44) into a voltage equation using the following one-to-one scale factors.

$$\left. \begin{array}{ll} T \to 1°F = 1 \text{ volt}, & \dfrac{d^2T}{d\tau^2} \to \dfrac{1°F}{\text{sec}^2} = 1 \text{ volt} \\[3mm] \dfrac{dT}{d\tau} \to \dfrac{1°F}{\text{sec}} = 1 \text{ volt}, & \dfrac{d^3T}{d\tau^3} \to \dfrac{1°F}{\text{sec}^3} = 1 \text{ volt} \end{array} \right\} \tag{6-45}$$

These one-to-one relations are not peculiar to this example; we *always* assume them in just this way. In writing the voltage equation we could define new symbols, say, V_T for a voltage representing temperature T, but we prefer to retain the actual physical system symbol and just remember that from here on, it means a voltage rather than a temperature. Since the conversion factors (6-45) are all one-to-one, Eq. (6-44) can be taken now to be a voltage equation. At this point it is necessary to estimate the maximum values of all terms in this equation. How good these estimates are depends upon how familiar or unfamiliar we are with this particular physical system. In Eq. (6–39), with Q constant at 150,000 it is clear that the final value (particular solution) of T will be 75°F. That this will be the *maximum* value of T is intuitively clear if one has some feeling for the response of thermal systems. The maximum values of the various derivatives of T are not nearly so obvious and we may have to decide our initial scaling based only on our knowledge of T, set up the computer on this basis and let the computer tell us if rescaling is needed. Treating Eq. (6–44) as a voltage equation (numerical parameters such as 1.87, 6.86, etc., are treated as pure numbers with no dimensions) and using scale factors (6–45), we see that the two terms that are known or can be estimated both have maximum values of 75 volts. If we are doing this problem on a 10-volt full scale computer, both of these terms would clearly overload it. The scaling procedure consists simply of dividing the whole equation through by some convenient number which will reduce the maximum voltages to acceptable ranges. Since the original equation was correct, dividing through by any constant will of course still give a correct equation. A convenient number here might be 10, since it gives maximum voltages of 7.5 volts, which are big enough to be accurate and small enough to not overload even if our estimated maximums are somewhat too low. Our final equation can thus be written in a form ready for computer setup as

$$0.187 \frac{d^3T}{d\tau^3} = 7.5 - 0.686 \frac{d^2T}{d\tau^2} - 0.504 \frac{dT}{d\tau} - 0.1T \qquad (6\text{–}46)$$

Figure 6–7a shows a computer diagram for this system; all initial conditions are zero since the system is in equilibrium at $T = 0°F$ when the heat is turned on at $t = 0$. A recorder connected to $0.1T$ shows the time history of T. Since the scale factors (6–45) tell us that, for the voltage T, one volt represents one °F, if at some time instant the signal $0.1T$ measures, say, 5 volts, the temperature at this instant would be 50°F. If in the computer this occurred at $\tau = 6$ sec, in the real system it would occur at $t = 6000$ sec $= 1.66$ hours. If at the same instant we record the signal $-0.748 d^2T/d\tau^2$ and get, say, -2 volts, this makes $d^2T/d\tau^2 = 2.68°F/sec^2$ and from (6–41) the real system d^2T/dt^2 is $2.68 \times 10^{-6}°F/sec^2 = 34.7°F/hr^2$. This problem was actually run, giving the response curves of Fig. 6–7b. We see there that none of the signals overload and all are large enough to be accurate, so the initial scaling need not be revised.

The accuracy of the best individual analog computer components may approach 0.01%, but when many of these are used together in a complete simulation, the overall accuracy deteriorates. However, careful attention to detail allows simulation of problems requiring several hundred components, with acceptable engineering accuracy on the order of 1 to 5%. Large modern analog installations may provide considerable automation of various operations. For example, the manual setting of several hundred

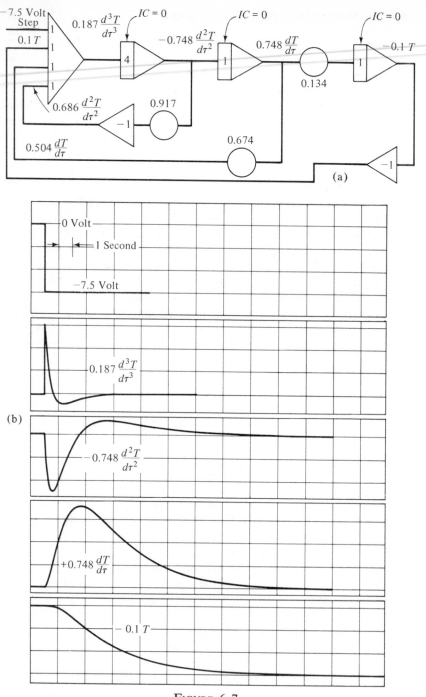

FIGURE 6–7

Analog Simulation of Heating System

potentiometers is quite time consuming, but can be made fast and automatic by use of servomechanisms commanded by numerical values fed in on punched paper tape. There is also an increasing use of combinations of analog and digital computer technology to perform needed operations, the ultimate realization of this being the *hybrid computer*, which shall be discussed briefly later.

6-5. DIGITAL SIMULATION.

While digital computers have from their earliest days been used to solve differential equations, the concept of digital simulation is of relatively recent origin. It is part of the general movement to make the power of digital computers more accessible to those who are not expert programmers. The basic concept behind this movement is the development of specialized computer languages to solve particular classes of problems. It takes a high degree of programming skill to create these languages, but they may then be used by those who may have neither time nor inclination to become expert programmers. The various digital simulation languages which have been developed are good examples of this approach to improved efficiency in the utilization of digital computers.

Many of the features of digital simulation programs amount to a digital implementation of analog computer methods; in fact, some of them actually are called "digital simulation of an analog computer." The reason for this is that the analog method of equation solution is basically simple and direct and appeals greatly to most engineers. Fundamental to all digital methods of differential equation solution is the process of numerical integration. Various algorithms for numerical integration with varying degrees of accuracy and speed are available, and most digital simulation programs give one a choice of several of these methods. Whatever method is used, the selection of a proper computing increment for the independent variable (usually time) is critical. That is, while an analog computer deals with time as a continuous variable, in digital analysis the independent variable varies in discrete steps. If the step size is too small, the total number of computations becomes large, using up expensive computer time. Also, since each calculation causes a roundoff error, the accumulated error may become excessive. Too large a step size causes inaccuracy due to extrapolation error and may even cause instability and divergent oscillations in the solution. Fortunately, most algorithms exhibit a fairly wide range of step sizes for which the solution is essentially the same, and this is, in fact, a practical criterion for judging the adequacy of a particular step size. The more sophisticated algorithms automatically adjust their step size to meet a prespecified error criterion and are thus able to use a large step size (and thus compute rapidly) when the solution is changing slowly, but will switch to a smaller step when rapidly changing variables require it.

To illustrate how a digital simulation language is used, let us employ the IBM CSMP (Continuous System Modeling Program)[1] to solve Eq. (6–39). A comprehensive digital simulation language suitable for large problems requires a significant fraction of the capabilities of a large and fast computer. Once the language has been implemented on the computer, however, the user can call it into action very simply without concern for all that is going on "behind the scenes." We first note that we can work with Eq. (6–39) *directly*, and that *no time or magnitude scaling is necessary*, an obvious advantage over the analog computer. This is due to the extremely wide range of numbers accepted by a digital computer and the fact that the time taken by the computer to run the problem is essentially unrelated to the time scale of the real physical system. Just as in the analog computer, the first step is to write an expression for the highest derivative in the equation:

$$\text{TD\overline{O}T3} = 0.0125*Q - 13.5*\text{TD\overline{O}T2} - 35.*\text{TD\overline{O}T1} - 25.*\text{T} \quad \text{(6–47)}$$

Note that this is just an ordinary Fortran statement in which we are using the symbols TDŌT3, TDŌT2 and TDŌT1 for the derivatives of T. The program allows us to make up any names we wish for variables as long as they contain no more than six alphanumeric characters and start with an alphabetic character. The names chosen above for the derivatives are merely convenient; any other names conforming to the rules could have been employed. The statement (6–47) is equivalent to the connection of inputs to an analog summer which will have TDŌT3 as its output; of course, the "digital summer" does not change the algebraic sign. The next step, just as in the analog computer, is to integrate TDŌT3 to get TDŌT2; however, now the integration is done numerically using some chosen algorithm. In CSMP seven different integration methods are available; if none is specified the program will use a fourth order Runge-Kutta with variable step size, a good general-purpose method. Whatever integration method is selected, the statement to actually perform the integration is the same. To get TDŌT2 from TDŌT3, we would write

$$\text{TD\overline{O}T2} = \text{INTGRL}(0.0, \text{TD\overline{O}T3}) \quad \text{(6–48)}$$

where 0.0 is the initial value of TDŌT2. Similarly, the other two integrations are specified by

$$\text{TD\overline{O}T1} = \text{INTGRL}(0.0, \text{TD\overline{O}T2}) \quad \text{(6–49)}$$

$$\text{T} = \text{INTGRL}(0.0, \text{TD\overline{O}T1}) \quad \text{(6–50)}$$

---

[1] IBM Publication H20-0367-3.

In Eq. (6–47) we could have inserted Q directly as the number 1.5E05, however, CSMP also has a statement for specifying step inputs of unit height which occur at any desired time TS. The statement for a variable Y would be Y = STEP(TS). To specify Q, we could thus write Q = 1.5E05*Q1 and Q1 = STEP(0.0). It is now necessary to tell the computer what time-step size to use, how far in time we wish to run, and the increments of time at which results are to be printed and/or plotted. We will use the Runge-Kutta variable step integration method so the computer will adjust the step size for optimum performance; however, it is still desirable to give an initial estimate of step size. A ballpark rule is to take a step size on the order of 1/2000 of the total time or, if an oscillatory solution is anticipated, a step size about 1/200 of a cycle of oscillation. We earlier estimated the time for the house to heat up at about two hours; thus, a computing increment of 0.001 hour would be reasonable, with results printed every 0.01 hour. Printer plots of selected variables against time may be obtained simply by asking for them; scaling to fit the plots on the printer page size is automatically provided. We can now show the complete set of statements which must be punched into cards to run this problem

```
TITLE    HOUSE HEATING TRANSIENT
         TDOT3 = 0.0125*Q−13.5*TDOT2−35.*TDOT1−25.*T
         TDOT2 = INTGRL(0.0,TDOT3)
         TDOT1 = INTGRL(0.0,TDOT2)
         T = INTGRL(0.0,TDOT1)
         Q1 = STEP(0.0)
         Q = 1.5E05*Q1
TIMER    DELT = 0.001, FINTIM = 2.0, PRDEL = 0.01,OUTDEL = 0.04
PRINT    T,TDOT1,TDOT2,TDOT3
PRTPLT   T,TDOT1,TDOT2,TDOT3
END
STOP
ENDJOB
```

On the TIMER card, DELT is the step size, FINTIM is the final time, PRDEL is the printing increment, and OUTDEL is the plotting increment. The PRINT card tells which variables are to be printed and the PRTPLT card which are to be plotted against time. The order in which the statements are given is largely immaterial, the computer sorts them into a proper sequence.

The above problem was run on an IBM 360–75 and gave numerical values in close agreement with the analog results of Fig. 6–7. As could have been noted from the analog results, the time to come to equilibrium is more like five hours than the two originally estimated. With FINTIM changed to 5.0 in the above program, so as to show the entire warmup period, the actual computing time in the central processing unit was 21.1

seconds. From Fig. 6–7, we see that the analog solution took about 20 seconds; however, we should point out that many modern analog machines could easily be time-scaled to run this problem in as little as 0.05 seconds if desired, whereas the speed of the digital solution *cannot* be significantly increased using the Runge-Kutta integrating algorithm and the IBM 360–75 computer. In fact, no matter what digital machine or algorithm is employed, no really large increase in computing speed for such a problem should be expected with the present generation of computers.

To compare some integration methods and also to demonstrate the effects of step size, this same problem was run using the rectangular method of numerical integration (perhaps the simplest algorithm) with several fixed step sizes. Figure 6–8 shows the results of these tests. (First note that with the Runge-Kutta variable-step-size technique which was originally used, the program ran with 0.01-hr steps, even though we specified 0.001. That is, the algorithm applied its accuracy criterion and found it could use the larger step with adequate accuracy.) The rectangular method merely computes the present value of the highest derivative and assumes this constant over the next time step. This allows one to calculate the change in the next lower derivative and so forth, until a new value of T is found. The process starts from the given initial conditions and progresses in the above fashion until FINTIM is reached. Figure 6–9 shows the first few steps. Returning to Fig. 6–8, we see that the rectangular method works well with step sizes of 0.1 or smaller, but begins to oscillate divergently for 0.2 and violently for 0.4.

The above example problem should clearly show the ease and speed with which the digital simulation language allows us to deal with differential equations. It does not, however, bring out the many other powerful features which are built into such an operating system and which are extremely useful in more complex problems. We cannot here explore these exhaustively, but a few of the more important should be mentioned. Nonlinear equations such as (6–35) are easily handled using ordinary Fortran statements to express the highest derivative and then the CSMP integrating statements to get the lower derivatives:

$$\text{XD}\overline{\text{O}}\text{T2} = (\text{F}-\text{B}*\text{XD}\overline{\text{O}}\text{T1}**3.-\text{K}*\text{SIN(X)})/(1.-\text{A}*\text{TIME})$$
$$\text{F} = 2. + 4.*\text{TIME}$$

PARAM $\text{B} = 3.6, \text{K} = 10.4, \text{A} = .05$
$$\text{XD}\overline{\text{O}}\text{T1} = \text{INTGRL(IC1,XD}\overline{\text{O}}\text{T2)}$$
$$\text{X} = \text{INTGRL(IC2,XD}\overline{\text{O}}\text{T1)}$$
INC$\overline{\text{O}}$N $\text{IC1} = -3.,\text{IC2} = +2.7$

This example brings out several other aspects of CSMP. The independent variable is always called TIME; we use it above in the 1−At term and also to express the driving force F, which has here been taken to be a step plus

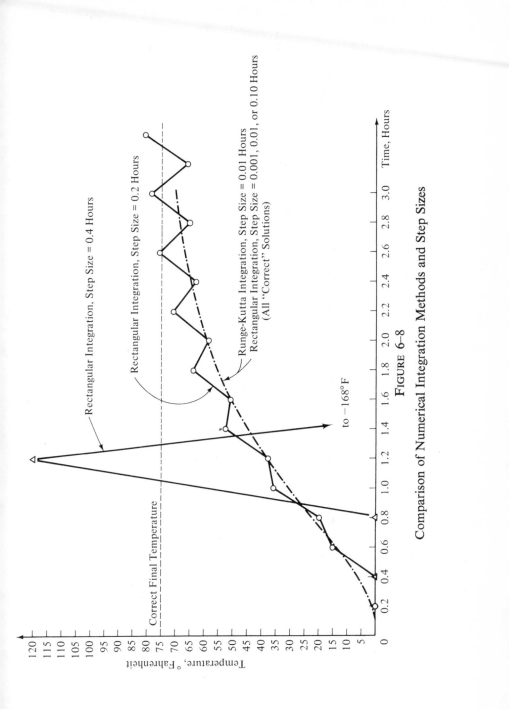

FIGURE 6–8

Comparison of Numerical Integration Methods and Step Sizes

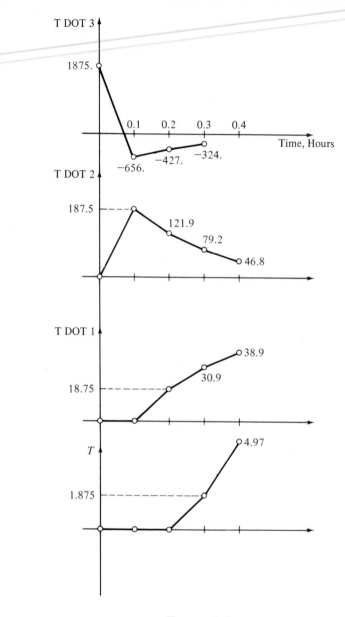

FIGURE 6–9

Rectangular Integration with 0.1 Step Size

266

a ramp, F = 2 + 4t. The $\dot{x}^3$ is formed by the usual Fortran statement as is the trigonometric function sin x. Numerical values of parameters B, K and A are in this example given on a separate card labelled PARAM. This is convenient if we wish to change parameters with a minimum of effort. In fact, if *several* values of a parameter, such as B, are given in the form

$$\text{PARAM} \quad B = (3.6, 4.0, 4.4), \quad K = 10.4, \quad A = .05$$

the system will *automatically* produce three runs, using in turn the three values of B given. Similarly, initial conditions IC1 $\triangleq \dot{x}(0)$ and IC2 $\triangleq x(0)$ are here given on a separate card labelled INC$\overline{\text{O}}$N to facilitate changes in initial conditions, should they be desired. Another extremely useful nonlinear feature is the arbitrary function generator capability. In an analog computer, arbitrary functions are fitted with 10 or 20 straight line segments of adjustable slope and intercept. In CSMP the same approach is used; however, the implementation is much quicker and easier and any number of segments (not limited to 10 or 20) may be used. In Fig. 6–10 a quantity y is expressed as a function of x by a graph of experimental data; no mathematical formula is known. To express this relation in CSMP, it is merely necessary to quote as many pairs of x, y data points as desired; the computer will then linearly interpolate between these given points to find the y value corresponding to any x which might occur during a differential equation solution. The needed statements are

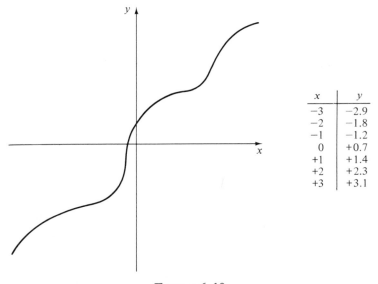

x	y
−3	−2.9
−2	−1.8
−1	−1.2
0	+0.7
+1	+1.4
+2	+2.3
+3	+3.1

FIGURE 6–10

Arbitrary Function

$$Y = AFGEN(FCN1,X)$$

AFGEN FCN1 $= -3.,-2.9,-2.,-1.8,-1.,-1.2,0.,.7,1.,1.4,2.,2.3,3.,3.1$

Many more powerful features of digital simulation languages could be enumerated if time and space permitted; hopefully, those given will convince the reader of the great utility of this approach to system analysis and he will explore further on his own.

6-6. HYBRID COMPUTERS, COMPARISON OF SOLUTION METHODS.

The concept of combining the best features of analog and digital computers into a *hybrid* computer (see Fig. 6–11) was born out of certain computing

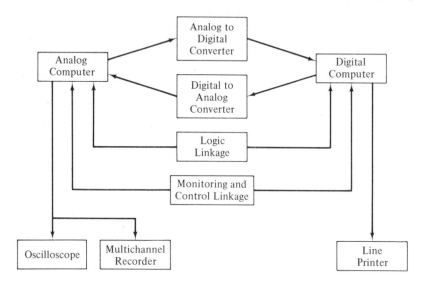

FIGURE 6–11

Hybrid Computer Block Diagram

needs in the aerospace industry about 1958. Problems such as the reentry of manned vehicles (Apollo, for instance) into the atmosphere require computation of trajectories with simultaneous consideration of vehicle attitude (rolling, pitching, yawing) motions. The trajectory calculations are relatively slow in time scale, but must be very accurate, whereas the attitude calculations involve rapid motions, but require less accuracy. It was found that there was an economic advantage in treating such problems with a hybrid approach. An all-digital solution was technically possible, but consumed excessive time. For those parts of the problem requiring extreme accuracy, an analog approach was technically not feasible. In some cases, where calculations had to be made in "real time" or faster, the digital methods were too slow. For complex analog simulations many of the auxiliary operations such as pot setting, searching for optimum solutions, and trial-and-error methods of solving boundary-value problems can be made automatic if put under the control of a digital computer. The aerospace experience in this field showed applications in other industries were also feasible and hybrid computing is now employed, wherever it is economical to do so, in many fields of engineering.

Over the years there has been considerable competition between proponents of analog and digital computers. Some devotees of the digital approach go so far as to predict the complete elimination of general-purpose analog machines in the near future. Such predictions have been made for about 10 years now, and analog machines still are being manufactured and used today; however, it is difficult to dispute that digital techniques have made significant inroads on problem areas earlier reserved for analog methods. Powerful and easy-to-use simulation languages such as CSMP are undoubtedly the main reason behind this development. Perhaps the main advantage (other than speed) still retained by analog methods is the close interaction between man and machine inherent in analog operation. Many analog computer installations operate as "open shops" where the engineer having the physical problem to solve actually participates in running the computer. He thus *immediately* observes results and can also immediately make changes based on his observations. This close man-machine synergism can be extremely important in the efficient use of computing facilities. For teaching purposes, a high-speed analog is unmatched in effectiveness since a large-screen oscilloscope can display results to an entire class and the effects of changes in system parameters appear instantly.

Proponents of digital methods are quite aware of these analog advantages and are busily engaged in duplicating them in digital form. The two main developments which allow this are interactive time-sharing terminals and computer graphics. An interactive terminal allows a user access to a large central digital computer as if it were his personal machine. The time-sharing system makes such service available to many users simultaneously. Computer graphics provides large-screen oscilloscope displays of system behavior ranging from simple curve plotting to sophisticated 3-dimensional views of moving objects such as space vehicles or vibrating structures. It should be clear that such capabilities essentially duplicate or even improve upon the man-machine interaction inherent in analog systems. The main drawback at present is the high cost of such interactive terminals with graphic displays; few installations can afford very many of them at present prices. Even without these sorts of facilities, however, the availability of a simulation language such as CSMP makes possible a very effective use of a computer by large numbers of people with ordinary batch processing. It is quite feasible to have several hundred students per day submitting fairly complex CSMP problems to a single machine (which is also doing many other jobs); this would be quite impossible with analog machines unless one had very many of them, which is again not economically feasible.

The above discussions should indicate that while digital methods are undoubtedly in the ascendancy, analog and hybrid techniques will certainly be with us for many years in general-purpose computers and probably always in special-purpose devices that are built into operating processes such as machine tools, vehicles, power plants, and chemical plants.

BIBLIOGRAPHY

1. Kaplan, W., *Elements of Differential Equations*. Reading, Mass.: Addison-Wesley Publishing Co., Inc., 1964.

2. Martin, W. T. and E. Reissner, *Elementary Differential Equations (2nd ed.)*. Reading, Mass.: Addison-Wesley Publishing Co., Inc., 1961.

3. Rekoff, M. G., *Analog Computer Programming*. Columbus: Charles E. Merrill Publishing Co., 1967.

4. Rogers, A. E. and T. W. Connolly, *Analog Computation in Engineering Design*. New York: McGraw-Hill Book Company, 1960.

5. Salvadori, M. G. and R. J. Schwarz, *Differential Equations in Engineering Problems*. Englewood Cliffs, N. J.: Prentice-Hall, Inc., 1954.

PROBLEMS

6-1. Find complementary solutions for the following equations. Digital computer root-solvers may be used.

a. $3\dfrac{dx}{dt} + 5x = 7 \qquad x(0) = -2$

b. $3\dfrac{dx}{dt} - 5x = 7t \qquad x(0) = 0$

c. $2\dfrac{d^2x}{dt^2} + 5x = e^{-5t} \qquad x(0) = 0, \quad \dfrac{dx}{dt}(0) = 0$

d. $2\dfrac{d^2x}{dt^2} + 0.1\dfrac{dx}{dt} + 5x = 4$ $\qquad x(0) = 1, \qquad \dfrac{dx}{dt}(0) = 0$

e. $2\dfrac{d^2x}{dt^2} + 20\dfrac{dx}{dt} + 5x = 4 + 3t$ $\qquad x(0) = 1, \qquad \dfrac{dx}{dt}(0) = -1$

f. $5\dfrac{d^4x}{dt^4} + 4\dfrac{d^3x}{dt^3} + 10\dfrac{d^2x}{dt^2} + 5\dfrac{dx}{dt} + 3x = 0$

All initial conditions are zero except $x(0) = 2$.

g. $4\dfrac{dx}{dt} + 3x - 7y = 2t \qquad x(0) = 1$

$2x + 7\dfrac{dy}{dt} - 8y = 0 \qquad y(0) = 0$

h. $6\dfrac{d^2x}{dt^2} + 3\dfrac{dx}{dt} + 5y = \sin 2t \qquad y(0) = -2$

$2\dfrac{dy}{dt} + 4y - 6x = 0 \qquad x(0) = 0, \qquad \dfrac{dx}{dt}(0) = 1$

i. $8\dfrac{d^2x}{dt^2} + 2\dfrac{dx}{dt} + 10x - 2\dfrac{dy}{dt} - 10y = 2t - 2$

$3\dfrac{d^2y}{dt^2} + 2\dfrac{dy}{dt} + 15y - 2\dfrac{dx}{dt} - 10x = 3\sin t$

All initial conditions are zero.

6-2. Find particular solutions only for the equations of problem 6-1.

6-3. Find complete (complementary plus particular) solutions for the equations of problem 6-1 but do not evaluate the constants of integration.

6-4. Get complete, specific solutions (including the constants of integration) for the equations of problem 6-1.

6-5. Set up analog computer diagrams for the equations of problem 6-1, but do not attempt to time scale or magnitude scale.

6-6. For the equations of problem 6-1a through e, set up analog computer diagrams complete with time and magnitude scaling, using the analytical solutions to estimate maximum amplitudes and rates. Treat the equations as voltage equations, take time t in minutes, assume computer is ±10 volts full scale, and try for a problem duration of about 10 seconds.

6-7. Using CSMP or whatever digital simulation program is available to you, write programs to solve the equations of problem 6-1 but do not run them.

6-8. Repeat problem 6-7, but now actually run the programs to get the solutions.

6-9. Using CSMP with the rectangular integration method (insert a card METHŌD RECT) and fixed step size, investigate for equation a of problem 6-1 the range of step sizes which gives "correct" solutions.

6–10.   Using CSMP, solve for x_1 and x_2 in Fig. 6–1a if $M_1 = M_2 = 1$ slug, $K_{s1} = 1000$ lb$_f$/ft, $K_{s2} = 2000$ lb$_f$/ft, $B = 10$ lb$_f$/(ft/sec), all initial conditions are zero, and f is a step input of 200 lb$_f$.

6–11.   Set up an analog computer diagram (no scaling) for the system of Fig. 6–1a.

6–12.   Set up an analog computer diagram (no scaling) for the system of Fig. 6–1b.

6–13.   Using CSMP, solve for i_1 and i_2 in Fig. 6–1b if $R = 10{,}000$ ohms, $L = 0.5$ H, $C = 0.01$ μF, and e is a 100-volt rectangular pulse of 0.001 sec duration. Both currents are initially zero.

6–14.   Set up analog computer diagrams (no scaling required) for the following equations.

a. $3\dfrac{dx}{dt} + 5\sqrt{x} = 20 + 2\sin 6.28t \qquad x(0) = 15$

b. $35\dfrac{d^2x}{dt^2} + 3\dfrac{\dot{x}}{|\dot{x}|}|\dot{x}|^{1.75} + x = 0, \qquad \dot{x}(0) = 0, \qquad x(0) = 5$

c. $\dfrac{d^2x}{dt^2} + 50\dfrac{dx}{dt} + 9760x^3 = 2000 \qquad \dot{x}(0) = 0, \qquad x(0) = 0$

d. $0.01\dfrac{d^2x}{dt^2} + 1.25(x - y)^3 = f$

$-1.25(x - y)^3 + 100y + 0.2\dfrac{dy}{dt} + 0.01\dfrac{d^2y}{dt^2} = 0$

For $0 \le t \le 2.0, \qquad f = 10 - 10t$
For $2.0 \le t < \infty, \qquad f = -10$

$x(0) = y(0) = \dfrac{dx}{dt}(0) = \dfrac{dy}{dt}(0) = 0$

6–15.   Using CSMP or whatever digital simulation program is available to you, write programs for the equations of problem 6–14 but do not run them.

6–16.   Repeat problem 6–15, but now actually run the programs.

7

FIRST-ORDER SYSTEMS

7-1. INTRODUCTION.

We now begin a detailed consideration of certain combinations of the basic system elements from the mechanical, electrical, fluid, and thermal areas. Since we found analogous behavior quite common when considering elements, we should not be surprised to encounter it also in systems. This commonality allows a great efficiency in our study of system response since knowledge of the characteristics of a particular *class* of systems is immediately applicable to any member of that class, irrespective of its physical nature. Two classes of systems, the so-called *first-order* and *second-order*, are found to be of fundamental importance. Many practical devices will be found to fit these two classes; thus they are important in their own right. Furthermore we shall find that more complex systems may profitably be considered in terms of combinations of the simple first- and second-order types.

7-2. MECHANICAL FIRST-ORDER SYSTEMS.

Figure 7–1 displays some examples of combinations of masses, springs, and dampers which will be found to be first-order systems. The usefulness of

273

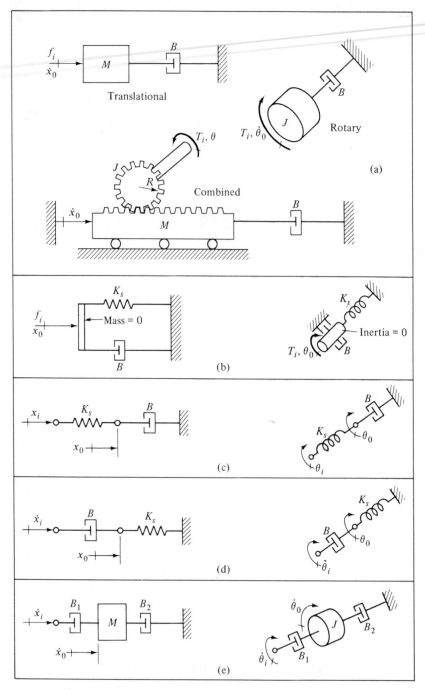

Figure 7-1

Some Mechanical First-Order Systems

such models in dealing with practical engineering problems can be illustrated by quoting some actual applications. The model of Fig. 7–1a is
directly applicable to the study of the motions of the slides and carriages
of machine tools such as milling machines and lathes. The driving force
f_i (or torque T_i) is provided by some kind of motor (electric or hydraulic)
which is to move the slide (whose mass is M) at some desired speed $\dot{x}_o$ or
to some desired position x_o. Slides on machines may be supported and
guided by lubricated bearing surfaces called "ways"; the damper B may
represent the viscous friction due to shearing the lubricating oil film. Since
both rotary and translational motions are found in machines, Fig. 7–1a
shows models for each. It also shows a *combined* rotary and translational
system, since it is quite common to use a rotary motor to power a translational machine motion.

The model of Fig. 7–1b is sometimes used to represent a real spring
which has significant energy dissipation. Since a pure spring element has
no losses, we must add a damper to our model to provide for this. Figure
7–1c might represent a delayed-action (or mechanical lag) mechanism.
Analysis will show that motions applied at x_i are reproduced at x_o if they
are slow, but will be "delayed" if they are rapid. This action has been
usefully applied in hydromechanical control systems to provide a stabi-
lizing effect. The rotary version of Fig. 7–1d is used as a speed-measuring
instrument in automobile speedometers. An angular velocity $\dot{\theta}_i$ (provided
by the speedometer cable) produces an angular deflection θ_o of the needle
which indicates speed. Figure 7–1e (rotary version) could represent a
mechanical drive system in which a power source running at speed $\dot{\theta}_i$ is
used to drive a load of inertia J and friction B_2 through a fluid coupling
modeled as damper B_1.

Preliminaries to Equation Setup. Let us use the translational system of
Fig. 7–1a as a vehicle for illustrating some general concepts useful in
mechanical system analysis. We recall first that the mass and damper
elements shown represent an idealized model of some real, practical
system and that it was necessary for the analyst to make a judgment, as
best he could, that such a model would be adequate for his purposes. The
ultimate test of such judgments is experimental testing of the real system.
If the predictions of the analysis are largely confirmed by such experiments,
the analyst will have had his judgment justified, and if a related problem
occurs in the future, he will feel much more confident about making
similar assumptions. On the other hand, if experiments do *not* agree with
predictions the analyst will carefully study the experimental data to try to
discover those aspects of system behavior which he did not sufficiently
understand. He may then change the model so that it will agree more
closely with reality.

It is necessary to be clear from the outset which physical variables will be taken as the system input and output. This decision generally requires consideration of the interfaces between the subsystem being studied and the overall system of which it is a part. In our present example the force f_i might be provided by a driving motor of some sort and used to control the speed $\dot{x}_o$ of the moving mass, which speed might be measured by some instrument which is part of the overall machine. It would thus be natural to consider the force f_i as an input which causes the velocity $\dot{x}_o$. We must next become more specific about the input and output signals with regard to such things as coordinate systems and sign conventions. Since both force and velocity are vector quantities they can in general take arbitrary directions in space. We here consider, as part of the definition of our model, that forces and motions are constrained to a single axis; that is, the model is one-dimensional. With this restriction it is still necessary to decide on an origin and positive direction for the single coordinate x_o. That is, if we find out later that $x_o = -5.3$ inches at some instant of time we have to know the answer to the questions "5.3 inches from where?" and "5.3 inches in which direction?" These choices may be made freely, but in some cases one choice is more convenient than another. In Fig. 7–2a we note that with no force applied there is no preferred position for the mass M—it will sit wherever it is put; thus the choice of an origin for coordinate x_o is arbitrary. In Fig. 7–1b, however, for $f_i = 0$ the spring

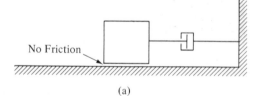

(a)

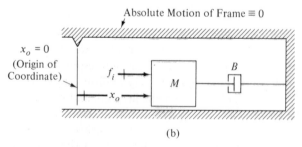

(b)

FIGURE 7–2

Coordinate Origins and Sign Conventions

will assume a definite position, which would make a good choice for the $x_o = 0$ location, since then the spring force is most simply written. The origin of x_o shown in Fig. 7–2b can thus correspond to any convenient location in the actual machine *but is assumed to have zero absolute motion;* that is, the displacement x_o (measured relative to its origin) is an *absolute* displacement. (Since any system resting on the earth must participate in the earth's own motions, the notion of a "fixed point" to use as a reference is theoretically somewhat elusive, but for the vast majority of problems it is conventional [and adequately accurate] to consider the earth "fixed.") The reason we are so concerned about the "fixity" of our reference points is that our analyses are rooted in Newton's law of motion $F = MA$ which requires that the acceleration A be the *absolute* acceleration. The choice of a positive direction for x_o is completely arbitrary and there is generally little if any advantage to choosing one direction or the other; however, a choice must be made, and it must be made at the beginning of the analysis, not at the end.

Once a positive direction for x_o has been chosen, the positive directions for $\dot{x}_o$, $\ddot{x}_o$, and forces *must* coincide with the choice for x_o. Velocity $\dot{x}_o$ is defined as $\lim(\Delta x/\Delta t)$ and since Δt is positive (time does not run backwards) a positive Δx must correspond to a positive velocity (and similarly for acceleration). Forces must be taken positive in the same direction as x_o because Newton's law $F = MA$ says that a positive force must cause a positive acceleration (mass is by definition positive). It is not necessary at this stage to be explicit about the exact nature of the driving force; we may carry it through as an arbitrary function of time $f_i(t)$. In Fig. 7–2b we have taken x_o positive to the right (symbol $+\longrightarrow$) so $\dot{x}_o$, $\ddot{x}_o$, and all forces are also positive to the right.

Writing the System Equation. Analysis of any system is based upon proper application of the appropriate physical laws. For mechanical systems Newton's law is fundamental, even though it may appear in various forms such as D'Alembert's principle or energy methods. It is generally best to first express the applicable law in *word* (rather than equation) form. For our present example we might say to ourselves, "At any instant of time, the summation of forces acting on the mass in the x_o direction must equal the product of the mass and its instantaneous acceleration $\ddot{x}_o$." The concept of the free-body diagram is widely useful in mechanics problems as an aid to properly enumerating and expressing the various forces which enter into the summation. One draws a sketch of the body to which Newton's law is to be applied and carefully indicates *all* of the forces which impinge on the body in the direction of motion being considered. Recall that fundamentally only two types of forces can act on a body; forces due to actual contact with another body, and the "mysterious" gravitational, magnetic, and electrostatic forces (so-called "action-at-a-distance" forces). In Fig. 7–3a all the forces acting on M are shown. The gravity force (weight of M) is just balanced by the reaction force of

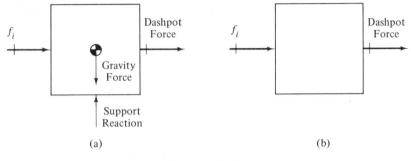

(a)   (b)

FIGURE 7–3

Free-Body Diagrams

the support; thus the net force in the vertical direction is zero, and no motion occurs in this direction. We can thus employ the simpler free-body diagram of Fig. 7–3b for the x_o direction only. The only forces acting in this direction are direct-contact forces due to the input force source and the force of the damper on the mass.

We can now write Newton's law in equation form as

$$\sum \text{Forces} = (\text{Mass})(\text{Acceleration}) \qquad (7\text{–}1)$$

$$f_i - B\dot{x}_o = f_i - Bv_o = M\ddot{x}_o = M\dot{v}_o \qquad (7\text{–}2)$$

It is important to be careful about the algebraic signs of the various forces. The sign of f_i is positive since it is defined as the force of the source *on* the mass; a positive force must cause a positive acceleration, a negative force a negative acceleration. We know from chapter two that the magnitude of the damper force is Bv_o and that it *opposes* the velocity; since our equation must agree with these known facts, let us check it. If v_o is positive (mass moving to the right) the damper opposes with a force to the left, a negative force; thus $-Bv_o$ should be negative, which it is. If v_o were negative (mass moving to the left) the force of damper on mass is to the right (positive) which again agrees with $-Bv_o$. Finally, if $v_o = 0$, the damper exerts no force on the mass. We see that, no matter whether the velocity is $+$, $-$, or 0, our damper force term in Eq. (7–2) is correct in magnitude and direction. We now put Eq. (7–2) in the conventional form of Eq. (6–1):

$$M\frac{dv_o}{dt} + Bv_o = f_i \qquad (7\text{–}3)$$

As another example let us analyze the rotational version of Fig. 7–1c. At the "junction" of the spring and damper (the θ_o location) the torque of spring and damper must be identical:

$$K_s(\theta_i - \theta_o) = B\dot{\theta}_o \tag{7-4}$$

$$B\frac{d\theta_o}{dt} + K_s\theta_o = K_s\theta_i \tag{7-5}$$

Note the similarity of form between Eqs. (7–5) and (7–3). Another way of analyzing this system places a fictitious inertia J at the spring/damper junction as in Fig. 7–4. This allows us to use J as a free body and write:

$$\sum \text{Torques} = (\text{Moment of Inertia})(\text{Angular Acceleration}) \tag{7-6}$$

$$+K_s(\theta_i - \theta_o) - B\dot{\theta}_o = J\ddot{\theta}_o \tag{7-7}$$

and since $J \equiv 0$,

$$B\frac{d\theta_o}{dt} + K_s\theta_o = K_s\theta_i$$

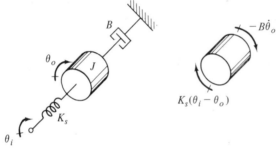

FIGURE 7–4

Use of Fictitious Inertia

The algebraic signs of the torque terms in Eq. (7–7) must again be justified. Positive directions for angles θ_i and θ_o may be arbitrarily chosen; in fact we could take θ_i positive clockwise and θ_o positive counterclockwise if we wished. However, since θ_i is the input which *produces* θ_o, it may be less confusing, once the positive direction for θ_i is chosen, to pick θ_o the same way. Then a positive θ_i will in steady state produce a positive θ_o, which, though not a necessity, may be convenient. That is, certain choices may be mathematically equivalent but one or the other may be practically preferable. Once a positive direction for θ_o is chosen, then $\dot{\theta}_o$, $\ddot{\theta}_o$, and torques on J *must* conform to this choice. If $\theta_i > \theta_o$, then the spring exerts a clockwise $(+)$ torque on J; if $\theta_i < \theta_o$, the torque is counterclockwise $(-)$; thus, the torque term $+K_s(\theta_i - \theta_o)$ correctly represents the physical facts. Similarly, $-B\dot{\theta}_o$ gives the correct damper torque at any instant. The origins for angular displacements θ_i and θ_o can be chosen at any desired locations; however, the spring torque can be written as $K_s(\theta_i - \theta_o)$ only if the origin for θ_i is identical with that for θ_o so that the torque goes to zero when $\theta_i = \theta_o$.

Response to Step Inputs. Having shown a large number of mechanical first-order systems and having derived the system equations of two of these, it is now appropriate to begin a study of the response of such systems to various inputs. As we have pointed out before, ideally one would use as input functions for analysis the actual input forces or motions encountered in the operation of the real system; however, these are generally unpredictable and we thus employ simple "standard" inputs such as steps and sine waves. Experience has shown this approach to be quite satisfactory in most cases. For a standard step input we assume that the system is initially in equilibrium with both input and output at the zero level when the input suddenly jumps up to some constant value at which it remains thereafter. For the system of Fig. 7–2b, we have f_i and v_o both initially zero when f_i suddenly jumps up to a value f_{is} and remains at this value thereafter. We wish to find the resulting motion of the mass; in particular the way that velocity v_o varies with time.

At this point we shall define the generalized first-order system with input q_i and output q_o as a system with equation

$$a_1 \frac{dq_o}{dt} + a_0 q_o = b_1 \frac{dq_i}{dt} + b_0 q_i \qquad (7\text{–}8)$$

Any system whose equation fits this pattern is, by definition, a first-order system. In this chapter we will concentrate on the most important form of this equation, that in which $b_1 \equiv 0$. This type of equation may be reduced to its simplest form by dividing through by a_0 so as to make the coefficient of q_o equal to 1.0:

$$\frac{a_1}{a_0} \frac{dq_o}{dt} + q_o = \frac{b_0}{a_0} q_i \qquad (7\text{–}9)$$

If we now define

$$\frac{a_1}{a_0} \triangleq \tau \triangleq \text{system time constant} \qquad (7\text{–}10)$$

$$\frac{b_0}{a_0} \triangleq K \triangleq \text{system steady-state gain or static sensitivity} \qquad (7\text{–}11)$$

then the equation assumes the compact *standard form*

$$\tau \frac{dq_o}{dt} + q_o = Kq_i \qquad (7\text{–}12)$$

It thus requires only two parameters, τ and K, to completely describe this first-order system. The physical significance of these parameters and the reasons for defining them as we did will become apparent shortly. Whenever one encounters a new physical device whose equation conforms to (7-8) he should *immediately* define K and τ and convert to the standard form (7-12). All the results which we are about to develop pertaining to first-order systems are then instantly available. The operational transfer function relating q_o to q_i is defined in the usual way from the operator form of the differential equation.

$$(\tau D + 1)q_o = Kq_i \qquad (7\text{-}13)$$

$$\frac{q_o}{q_i}(D) = \frac{K}{\tau D + 1} \qquad (7\text{-}14)$$

Applying these procedures to Eq. (7-3) we have

$$\frac{M}{B}\frac{dv_o}{dt} + v_o = \frac{1}{B}f_i \qquad (7\text{-}15)$$

$$\tau \triangleq \frac{M}{B} = \frac{\text{lb}_f\text{-sec}^2}{\text{inch}}\frac{\text{inch/sec}}{\text{lb}_f} = \text{sec}, \qquad K \triangleq \frac{1}{B} = \frac{\text{inch/sec}}{\text{lb}_f} \qquad (7\text{-}16)$$

and thus

$$(\tau D + 1)v_o = Kf_i \qquad (7\text{-}17)$$

$$\frac{v_o}{f_i}(D) = \frac{K}{\tau D + 1} \qquad (7\text{-}18)$$

For a step input of size f_{is} Eq. (7-17) becomes, for $t > 0$,

$$(\tau D + 1)v_o = Kf_{is} \qquad (7\text{-}19)$$

By inspection we see that the particular solution is $v_{op} = Kf_{is}$. The characteristic equation $\tau D + 1 = 0$ has the single root $s = -1/\tau$ so the complementary solution is $v_{oc} = Ce^{-t/\tau}$ and the complete solution is

$$v_o = Ce^{-t/\tau} + Kf_{is} \qquad (7\text{-}20)$$

To find the constant of integration C the initial condition, $v_o(0^+)$, the velocity just an instant *after* the step force is applied, must be known. (For the general first-order system $q_o(0^+)$ must be known.) We know that $v_o = 0$ just before the step is applied. A step force of finite size f_{is} causes a sudden (step) increase in acceleration of f_{is}/M, since $A = F/M$ from Newton's law. This finite acceleration *cannot*, however, produce a finite change in velocity during the infinitesimal time interval from $t = 0$ to $t = 0^+$, since $v = \int A\, dt$ and $\int_0^{0^+} (f_{is}/M)\, dt = 0$. Thus, we conclude that v_o is still zero at $t = 0^+$. This will be the case in *all* first-order systems as given by Eq. (7-12); that is, $q_o(0^+)$ is always 0 for a step input. This can be proven in each physical system that might come up by reasoning from the appropriate physical law, or can be shown once and for all using a strictly mathematical approach as we now demonstrate.

The basic equation

$$\tau \frac{dq_o}{dt} + q_o = Kq_i$$

may be multiplied through by dt and integrated term by term between limits $t = 0$ and $t = 0^+$ to give

$$\int_0^{0^+} \tau \, dq_o + \int_0^{0^+} q_o \, dt = K \int_0^{0^+} q_{is} \, dt \qquad (7\text{--}21)$$

$$\tau q_o \Big|_0^{0^+} + 0 = 0 \qquad (7\text{--}22)$$

$$\tau[q_o(0^+) - q_o(0)] = \tau q_o(0^+) = 0 \qquad (7\text{--}23)$$

The term $\int_0^{0^+} q_o \, dt$ is zero no matter what q_o is, as long as it is not infinite, like an impulse. We can show that q_o cannot be an impulse by integrating the equation *twice:*

$$\tau \int_0^{0^+} q_o \, dt + \int_0^{0^+} \left(\int_0^{0^+} q_o \, dt \right) dt = K \int_0^{0^+} \left(\int_0^{0^+} q_{is} \, dt \right) dt \quad (7\text{--}24)$$

If q_o *were* an impulse, this equation would say

$$\tau \, (\text{something not zero}) + (\text{zero}) = (\text{zero}) \qquad (7\text{--}25)$$

which obviously cannot be true; thus q_o cannot be an impulse. We conclude from Eq. (7–23), then, that $q_o(0^+)$ is always zero.

We can now proceed with the solution for the constant of integration in Eq. (7–20):

$$v_o(0^+) = 0 = C + Kf_{is}, \qquad C = -Kf_{is} \qquad (7\text{--}26)$$

and thus

$$v_o = Kf_{is}[1 - e^{-t/\tau}] \qquad (7\text{--}27)$$

Figure 7–5a shows a plot of v_o versus time. A universal nondimensional plot may be achieved by graphing v_o/Kf_{is} against t/τ as in Fig. 7–5b. This graph is usable for any first-order system with any value of K and τ and for a step input of any size f_{is}. Let us now point out some characteristic features of the step response of a first-order system. As $t \to \infty$ the output v_o approaches asymptotically the value Kf_{is}. While this final value of v_o theoretically requires an infinite time to achieve, in actual practice after a time equal to 4 or 5 time constants has elapsed, the difference between the actual velocity and its ideal final value is so small as to be unmeasurable. We then say the system has achieved its *steady state* for the particular input applied. The ratio of output velocity to input force for this steady-state condition is seen to be $Kf_{is}/f_{is} = K$, the system steady-state gain or static sensitivity. *We thus interpret K to be the amount of steady-state output produced for each unit of input applied.*

In our example $K = 1/B$ and this would be inches/sec of output velocity per pound of input force. Note that the numerical value of K has no effect

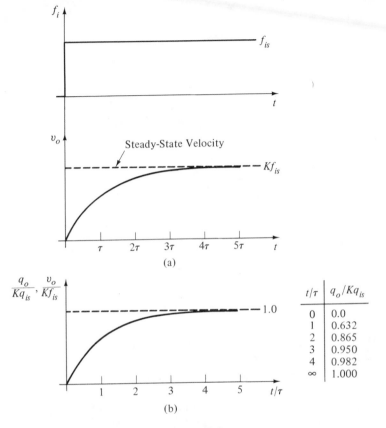

FIGURE 7–5

Specific and Generalized Step Response of First-Order Systems

whatever on *how rapidly* the steady state is reached; the time constant τ serves this function. In linear systems, speed of response is generally defined in terms of how long the system takes to reach some given *percentage* of its steady-state output. This is possible because the response is proportional to the stimulus; doubling the input, for example, also doubles the output. Thus the time to reach the same *actual value* of output may be different for a large step input than for a small, but the time to reach the same *percentage* of final value will be the same for all sizes of steps (see Fig. 7–6). Using this viewpoint, the speed of response for a first-order system is determined entirely by the numerical value of its time constant, since the percentage of steady-state response is given by

$$\frac{v_o}{Kf_{is}} = 1 - e^{-t/\tau} \qquad (7\text{–}28)$$

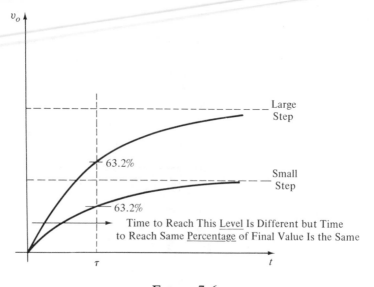

FIGURE 7–6

Definition of Speed of Response

Clearly, if τ is, for example, cut in half, identical values of v_o/Kf_{is} will be achieved in one half the time; thus, *the speed of response is inversely proportional to* τ. We can thus now summarize the significance of the two basic first-order system parameters by saying that K is an indication of *how much* steady-state output will be produced for each unit of input, and that τ will determine *how fast* the steady state will be reached.

Let us compute the initial slope of the curve of v_o versus t:

$$\left.\frac{dv_o}{dt}\right|_{t=0^+} = \left.\frac{Kf_{is}}{\tau}e^{-t/\tau}\right|_{t=0^+} = \frac{Kf_{is}}{\tau} \qquad (7\text{–}29)$$

If the slope stayed constant at this initial value (rather than decreasing according to $(Kf_{is}/\tau)e^{-t/\tau}$) the steady state would be reached in τ seconds (see Fig. 7–7). This fact is useful in sketching such curves and also in checking experimentally measured step responses to see whether they fit the first-order model. For $t = \tau$ in Eq. (7–27), we have $v_o = Kf_{is}(1 - e^{-1}) = 0.632Kf_{is}$; thus 63.2% of the final value is achieved in one time constant. This is a quick way to estimate τ from a measured step response curve. A *critical* check of how well a real system fits the first-order model is obtained by replotting its measured step response semilogarithmically as follows:

$$Z \triangleq \log_e\left(1 - \frac{v_o}{Kf_{is}}\right) = \log_e\left(e^{-t/\tau}\right) = -\frac{t}{\tau} \qquad (7\text{–}30)$$

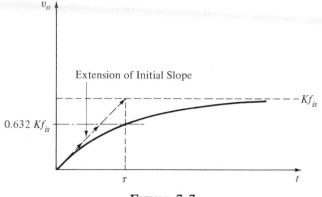

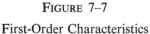

FIGURE 7–7

First-Order Characteristics

A plot of Z versus time t is thus a straight line whose slope is $-1/\tau$. Figure 7–8 shows how this procedure is applied to a general first-order system with input q_i and output q_o. Since q_o at any instant, and its final value Kq_{is}, are known, we can compute as many z values as we wish from

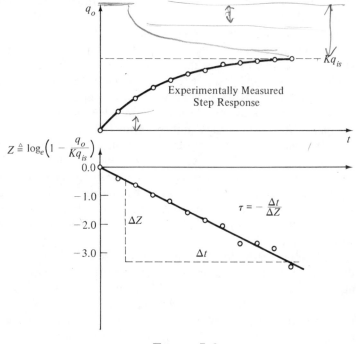

FIGURE 7–8

Experimental Modeling by Step Testing

$Z = \log_e (1 - q_o/Kq_{is})$ and plot against t as shown. If the data points can be well fitted with a straight line, then the real system is behaving very nearly as a first-order system and an accurate numerical value of τ is obtained from the slope of the line. The numerical value of K is obtained simply by dividing the final value of q_o by the known value of input q_{is}.

In Fig. 7–9, a complete picture of the behavior of all the pertinent physical variables in the example system is given. The displacement x_o of the mass is obtained from

$$x_o = x_o(0) + \int_0^t v_o \, dt = Kf_{is}t + Kf_{is}\tau(e^{-t/\tau} - 1) \qquad (7\text{–}31)$$

where we have taken the initial displacement $x_o(0)$ to be zero. Note that after a transient period during which $e^{-t/\tau}$ is dying out, the displacement becomes asymptotic to the straight line $Kf_{is}(t - \tau)$. (If the force f_{is} is left on, a translational damper must sooner or later encounter mechanical stops and cause the mass to stall. However, the rotational version allows continuous unimpeded motion such that the output angular displacement can actually "approach infinity" as indicated by the equations.)

A mathematical description of system behavior as given by the graphs of Fig. 7–9 should generally be reinterpreted in physical terms as a means of checking the plausibility of the results and reinforcing our intuitive feelings about the system. We might put it this way: the suddenly applied force f_{is} causes a sudden acceleration of the mass M; however, as the acceleration acts over time and produces some velocity, the dashpot develops an opposing force which reduces the net accelerating force available. As the velocity builds up, the dashpot force approaches the external driving force more and more closely, and the acceleration approaches zero. The system thus approaches asymptotically a terminal velocity given by the ratio of applied force f_i and the damper constant B. After three or four time constants have gone by, transient effects in all the variables have practically disappeared and the system is in steady-state operation. If we define speed of response as the speed with which the system gets into steady state, system speed can be increased only by reducing the time constant $\tau = M/B$. We may thus try to reduce M, increase B, or both. If we reduce M, the system steady-state gain $K = 1/B$ is unaffected; an increase in B will, however, reduce the gain.

Perhaps the main reason for defining speed of response as we have done is that the inputs to our systems are often in the nature of *commands* which the output quantity is to reproduce or follow. In a differential equation, the command input appears on the right-hand side and is the element which produces the particular solution portion of the output. This portion of the response is sometimes called the *forced* part, while the complementary solution (which is present whether there is a driving input or not) is called the *natural* part of the response. If the forced portion of the re-

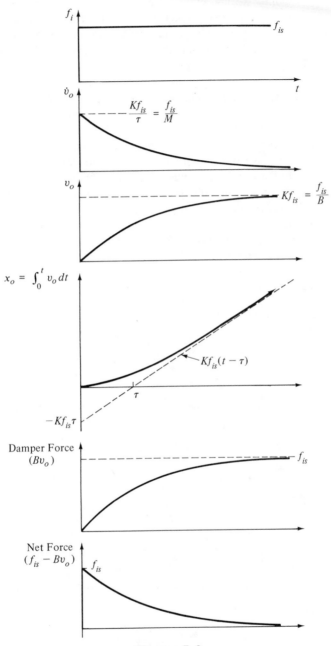

FIGURE 7–9

Mass/Damper System Step Response

sponse corresponds to the command input, and the natural part is of a form that eventually disappears [such as $e^{-t/\tau}$ in Eq. (7–27)], then the faster the natural part disappears the faster the system response conforms to the command input. In a first-order system this leads to the conclusion that a small time constant denotes a fast system.

Computer Simulation and Nonlinearities. An analog simulation of Eq. (7–19) is shown in Fig. 7–10a. If individual adjustment of M and B is desired, Eq. (7–3) may be set up as in Fig. 7–10b. This latter diagram also gives a particularly good view of the physical happenings in the system. The input force f_{is} is summed with the dashpot force $-Bv_o$ to give the net force which, by Newton's law, must equal $M\dot{v}_o$. Division by M gives the acceleration $\dot{v}_o$ which is then integrated to get v_o; multiplication by B gives the damper force. Furthermore, if f_{is} suddenly jumps up, the summer output $-M\dot{v}_o$ and the pot output $-\dot{v}_o$ also suddenly jump up, while the integrator output v_o comes up gradually. Using digital simulation the CSMP statements needed would be

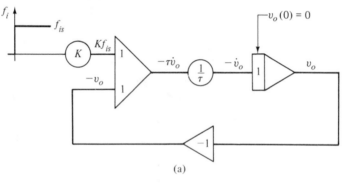

(a)

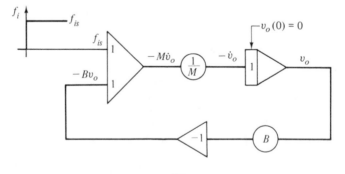

(b)

FIGURE 7–10

Analog Simulations of First-Order Systems

$$\overline{VO}\dot{OT} = (FIS*STEP(0.0) - B*\overline{VO})/M$$
$$\overline{VO} = INTGRL(0.0, \overline{VODOT})$$

where step size FIS, damper coefficient B and mass M would be entered as numbers. Since first-order systems occur frequently in practice, the CSMP language also provides a special statement to handle them with a minimum of effort. For a system as in Fig. 7–11 the statements are

$$QIK = QI*K$$
$$\overline{QO} = REALPL(IC, TAU, QIK)$$

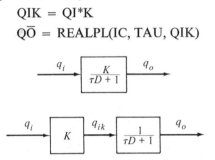

FIGURE 7–11

CSMP Simulation of First-Order System

The second statement is the basic CSMP statement for any first-order system with steady-state gain of 1.0, input QIK, output $\overline{QO}$, time constant TAU and initial value of $\overline{QO}$ equal to IC. The names $\overline{QO}$ and QIK and the numerical values IC and TAU change from problem to problem; the name REALPL (stands for real pole or real root) tells the computer that a first-order system is wanted.

Since real dampers always exhibit at least some nonlinearity, it may be instructive to study the step response of a system with such a damper using both linearizing approximations and "exact" nonlinear computer simulations. Let us take a system with nonlinear damping Cv_o^2 and described by the equation

$$f_i - Cv_o^2 = M\dot{v}_o = f_i - v_o^2 = \dot{v}_o \qquad (7\text{--}32)$$

Recall that the linearizing approximations assume the system to be in equilibrium at some operating point when a perturbation in the input occurs. We wish to get some idea as to the allowable size of such perturbations consistent with reasonable accuracy of the linearized model, so we shall analyze the response for both a small and a large change in input force. For the small change we assume f_i steady at 0.81 lb$_f$ and v_o steady at 0.9 ft/sec when f_i suddenly jumps to 1.21, causing v_o to level off eventually at 1.1 ft/sec. This thus represents a velocity change of $\pm 10\%$ about a base value of 1.0 ft/sec. For the large change, f_i jumps from 0.25 to 2.25 (corresponding to a velocity change from 0.5 to 1.5 ft/sec, a $\pm 50\%$ change about 1.0 ft/sec). Since the mean value of v_o is the same for both cases, the linearized model will be identical for each and is given by

$$v_o^2 \approx v_{o,0}^2 + 2v_{o,0}(v_o - v_{o,0}) = -1.0 + 2v_o \qquad (7\text{-}33)$$

Substituting this into (7–32) gives

$$\dot{v}_o + 2v_o = f_i + 1.0 \qquad (7\text{-}34)$$

If f_i is steady at 0.81, then v_o levels off at 0.905. A step change of f_i to 1.21 then gives

$$v_o = 0.905 + 0.2(1 - e^{-2t}) \qquad (7\text{-}35)$$

For the large step change, the response is

$$v_o = 0.625 + 1.0(1 - e^{-2t}) \qquad (7\text{-}36)$$

To compare these approximations with the "exact" nonlinear responses, we can use CSMP to "solve" the nonlinear equations with the statements

```
           VODOT = FIO + FIS*STEP(0.0) − VO*VO
           VO = INTGRL(IC, VODOT)
PARAM      FIO = 0.81, FIS = 0.4, IC = 0.9
TIMER      FINTIM = 3.0, DELT = 0.01, PRDEL = .05
PRINT      DELT, VO, VODOT
END
PARAM      FIO = .25, FIS = 2.0, IC = 0.5
END
STOP
ENDJOB
```

Figure 7–12 shows the results of these studies. It is clear that for the small input change the linearized model is quite accurate, but this is not the case for the ±50% change. Thus, whether we are allowed the use of a linearized model depends both on the nature of the nonlinearity and also on the magnitude of the excursions of the system variables. In a practical application the expected ranges of variables can often be estimated and a judgment reached on this question.

In these days when simple and quick "solutions" of nonlinear models are often obtainable by computer simulation, there is sometimes a question as to the continued need for linearized models. It turns out that these are still useful in the early stages of system development where the overall view of the effect of system parameters provided by analytical (rather than computer) methods is important. In our example above we could have treated Eq. (7–32) in terms of letter (rather than number) values for C and M, giving

$$M\dot{v}_o + C[v_{o,0}^2 + 2v_{o,0}(v_o - v_{o,0})] = f_i \qquad (7\text{-}37)$$

$$M\dot{v}_o + (2Cv_{o,0})v_o = f_i + Cv_{o,0}^2 \qquad (7\text{-}38)$$

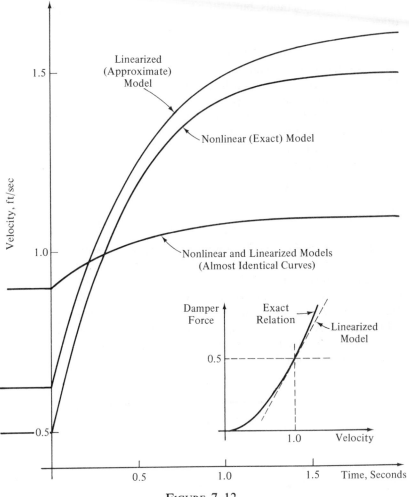

FIGURE 7–12

Effect of Step Size on Linearization Accuracy

The time constant of this linearized model is

$$\tau = \frac{M}{2Cv_{o,0}} \qquad (7\text{--}39)$$

which clearly indicates the effect of M, C, and the operating point $v_{o,0}$ on the system's speed of response. Many computer runs with different values of M, C, and $v_{o,0}$ would be needed to discover the general nature of this relation from the *nonlinear* model and, without careful data analysis, it might still remain obscure; thus, linearized analysis has not been *replaced* by the newer computer methods, but rather the two techniques *complement* each other.

Application of Generalized Results to Specific Cases. Most of our work on first-order systems so far has been presented in terms of the example of Fig. 7–2b and the generalized first-order system of Eq. (7–12). Any new examples of first-order systems (such as those in Fig. 7–1) could be worked out "from scratch" or, more efficiently, by analogy with the generalized system. Since we already have the basic equation for the rotational system of Fig. 7–1c, let us use it as an example. We first put Eq. (7–5) into standard form by defining

$$K \triangleq \frac{K_s}{K_s} = 1.0 \frac{\text{rad}}{\text{rad}} \qquad \tau \triangleq \frac{B}{K_s} \text{ sec} \qquad \text{(7–40)}$$

thus giving

$$\tau \dot{\theta}_o + \theta_o = K\theta_i \qquad \text{(7–41)}$$

Comparison with Eq. (7–12) shows that θ_i plays the role of q_i and θ_o the role of q_o, thus for a step input of θ_i the response of θ_o is given by the generalized response curve of Fig. 7–5, which we may now interpret as in Fig. 7–13 if we wish. Furthermore, since we know that fast response requires a small value of τ, Eq. (7–40) tells us that small B and/or large K_s will give this. Any first-order system (including non-mechanical ones) can be treated in exactly this fashion once its basic equation is derived from fundamental physical laws.

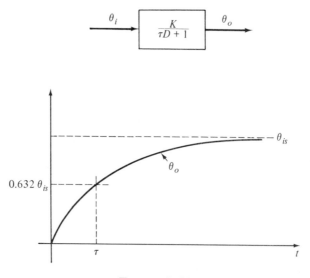

FIGURE 7–13

Step Response

Response to Ramp Input. Further insight into the behavior of first-order systems is obtained by consideration of their response to an input which changes uniformly with time, the so-called ramp input. In Fig. 7–2 this implies the system is initially at rest when at $t = 0$ a force $f_i = \dot{f}_i t$ is applied, where $\dot{f}_i$ is a constant and equal to the time rate of change of the force, lb_f/sec. Equation (7–17) now becomes

$$(\tau D + 1)v_o = K\dot{f}_i t \qquad (7\text{–}42)$$

whereas the generalized form would be

$$(\tau D + 1)q_o = K\dot{q}_i t \qquad (7\text{–}43)$$

We again need the initial condition $q_o(0^+)$ and it is again zero, since if a *step* change in q_i was unable to cause a sudden change in q_o from $t = 0$ to $t = 0^+$, then a *gradual* (ramp) change in q_i certainly would also be unable to cause any change. The particular solution (forced response to q_i) will be of the form

$$q_{op} = At + B \qquad (7\text{–}44)$$

giving

$$\tau A + At + B \equiv K\dot{q}_i t \qquad (7\text{–}45)$$

$$A = K\dot{q}_i \qquad B = -K\dot{q}_i \tau \qquad (7\text{–}46)$$

and thus

$$q_o = Ce^{-t/\tau} + K\dot{q}_i(t - \tau) \qquad (7\text{–}47)$$

Using the initial condition

$$0 = C + K\dot{q}_i(-\tau) \qquad (7\text{–}48)$$

we finally get

$$q_o = K\dot{q}_i \tau e^{-t/\tau} + K\dot{q}_i(t - \tau) \qquad (7\text{–}49)$$

or for our example system

$$v_o = K\dot{f}_i \tau e^{-t/\tau} + K\dot{f}_i(t - \tau) \qquad (7\text{–}50)$$

The natural part $K\dot{q}_i \tau e^{-t/\tau}$ of the motion dies out at a rate determined by τ, and then only the forced part $K\dot{q}_i(t - \tau)$ remains. A universal non-dimensional response curve can be obtained by plotting

$$\frac{q_o}{K\dot{q}_i \tau} = (e^{-t/\tau} - 1) + \frac{t}{\tau} \qquad (7\text{–}51)$$

Figure 7–14 shows the above ramp response curves.

For the mass-damper system we see that after an initial transient period the velocity v_o increases at a constant rate; that is, the motion is one of constant acceleration. The acceleration $\dot{v}_o$ can of course be calculated directly from Eq. (7–50) as

$$\dot{v}_o = K\dot{f}_i(1 - e^{-t/\tau}) \qquad (7\text{–}52)$$

which shows clearly how the acceleration builds up from zero to its constant value in steady state. In a real system a ramp input clearly cannot

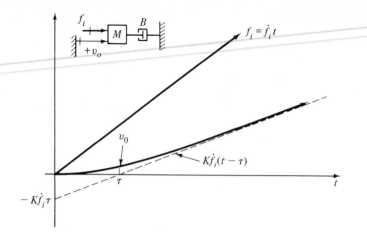

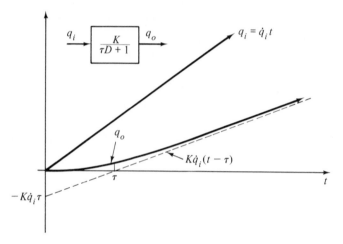

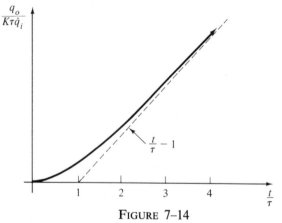

FIGURE 7-14

Ramp Input Response Curves

persist indefinitely since the input force and output velocity both approach infinity; however, a ramp input may be a reasonable model for a real input over a restricted time interval.

The ramp response of the rotational system of Fig. 7–1c is given by

$$\theta_o = K\dot{\theta}_i\tau e^{-t/\tau} + K\dot{\theta}_i(t - \tau) = \dot{\theta}_i\tau e^{-t/\tau} + \dot{\theta}_i(t - \tau) \qquad (7\text{–}53)$$

where the ramp input in this case consists of the θ_i shaft rotating at a constant velocity $\dot{\theta}_i$. Figure 7–15 shows the relation between input and output

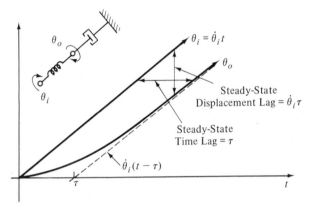

FIGURE 7–15

Ramp Response

for this system. In steady state, the output shaft turns at the same speed $\dot{\theta}_i$ as the input shaft; however, its angular displacement lags behind by an amount $\dot{\theta}_i\tau$, just sufficient that the spring torque due to $(\theta_i - \theta_o)$ will provide the torque $\dot{\theta}_iB$ to drive the damper at constant speed. One could also say that the displacement θ_o lags θ_i by τ seconds, since at any time (when in steady state) the output displacement equals what θ_i *was* τ seconds ago. This is another useful interpretation of the significance of the time constant.

Frequency Response. When the input of a first-order system is made sinusoidal at frequency ω and with amplitude q_{i0} the system equation becomes

$$(\tau D + 1)q_o = Kq_{i0} \sin \omega t \qquad (7\text{–}54)$$

Recall that for frequency response we are interested only in the sinusoidal steady state which is achieved after transients die out; that is, we want the *forced*, not the natural, part of the response. The particular solution must be of form

$$q_{op} = A \sin \omega t + B \cos \omega t \qquad (7\text{–}55)$$

giving

$$\tau(\omega A \cos \omega t - \omega B \sin \omega t) + A \sin \omega t + B \cos \omega t = Kq_{i0} \sin \omega t \quad (7\text{--}56)$$

$$-\omega B\tau + A = Kq_{i0} \qquad \omega A\tau + B = 0 \qquad (7\text{--}57)$$

$$A = Kq_{i0}/(\omega^2\tau^2 + 1) \qquad B = -\omega\tau Kq_{i0}/(\omega^2\tau^2 + 1) \quad (7\text{--}58)$$

and thus

$$q_{op} = \frac{Kq_{i0}}{\omega^2\tau^2 + 1} \sin \omega t - \frac{\omega\tau Kq_{i0}}{\omega^2\tau^2 + 1} \cos \omega t \qquad (7\text{--}59)$$

This may be simplified by using the trig identity

$$A \sin \alpha + B \cos \alpha = \sqrt{A^2 + B^2} \sin \left(\alpha + \tan^{-1}\frac{B}{A} \right)$$

to give finally

$$q_{op} = \frac{Kq_{i0}}{\sqrt{\omega^2\tau^2 + 1}} \sin \left[\omega t + \tan^{-1}(-\omega\tau) \right] \qquad (7\text{--}60)$$

The same result can be obtained much more quickly using the sinusoidal transfer function:

$$\frac{q_o}{q_i}(i\omega) = \frac{K}{i\omega\tau + 1} = \frac{K}{\sqrt{\omega^2\tau^2 + 1}} \underline{/\tan^{-1}(-\omega\tau)} \qquad (7\text{--}61)$$

That is, the amplitude ratio is $K/\sqrt{\omega^2\tau^2 + 1}$ and the phase angle ϕ is $\tan^{-1} - \omega\tau$, just as seen from Eq. (7–60).

For a given K and τ the amplitude ratio and phase angle curves appear as in Fig. 7–16a. As frequency ω approaches zero, the amplitude ratio approaches the steady-state gain K and the phase angle approaches zero. For increasing frequency the amplitude ratio decreases toward zero while the phase angle approaches $-90°$ asymptotically. If the system time constant is decreased, at any given frequency the amplitude ratio will be larger and the phase angle less lagging. The universal curves of Fig. 7–16b are obtained by plotting the amplitude ratio of q_o/Kq_i (rather than q_o/q_i) against $\omega\tau$ (rather than ω):

$$\frac{q_o}{Kq_i}(i\omega) = \frac{1}{\sqrt{(\omega\tau)^2 + 1}} \underline{/\tan^{-1} - \omega\tau} \qquad (7\text{--}62)$$

We introduce at this point the widely used logarithmic methods of plotting frequency-response curves. Here the amplitude ratio is expressed in *decibels*, where the decibel equivalent of a number N is given by

$$\text{decibel value of } N \triangleq dB \triangleq 20 \log_{10} N \qquad (7\text{--}63)$$

Thus, if at some frequency, the amplitude ratio is, say, 2.4, it could also be given as $20 \log_{10} 2.4 = 7.60$ decibels or 7.60 dB. Note that an amplitude ratio of 1.0 is 0 dB and ratios less than 1.0 will have negative dB values. For a first-order system with a steady-state gain of 1.0 we have

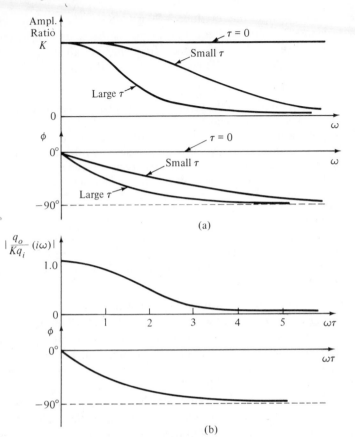

FIGURE 7–16

Frequency Response of First-Order System

$$\text{amplitude ratio in dB} = 20 \log_{10} \frac{1}{\sqrt{(\omega\tau)^2 + 1}} = -20 \log_{10} \sqrt{(\omega\tau)^2 + 1}$$

$$(7\text{–}64)$$

For very low frequencies $(\omega\tau)^2 \ll 1$ and

$$\text{dB} \approx -20 \log_{10} 1 = 0 \qquad (7\text{–}65)$$

while for high frequencies $(\omega\tau)^2 \gg 1$ and

$$\text{dB} \approx -20 \log_{10} \omega\tau = -20 \log_{10} \tau - 20 \log_{10} \omega \qquad (7\text{–}66)$$

When amplitude ratio in dB is plotted on a linear scale against frequency ω on a logarithmic scale, we get a curve which is asymptotic at low frequencies to a straight horizontal line at 0 dB [Eq. (7–65)] and at high frequencies to a straight line of slope -20 dB/decade [Eq. (7–66)], where a *decade* denotes any 10-to-1 frequency range. These low- and high-frequency asymptotes meet at $\omega = 1/\tau$, which frequency is called the *breakpoint* frequency. The plotting method consists of the following steps, using semi-logarithmic paper:

1. Lay out a dB scale on the vertical (linearly divided) graduations, locating the 0 dB line and choosing a convenient scale.
2. Lay out a frequency scale on the horizontal (logarithmic) axis. Three-cycle (3 decade) semi-log paper gives a sufficient frequency range for most practical problems. The actual frequency range, i.e., 0.1–100, 1–1000, 10–10,000, etc., depends on the system being studied.
3. Locate the breakpoint $\omega = 1/\tau$ on your frequency scale and at 0 dB. Draw the high-frequency asymptote sloping downward to the right at -20 dB/decade slope.
4. Correct the asymptotic approximation at the breakpoint (-3 dB), at one octave above and below the breakpoint (-1 dB), and at two octaves above and below the breakpoint ($-\frac{1}{4}$ dB). (An octave is a two-to-one frequency interval.)

The above procedure is illustrated for a first-order system with $\tau = 0.05$ seconds in Fig. 7–17. The phase angle is generally plotted on the same sheet as shown. Such curves can be constructed for any first-order system with a given value of τ in just a few minutes, *much* more quickly than they could be *calculated* from Eq. (7–61). To interconvert dB values and ordinary amplitude ratio, the table of Fig. 7–18 is provided.

In addition to facilitating rapid plotting of theoretical frequency-response curves, the logarithmic scheme is an aid in the interpretation of experimental test data. While step function testing is relatively quick and easy and is thus often used, sinusoidal testing generally gives a more complete picture of system dynamics, though with considerably greater experimental effort. The main practical problem is that of devising suitable apparatus to vary the input quantity sinusoidally over the required frequency range and at the desired amplitude. For systems with electrical input, inexpensive and widely available electronic oscillators provide this function. When the system input signal is a fluid pressure or flow rate, temperature, force or motion, the required apparatus becomes more complex and may have to be designed and built for the particular application. Assuming that a sinusoidal test signal can be produced, the system input and output are measured with electrical transducers whose voltage signals are recorded on a two-channel recorder. Measurements on the recorder chart then give the desired amplitude ratio and phase angle information. When much frequency response testing is to be done, commercially available *frequency response analyzers* may be economically justifiable. These instruments can rapidly and accurately step or sweep through a desired

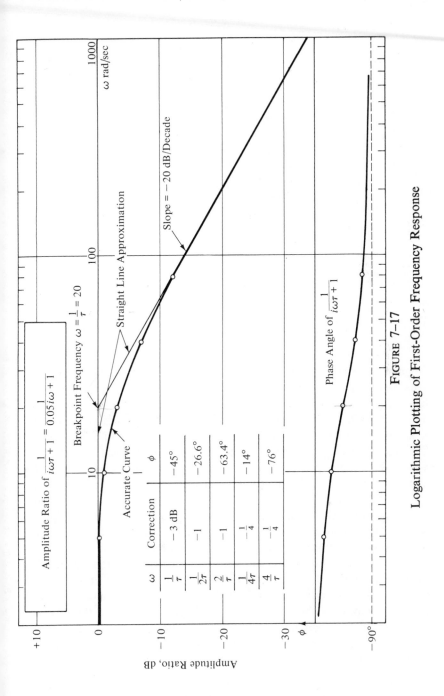

FIGURE 7-17

Logarithmic Plotting of First-Order Frequency Response

Number A	0	1	2	3	4	5	6	7	8	9
0		−40.00	−33.98	−30.46	−27.96	−26.02	−24.44	−23.10	−21.94	−20.92
0.1	−20.00	−19.17	−18.42	−17.72	−17.08	−16.48	−15.92	−15.39	−14.89	−14.42
0.2	−13.98	−13.56	−13.15	−12.77	−12.40	−12.04	−11.70	−11.37	−11.06	−10.76
0.3	−10.46	−10.16	−9.90	−9.63	−9.37	−9.12	−8.87	−8.64	−8.40	−8.18
0.4	−7.96	−7.74	−7.54	−7.33	−7.13	−6.94	−6.74	−6.56	−6.38	−6.20
0.5	−6.02	−5.85	−5.68	−5.51	−5.35	−5.19	−5.04	−4.88	−4.73	−4.58
0.6	−4.44	−4.29	−4.15	−4.01	−3.88	−3.74	−3.61	−3.48	−3.35	−3.22
0.7	−3.10	−2.97	−2.85	−2.73	−2.62	−2.50	−2.38	−2.27	−2.16	−2.05
0.8	−1.94	−1.83	−1.72	−1.62	−1.51	−1.41	−1.31	−1.21	−1.11	−1.01
0.9	−0.92	−0.82	−0.72	−0.63	−0.54	−0.45	−0.35	−0.26	−0.18	−0.09
1.0	0.00	0.09	0.17	0.26	0.34	0.42	0.51	0.59	0.67	0.75
1.1	0.83	0.91	0.98	1.06	1.14	1.21	1.29	1.36	1.44	1.51
1.2	1.58	1.66	1.73	1.80	1.87	1.94	2.01	2.08	2.14	2.21
1.3	2.28	2.35	2.41	2.48	2.54	2.61	2.67	2.73	2.80	2.86
1.4	2.92	2.98	3.05	3.11	3.17	3.23	3.29	3.35	3.41	3.46
1.5	3.52	3.58	3.64	3.69	3.75	3.81	3.86	3.92	3.97	4.03
1.6	4.08	4.14	4.19	4.24	4.30	4.35	4.40	4.45	4.51	4.56
1.7	4.61	4.66	4.71	4.76	4.81	4.86	4.91	4.96	5.01	5.06
1.8	5.11	5.15	5.20	5.25	5.30	5.34	5.39	5.44	5.48	5.53
1.9	5.58	5.62	5.67	5.71	5.76	5.80	5.85	5.89	5.93	5.98
2.0	6.02	6.44	6.85	7.23	7.60	7.96	8.30	8.63	8.94	9.25
3.0	9.54	9.83	10.10	10.37	10.63	10.88	11.13	11.36	11.60	11.82
4.0	12.04	12.26	12.46	12.67	12.87	13.06	13.26	13.44	13.62	13.80
5.0	13.98	14.15	14.32	14.49	14.65	14.81	14.96	15.12	15.27	15.42
6.0	15.56	15.71	15.85	15.99	16.12	16.26	16.39	16.52	16.65	16.78
7.0	16.90	17.03	17.15	17.27	17.38	17.50	17.62	17.73	17.84	17.95
8.0	18.06	18.17	18.28	18.38	18.49	18.59	18.69	18.79	18.89	18.99
9.0	19.08	19.18	19.28	19.37	19.46	19.55	19.65	19.74	19.82	19.91
10.0	20.00									
100.0	40.00									

dB value corresponding to A

FIGURE 7–18

Decibel Conversion Chart

range of frequencies and some even automatically produce graphical plots —usually the logarithmic decibel and phase angle curves described above. From these curves we can tell whether the tested device is adequately modeled as a first-order system and, if it is, what the numerical values of K and τ are. Figure 7–19 shows some typical experimental data. Note first that frequency is given in Hz (cycles/sec) rather than radians/sec. This changes nothing except that now, at the breakpoint, $\omega = 2\pi f = 1/\tau$ and thus $\tau = 1/2\pi f$. After plotting the experimental points for amplitude ratio in dB, one tries to fit straight lines to the low-frequency and high-frequency portions. For the example shown, these lines seem to fit quite well; the low-frequency line being horizontal and the high having a slope quite close to -20 dB/decade, thus the system appears to be close to a first-order system. The intersection of the straight line asymptotes locates the breakpoint at $f = 34.5$ Hz; thus the system time constant $\tau = 0.0046$ sec. From Fig. 7–16a, it is clear that the amplitude ratio of the low-frequency asymptote is the system steady-state gain K, which in this case is 12 dB or 4.0. The phase angle data further confirm the first-order nature of the system; thus we are justified in taking as a mathematical model for the tested system

$$\frac{q_o}{q_i}(D) = \frac{K}{\tau D + 1} = \frac{4.0}{0.0046D + 1} \tag{7–67}$$

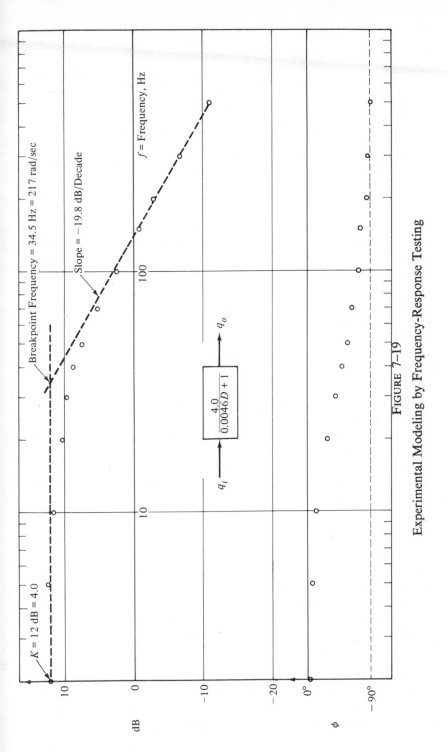

FIGURE 7-19

Experimental Modeling by Frequency-Response Testing

When the data are *not* well fitted by two straight lines of slope 0 and −20 dB/decade, this is an indication that the system has more complex dynamics and other models must be tried. We shall see later that the logarithmic plotting schemes continue to be helpful in finding these more complex models.

Up to this point we have treated the first-order system frequency response entirely in terms of a generalized system with input q_i and output q_o. Let us apply these results to the rotational system of Fig. 7–1c for which

$$\frac{q_o}{q_i}(i\omega) = \frac{K}{i\omega\tau + 1} = \frac{1}{i\omega B/K_s + 1} = \frac{\theta_o}{\theta_i}(i\omega) \qquad (7\text{–}68)$$

Suppose this model represents a torsional vibration isolator whose function is to isolate the motion θ_o from rotational oscillations $\theta_i = 0.1 \sin 200t$ known to be present at the input location. Suppose also that our design criterion is that the θ_o motion is to be no more than 1% of the θ_i motion and that the design problem is to find the spring and damper which will give these results. Since $K = 1$ for this system, the above requirements translate into the requirement that the amplitude ratio of $(\theta_o/\theta_i)(i\omega)$ be equal to $0.01 = -40$ dB at $\omega = 200$ rad/sec. From Fig. 7–17 it is clear that for an amplitude ratio of −40 dB the straight line asymptote will be *extremely* close to the exact curve and that the frequency for which dB = −40 will be very nearly $(100)(1/\tau)$. That is, if the slope is −20 dB/decade, it will take 2 decades (a 100-to-1 frequency change) above $1/\tau$ to reach −40 dB. We may thus solve for τ from

$$\frac{100}{\tau} = 200 \text{ rad/sec}, \qquad \tau = 0.5 \text{ seconds} \qquad (7\text{–}69)$$

As a check

$$\left|\frac{\theta_o}{\theta_i}(i200)\right| = \frac{1}{\sqrt{[(0.5)(200)]^2 + 1}} = 0.010 \qquad (7\text{–}70)$$

We see thus that any combination of K_s and B which gives $B/K_s = \tau = 0.5$ sec will meet the vibration isolation requirements. In a practical problem there will generally be other constraints which will give a value for either B or K_s and thus specific values of both can be found. For example, suppose that the motion source providing θ_i must not feel a torque greater than 1.5 inch-lb$_f$ when $\theta_i = 0.1 \sin 200t$ radians. Since for $\omega = 200$ the output shaft θ_o is practically motionless relative to θ_i, the spring deflection will be very nearly equal to θ_i alone. Thus the peak spring torque will be given by

$$\text{spring torque} = K_s(\theta_i - \theta_o) \approx K_s(\theta_i) = 0.1 \, K_s = 1.5 \text{ inch-lb}_f \qquad (7\text{–}71)$$

This makes $K_s = 15$ inch-lb$_f$/rad and thus $B = \tau K_s = 7.5$ inch-lb$_f$/(rad/sec). The next step in design would be to decide the detailed form of spring and damper to be used and then use formulas such as those in chapter two to actually calculate dimensions and material properties. This example is typical of the way in which system dynamics methods are useful in the overall design of a practical device.

Impulse Response.   We have earlier encountered the concept of impulse functions when considering the response of an inductor to a step input of current and a capacitor to a step voltage input. In these cases the impulse appeared as the *output* of the system. We now consider the response of a system to an impulse applied as an *input*. Recall that an ideal impulse has infinite "height," infinitesimal duration, but a finite area. No real physical variable can behave in precisely this fashion; however, an approximation sufficiently close for many practical purposes is often possible. Furthermore, the *theoretical* aspects of system response to impulsive inputs are of considerable importance. Let us consider again the translational system of Fig. 7–1a, whose analog computer diagram was given in Fig. 7–10b. In Fig. 7–20 this system is driven by a force input which is an impulse of area A_i lb$_f$-sec. The system equation would be

$$M\dot{v}_o + Bv_o = f_i = A_i\delta(t) \qquad (7\text{–}72)$$

where $\delta(t)$ is the symbol for a unit impulse function (an impulse of area

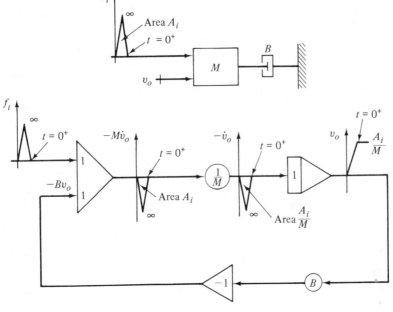

FIGURE 7–20

First-Order System Impulse Input

1.0). For $t > 0$, the right-hand side is zero, since the impulse is over in an infinitesimal time; thus the equation is

$$M\dot{v}_o + Bv_o = 0 \qquad (7\text{–}73)$$

with solution

$$v_o = Ce^{-t/\tau} \qquad (7\text{–}74)$$

To find C we need the initial condition $v_o(0^+)$, which is the velocity *just after the impulse has occurred*. The analog computer diagram is helpful in finding $v_o(0^+)$; in fact, such *diagrams* would be helpful in understanding the behavior of differential equations even if the computer itself had never been invented. We "track" the propagation of the input force through the diagram and see that $-M\dot{v}_o$ will just be the negative of f_i; the "feedback" signal $-Bv_o$ is of finite size and can thus, at the summer input, be neglected relative to f_i. Thus $-M\dot{v}_o$ and the acceleration $\dot{v}_o$ are both impulses, although of different areas. Since an integrator produces at its output the *area* of its input signal, the signal v_o must at time $t = 0^+$ be given by

$$v_o(0^+) = \int_0^{0^+} (f_i/M)\, dt = \int_0^{0^+} A_i\delta(t)/M\, dt = A_i/M \qquad (7\text{–}75)$$

The complete solution of Eq. (7–74) is thus

$$v_o = \frac{A_i}{M} e^{-t/\tau} = \frac{KA_i}{\tau} e^{-t/\tau} \qquad (7\text{–}76)$$

which is graphed in Fig. 7–21 for both this specific system and the general first-order. *It is clear that the response of an initially motionless system to an impulsive force of area A_i is identical to the response of an unforced system with an initial velocity of magnitude A_i/M.*

To see how the conditions of an impulsive input may arise (approximately) in a real-world situation, consider the force input shown in Fig. 7–22. If its area FT were kept constant as T approached zero, F would approach infinity and the force itself would approach an impulse. For $0 \le t \le T$ the input is a step of size F, thus the response is

$$v_o = KF(1 - e^{-t/\tau}) \qquad (7\text{–}77)$$

At $t = T$ we have

$$v_o = KF(1 - e^{-T/\tau}) \qquad (7\text{–}78)$$

For $t > T$ the system equation is

$$(\tau D + 1)v_o = 0 \qquad (7\text{–}79)$$

with initial condition

$$v_o(T) = KF(1 - e^{-T/\tau}) \qquad (7\text{–}80)$$

Thus

$$v_o = Ce^{-t/\tau})$$

$$KF(1 - e^{-T/\tau}) = Ce^{-T/\tau}$$

$$C = \frac{KF(1 - e^{-T/\tau})}{e^{-T/\tau}} = KFe^{T/\tau} - KF$$

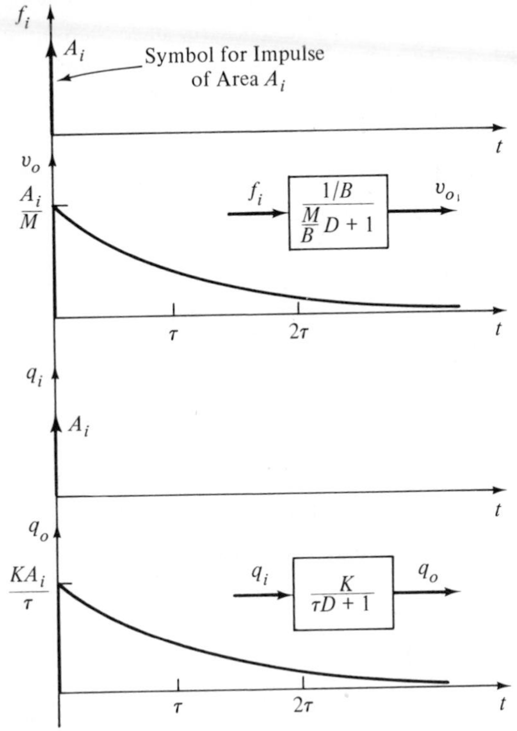

FIGURE 7–21

First-Order System Impulse Response

and

$$v_o = KF(e^{-(t-T)/\tau} - e^{-t/\tau}) \tag{7–81}$$

If we now let $FT = A_i$ and keep A_i constant as $T \to 0$, Eq. (7–81) becomes

$$v_o = \frac{KA_i}{T}(e^{-(t-T)/\tau} - e^{-t/\tau}) \tag{7–82}$$

As $T \to 0$, we get $v_o \to 0/0$, an indeterminate form, so we apply L'Hospital's rule to get

$$v_o = \frac{KA_i}{\tau} e^{-t/\tau} \tag{7–83}$$

which we see agrees with Eq. (7–76). The actual response (7–82) approaches the perfect-impulse response (7–83) more closely as T becomes small *compared to* τ. This can be seen by doing some numerical examples or, in general terms, by noting that the Taylor Series expansion for $e^{T/\tau}$ is

$$e^{T/\tau} = 1 + \frac{T}{\tau} + \frac{1}{2}\left(\frac{T}{\tau}\right)^2 + \frac{1}{6}\left(\frac{T}{\tau}\right)^3 + \ldots \tag{7–84}$$

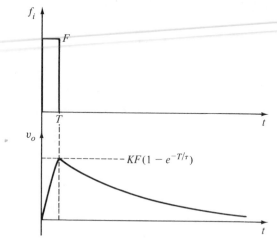

FIGURE 7–22

First-Order Response to Rectangular Pulse

For $T/\tau = 0.1$, for example,

$$e^{T/\tau} = 1 + 0.1 + 0.005 + 0.00017 + \ldots$$

Thus we might neglect all but the first two terms, giving in Eq. (7–82)

$$v_o = KA_i \frac{e^{-t/\tau}(e^{T/\tau} - 1)}{T} \approx KA_i \frac{e^{-t/\tau}\left(1 + \dfrac{T}{\tau} - 1\right)}{T} = \frac{KA_i}{\tau} e^{-t/\tau} \quad (7\text{–}85)$$

which we again see to be the perfect-impulse response. Thus for a rectangular pulse of duration the order of $\tau/10$ or less, the system acts nearly as if driven by a perfect impulse with the same area as the actual pulse. It can be shown that this is true for *any shape of pulse;* if the pulse is short enough, only its net area (*not* its shape) is of any consequence.

7–3. ELECTRICAL FIRST-ORDER SYSTEMS: GENERAL CIRCUIT LAWS AND SIGN CONVENTIONS.

The basic physical laws which govern the behavior of electrical circuits are *Kirchhoff's voltage loop law and current node law*. The voltage loop law is merely a statement of the following intuitive truth. If, at any instant of time, one chooses some point in a circuit and then traces out a loop along any chosen path which returns to the original point, keeping account of all voltage drops or rises encountered along the path, the net potential difference must be zero, since we have returned to the very same point.

For actual application to circuit analysis this law can be stated in several ways, the most common being:

1. The summation of the voltage drops around a closed loop must be zero at every instant.
2. The summation of the voltage rises around a closed loop must be zero at every instant.
3. The summation of the voltage drops around a closed loop must equal the summation of the voltage rises at every instant.

None of these statements has any particular relative advantage; however, most people tend to choose one and then stick with it. We shall generally use the first of the above three. The current node law is based on the fact that at any *point* (node) in a circuit there can be no accumulation of electrical charge. Since current is defined as the "flow of charge" (time rate of change of charge) we may say that, at every instant of time,

1. The summation of currents into a node must be zero. Or, alternatively,
2. The summation of currents out of a node must be zero. Or, still another way,
3. The summation of currents into a node must equal the summation of currents out.

Just as with the voltage loop law, any of the above three forms of the current node law may be employed in circuit analysis.

Just as sign conventions are needed for forces and motions in mechanical systems, so also must we specify the positive senses of voltage and current in electrical circuits. The assumed positive direction of a current is indicated by the symbol $+\!\!\longrightarrow$ and may in general be chosen arbitrarily at the beginning of the analysis. Later, when the solution for the unknown current has been obtained, for example $i = 2.8 \sin 377t$, at any instant when i is a positive number we know the current is actually in the direction given by the sign-convention arrow. If i should be a negative number, the current is *opposite* to the arrow. If the assumed positive direction of current has not been specified at the *beginning* of a problem, an orderly analysis is quite impossible, and any results which might be obtained cannot be properly interpreted, since the meaning of positive and negative currents is undefined. For voltages, the sign convention consists of $+$ and $-$ marks at the terminals where the voltage exists. Which terminal receives the $+$ mark is again an arbitrary choice made at the beginning of an analysis. When the solution is obtained, if the voltage is at some instant a positive number, then the actual polarity is the same as that shown by the sign-convention polarity marks. If the voltage is negative, the actual polarity is *opposite* to that shown by the polarity marks.

Once sign conventions for all currents and voltages have been chosen, combination of Kirchhoff's laws with the known voltage/current relations which describe the circuit elements leads us directly to the system differential equations. While practitioners of circuit analysis have developed many systematic and specialized techniques to speed the analysis of complicated circuits, these are beyond the scope of this text and are not really necessary or desirable for the relatively simple systems which are our main concern. We should also emphasize that everything said so far in this section is of course quite general and *not* restricted to the first-order type of system on which we are about to concentrate. Figure 7–23 shows some examples of circuits which will be of the first-order form $(\tau D + 1)q_o = Kq_i$ and which might arise in various practical applications. The circuit of Fig. 7–23a is widely used as an approximate integrating device and also as a low-pass filter (a filter which passes low-frequency components of e_i but rejects high frequencies). Piezoelectric measurement transducers (pressure gauges, load cells, accelerometers) are often modeled with the circuit of Fig. 7–23b, where the current source i_i represents the flow of charge produced by the conversion of mechanical into electrical energy. In the control of machines using DC motors, the circuit of Fig. 7–23c is useful as a model for the field circuit. The voltage e_i might come from an electronic amplifier; the current is the output of interest since field strength (and thus motor torque) depends on current, not voltage.

Let us illustrate the application of the general concepts of circuit analysis outlined above by consideration of the circuit of Fig. 7–23a. We might begin by showing, in Fig. 7–24, this circuit imbedded in a larger system, as it might ordinarily arise in practice. The input comes from an amplifier whose output circuit is modeled as an ideal voltage source in series with a 10-ohm resistance. A digital voltmeter with a very high input resistance (10^7 ohms) is connected to our circuit's output. While it is technically possible to analyze this entire assemblage of hardware as one system, it is often advantageous to attempt some initial simplification based on judgment and experience. Since the beginner in circuit analysis (as in any other field) has little of either of these two attributes to draw upon, he may tend to make few simplifications and instead choose to deal with a rather complex model. In these days of computer-aided analysis it might seem that a complex model should not cause too much worry; however, a model *more complex than really necessary* will always be undesirable, since it obscures basic understanding and also is expensive in time and money. In our present example we simplify the circuit by combining the source resistance R_s with R_1 into a total resistance R, and by treating the output terminals across C as an open circuit (infinite resistance) rather than, as is actually the case, shunted by a 10-megohm resistor. The latter simplification is also an approximation, but will be quite accurate because the "load" resistance R_L is so large *relative* to R_1 and R_s that the current it draws from the circuit may be neglected. The judgment necessary to make such decisions "on-

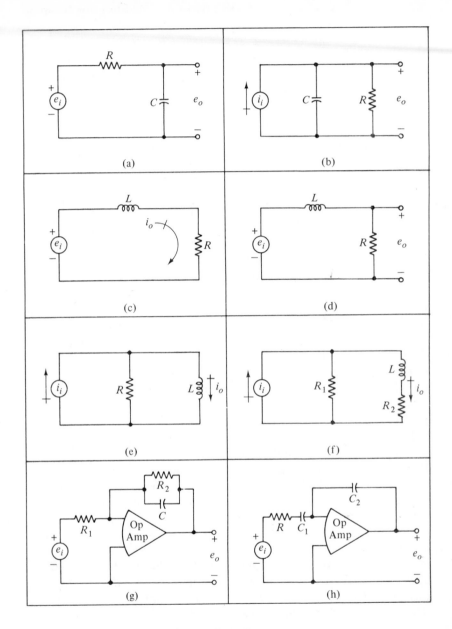

FIGURE 7–23

Some Electrical First-Order Systems

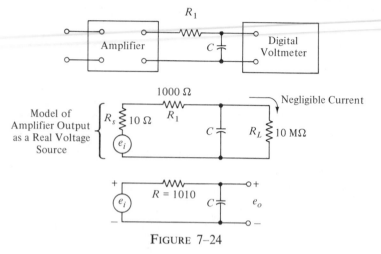

FIGURE 7–24

Derivation of Simplified Circuit Model

the-spot" comes from experience, both with *experimental tests* on real systems and also with *analysis* of similar systems in the past.

With the above background in mind we return to analysis of our circuit as an "isolated" entity. In Fig. 7–25a we initially choose the positive sense of e_i as shown; we could have chosen the reverse. For e_o we also have a free choice; however, once e_i's positive direction has been fixed there may be some incentive to choose e_o's such that a positive, constant e_i causes a positive e_o. This is not *necessary* but may be desirable and is what was done in Fig. 7–25a. Since we assume no current flowing at the output terminals, there is only one current in the system and we call it i. Its positive direction is open to choice but we are again influenced by a desire to have a positive e_i cause a positive i and thus choose i positive in the direction shown. The voltage across the resistor is called e_R and is chosen positive as shown (again the reverse choice could have been made) so that a positive i causes a positive e_R. Since the capacitor voltage is also e_o there is no need to define it again. Since this circuit has only one current, Kirchhoff's voltage loop law would seem to be more appropriate than the current node law. In applying this law one must choose a *point* in the circuit at which to start, the *loop path* to be followed, and the *direction* of traversing the loop. The starting point and the path to be followed depend on the circuit being analyzed; the direction of travel is arbitrary. In our example there is only one loop possible and the starting point may be chosen at will; we choose point x. The direction of travel is taken the same as that of positive current, since then the voltage drops for the passive (R, L, C) circuit elements will all be positive; $+iR$, $+L(di/dt)$, and $+(1/C) \int i \, dt$. This choice is not

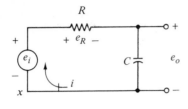

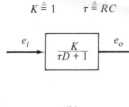

(a)

(b)

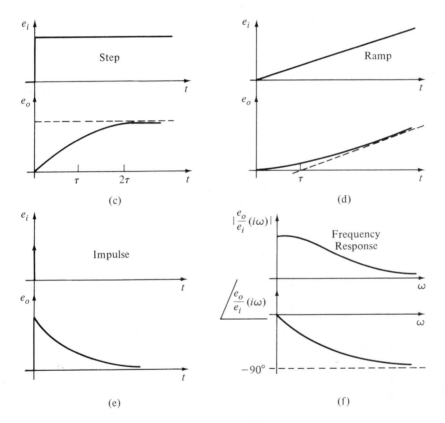

(c)

(d)

(e)

(f)

FIGURE 7-25

Electrical First-Order System Responses

necessary, but we prefer it because we earlier stated a preference for the "voltage drop version" of Kirchhoff's voltage loop law. Now, recalling this law we write the summation of the voltage drops around the loop as

$$-e_i + iR + \frac{1}{C} \int i \, dt = 0 \qquad (7\text{-}86)$$

Our objective is to find the relation between e_o and e_i; we achieve it indirectly by first finding i in terms of e_i from Eq. (7–86) and then recalling that $e_o = e_C = (1/C) \int i \, dt$. We have, in operator form

$$\left[R + \frac{1}{CD} \right] i = e_i, \qquad i = \left[\frac{CD}{RCD + 1} \right] e_i \qquad (7\text{–}87)$$

$$e_o = \left[\frac{1}{CD} \right] i = \left[\frac{1}{CD} \right] \left[\frac{CD}{RCD + 1} \right] e_i \qquad (7\text{–}88)$$

$$\frac{e_o}{e_i}(D) = \frac{1}{RCD + 1} = \frac{K}{\tau D + 1} \qquad (7\text{–}89)$$

$$K \triangleq 1 \qquad \tau \triangleq RC \qquad (7\text{–}90)$$

We see that this circuit clearly fits our definition of a first-order system and thus all the standard responses of Fig. 7–25 are immediately available.

Elementary AC Circuit Analysis. The frequency response of electric circuits is of particular interest since much of our electrical power is generated, transmitted, and utilized in the form of sinusoidal waves; that is, alternating current and voltage (AC). Commercial power in the United States is at the frequency 60 Hz, whereas some other countries use 50 Hz. In aircraft AC systems, 400 Hz is common since performance is improved while size and weight of equipment is reduced as frequency is increased. In many practical problems only the sinusoidal steady-state behavior is of interest, thus frequency-response methods and the sinusoidal transfer function apply directly. We have in chapter three defined electrical impedance; its sinusoidal version is widely used in AC circuit analysis. Let us develop a few of the standard techniques which utilize the impedance concept. Figure 7–26 shows how series and parallel combinations of impedances are combined to form the total impedance of the combination. For the series combination, both Z_1 and Z_2 must carry the same current i; thus

$$e_{\text{total}} = iZ_1(D) + iZ_2(D) = [Z_1(D) + Z_2(D)]i \qquad (7\text{–}91)$$

and since impedance is always the ratio voltage/current,

$$Z_{\text{total}}(D) = \frac{e_{\text{total}}}{i} = Z_1(D) + Z_2(D) \qquad (7\text{–}92)$$

Thus the rule for combining impedances in series is to simply add them, just as pure resistances add up in DC circuits. Note that this may be done with either operational ($Z(D)$) or sinusoidal ($Z(i\omega)$) impedances. Applying this method to Fig. 7–25a, the impedance "seen" by the voltage source e_i is the series combination of R and C:

$$Z(D) = R + \frac{1}{CD} = \frac{RCD + 1}{CD} = \frac{e_i}{i}(D) \qquad (7\text{–}93)$$

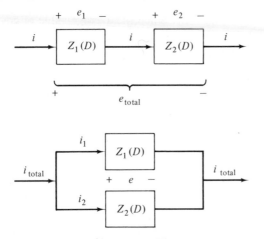

FIGURE 7–26

Parallel and Series Impedances

Note that this agrees with Eq. (7–87). For the parallel combination in Fig. 7–26, both Z_1 and Z_2 have the same voltage drop; thus

$$i_{\text{total}} = i_1 + i_2 = \frac{e}{Z_1(D)} + \frac{e}{Z_2(D)} = \left[\frac{1}{Z_1(D)} + \frac{1}{Z_2(D)}\right] e \quad (7\text{–}94)$$

$$Z_{\text{total}}(D) = \frac{e}{i_{\text{total}}}(D) = \frac{Z_1(D)Z_2(D)}{Z_1(D) + Z_2(D)} \quad (7\text{–}95)$$

The parallel combination is thus the product over the sum of the individual impedances, again just as for resistances in DC circuits.

In Eq. (7–93) let us consider voltage e_i as an input which produces current i as output. The quantity $(i/e_i)(D)$ is in general called the *admittance* and is clearly just the reciprocal of impedance. For the sinusoidal steady state we have

$$\frac{i}{e_i}(i\omega) = \frac{i\omega C}{i\omega RC + 1} = \frac{\omega C}{\sqrt{(\omega RC)^2 + 1}} \, \underline{/90° - \tan^{-1}\omega RC} \quad (7\text{–}96)$$

If $e_i = E \sin \omega t$,

$$i = \frac{E\omega C}{\sqrt{(\omega RC)^2 + 1}} \sin(\omega t + 90° - \tan^{-1}\omega RC) \triangleq I \sin(\omega t + \phi) \quad (7\text{–}97)$$

We wish to calculate the instantaneous power p supplied by the source to the circuit. Recall the definition of power,

$$p \triangleq e_i i = EI(\sin \omega t)(\sin(\omega t + \phi)) \quad (7\text{–}98)$$

Using appropriate trig identities this leads to

$$p = \frac{EI}{2}(\cos \phi - \cos(2\omega t + \phi)) \quad (7\text{–}99)$$

which is plotted in Fig. 7–27. Since the average value of $\cos(2\omega t + \phi)$ is zero, the average power into the circuit is

$$p_{\text{avg}} = \frac{EI\cos\phi}{2} \qquad (7\text{--}100)$$

The instantaneous power varies cosinusoidally around its average value at a frequency 2ω, just *twice* the frequency of the impressed voltage (and the

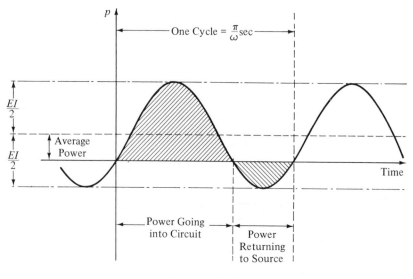

FIGURE 7–27

Power in AC Circuits

resulting current). During any one cycle, power flows into the circuit from the source for a portion of the time and is returned to the source from the circuit the rest of the time. Average power is that which is actually "used up" by the circuit and is what the electric company charges for. The angle ϕ by which the current leads the voltage is called the *power factor angle* and $\cos\phi$ is called the *power factor*. In general the angle ϕ may be "leading" (between $0°$ and $+90°$) or "lagging" (between $0°$ and $-90°$) and thus the power factor is between 0 and 1.

Many measurements and calculations in AC systems employ the so-called *effective values* of current and voltage. The effective value of a current or voltage is defined as that *constant* value which would produce the same average power dissipation in a resistor as would the actual *time-varying* voltage or current. An effective value exists for any waveform whatever, including random voltages and currents. For sinusoidal waveforms we can easily calculate effective values as follows.

$$p_{R,\text{avg}} = \frac{1}{\pi/\omega} \int_0^{\pi/\omega} ei\, dt = \frac{1}{\pi/\omega} \int_0^{\pi/\omega} \frac{e^2}{R}\, dt = \frac{\omega}{\pi R} \int_0^{\pi/\omega} E^2 \sin^2 \omega t\, dt = \frac{E^2}{2R}$$

$$(7\text{--}101)$$

(This same result follows from Eq. (7–100) when we note that $\phi = 0$ for a pure resistance.) Now since the average power produced by a constant voltage E_{eff} is E_{eff}^2/R, we have

$$\frac{E_{\text{eff}}^2}{R} \triangleq \frac{E^2}{2R}, \qquad E_{\text{eff}} = \frac{E}{\sqrt{2}} = 0.707E \qquad (7\text{--}102)$$

Thus the effective value of a sinusoidal voltage (also called the *root-mean-square* (RMS) value) is 0.707 times its peak value. Since Eq. (7–101) will work out exactly the same way if current I is used instead of voltage, the same relation, $I_{\text{eff}} = 0.707I$, holds for currents. Since most AC voltmeters and ammeters are calibrated to read effective values, we must be careful to multiply by 1.414 if we wish to convert their readings to peak values.

Let us apply some of the above methods to the circuit of Fig. 7–23a with $R = 10{,}000\ \Omega$, $C = 0.1\ \mu\text{F} = 10^{-7}\ \text{F}$ and e_i taken as the 115 V, 60 Hz available at a standard wall plug. First we must convert the 115 V (which is E_{RMS}) to the peak value $E = (1.414)(115) = 163$ V to get $e_i = 163 \sin 377t$. We can get e_o from

$$\frac{e_o}{e_i}(i377) = \frac{1}{\sqrt{(0.377)^2 + 1}} \bigm/ \tan^{-1} - 0.377 = 0.935 \bigm/ -20.6° \quad (7\text{--}103)$$

and thus

$$e_o = (163)(0.935) \sin (377t - 20.6°) = 152 \sin (377t - 20.6°) \quad (7\text{--}104)$$

The effective value of e_o would be given as $(0.707)(152) = 108$ V. To find the current,

$$\frac{i}{e_i}(i377) = \frac{(377)(10^{-7})}{\sqrt{(0.377)^2 + 1}} \bigm/ 90° - \tan^{-1} 0.377$$
$$= 0.0000352 \bigm/ 69.4° \qquad (7\text{--}105)$$

and thus,

$$i = (163)(0.0000352) \sin (377t + 69.4°)$$
$$= 0.00575 \sin (377t + 69.4°) \text{ amps} \qquad (7\text{--}106)$$

The effective value of i would be 4.06 mA. The power factor angle is $+69.4°$, giving a power factor of 0.352 leading and an average power of $(163)(0.00575)(0.352)/2 = 0.165$ W. To find the impedance $Z = (e_i/i)(i\omega)$ at $\omega = 377$ rad/sec we just take the reciprocal from Eq. (7–105),

$$Z(i\omega) = \frac{e_i}{i}(i\omega) = \frac{1}{0.0000352 \bigm/ 69.4} = 28400 \bigm/ -69.4°, \text{ ohms} \quad (7\text{--}107)$$

Note that impedances (being the ratio of voltage to current) *always* have the dimensions of ohms, irrespective of the combination of R, C and L they may represent.

Circuits with Current Sources, Op-Amp Circuits. When a circuit has elements in parallel, the current node law is often the most direct approach to analysis. The circuit of Fig. 7–23b has this feature and also is driven by a current source, so let us analyze it to gain some experience with both of these features. In Fig. 7–28 we have defined currents i_C and i_R as shown;

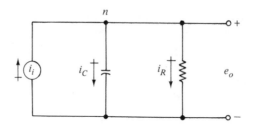

FIGURE 7–28

Circuit with Current Source

their positive directions may be freely chosen, the ones shown are merely convenient since they give positive i_C and i_R for a positive i_i and also give a positive e_o for a positive i_R. Applying the current law at the node n:

$$\text{sum of currents into } n = 0$$

$$i_i - i_C - i_R = 0 \qquad i_i = C\frac{de_o}{dt} + \frac{e_o}{R} \qquad \text{(7–108)}$$

$$(RCD + 1)e_o = Ri_i \qquad \frac{e_o}{i_i}(D) = \frac{K}{\tau D + 1} \qquad \text{(7–109)}$$

$$K \triangleq R \qquad \tau \triangleq RC \qquad \text{(7–110)}$$

The same result is achieved by the impedance approach:

$$i_i = \frac{e_o}{Z_{\text{total}}} = \frac{e_o}{\dfrac{(R)(1/CD)}{R + 1/CD}} = \frac{e_o}{R/(RCD + 1)}, \qquad \frac{e_o}{i_i}(D) = \frac{R}{RCD + 1}$$

$$\text{(7–111)}$$

For a sinusoidal input current the frequency response

$$\frac{e_o}{i_i}(i\omega) = \frac{R}{\sqrt{(RC\omega)^2 + 1}} \; \underline{/-\tan^{-1} RC\omega} \qquad \text{(7–112)}$$

shows that the voltage *lags* the current, which is the same as the current *leading* the voltage, just as we found in the circuit of Fig. 7–23a.

In the operational amplifier circuit of Fig. 7-29, recall that the node n is essentially at ground potential and that the current into the amplifier is negligible. Applying the current node law at n:

$$i_{R1} + i_{R2} + i_C = 0, \qquad \frac{e_i}{R_1} + \frac{e_o}{R_2} + C\frac{de_o}{dt} = 0 \qquad (7\text{-}113)$$

$$(R_2CD + 1)e_o = (-R_2/R_1)e_i, \qquad \frac{e_o}{e_i}(D) = \frac{K}{\tau D + 1} \qquad (7\text{-}114)$$

$$K \triangleq -\frac{R_2}{R_1}, \qquad \tau \triangleq R_2C \qquad (7\text{-}115)$$

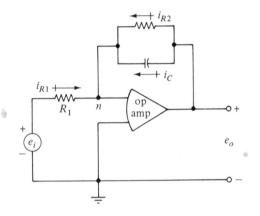

FIGURE 7-29

Op-Amp Circuit

In chapter three we showed that for op-amp circuits in general, $(e_o/e_i)(D) = -(Z_o/Z_i)$; thus we may also use this approach to obtain the above result:

$$\frac{e_o}{e_i}(D) = -\frac{(R_2)\left(\dfrac{1}{CD}\right)}{R_1\left(R_2 + \dfrac{1}{CD}\right)} = -\frac{R_2}{R_1}\frac{1}{(R_2CD + 1)} \qquad (7\text{-}116)$$

While we also showed in chapter three that "exact" integrating circuits could be achieved with op-amps, the present circuit may be used in practice as an approximate integrator under proper conditions. Its freedom from drift (compared to that of the "exact" circuit) may be important in some applications. Suppose we wish to use the circuit to integrate voltage signals produced by a vibration-measuring instrument sensitive to velocity. If we obtain the time integral of such a signal it would represent the *displacement* associated with the vibration. Vibration instruments are generally designed to operate over a specific range of frequencies, let us assume

5 to 500 Hz. To see how this circuit approximates an integrator we examine the frequency response

$$\frac{e_o}{e_i}(i\omega) = \frac{K}{i\omega\tau + 1} \qquad (7\text{--}117)$$

which, for $\omega\tau \gg 1$, becomes

$$\frac{e_o}{e_i}(i\omega) \approx \frac{K}{i\omega\tau}, \qquad \frac{e_o}{e_i}(D) \approx \frac{K}{\tau D}, \qquad e_o \approx \left(\frac{K}{\tau D}\right)e_i = \frac{K}{\tau}\int e_i\,dt \qquad (7\text{--}118)$$

We see thus that as long as $\omega\tau \gg 1$, we have an approximate integrator, the quality of the approximation depending on how much greater $\omega\tau$ is than 1.0. Since for a given circuit τ is fixed, it is clear that the approximation is least accurate for the lowest frequency; thus if we design for the lowest frequency the integration will be more nearly perfect at all higher frequencies. Suppose we decide that at $\omega = (5)(2\pi) = 31.4$ rad/sec we want $\omega\tau = 20$, and thus $\tau = 20/31.4 = 0.64$ sec. Any combination of R_2 and C giving this τ is theoretically satisfactory, although certain practical constraints limit the range of both R_2 and C; typical values might be $R_2 = 640,000\ \Omega$ and $C = 1.0\ \mu$F. Choice of R_1 will now set the steady-state gain $-R_2/R_1$. Since this gain is essentially the sensitivity of the integrator, its choice depends on the magnitude of the expected input signal e_i and the desired level of the output e_o. Suppose that at 5 Hz we expect a maximum input of 5 V, and desire that this produce full-scale reading on a 5-volt meter connected at e_o. This means that

$$\left|\frac{e_o}{e_i}(i31.4)\right| \approx \frac{R_2}{(R_1R_2C)(31.4)} = \frac{5}{5} = 1.0 \qquad (7\text{--}119)$$

and thus $R_1 = 31,800\ \Omega$. Our final design thus gives us

$$\frac{e_o}{e_i}(i\omega) = \frac{-20.1}{0.64i\omega + 1} \approx \frac{-31.4}{i\omega} \qquad (7\text{--}120)$$

Figure 7–30 compares the performance of the actual circuit and an equivalent ideal integrator. Note that the decibel plot of an integrator is a perfectly straight line with slope -20 dB/decade and requires no corrections at all.

7–4. FLUID FIRST-ORDER SYSTEMS.

Figure 7–31 shows simplified models of various fluid systems which arise in practical applications and which might reasonably be considered first-order systems. The first four are tanks (3 open and 1 pressurized) in which the liquid level is determined by a pressure or flow rate input. Such arrangements arise in many process plants (refineries, chemical plants, power

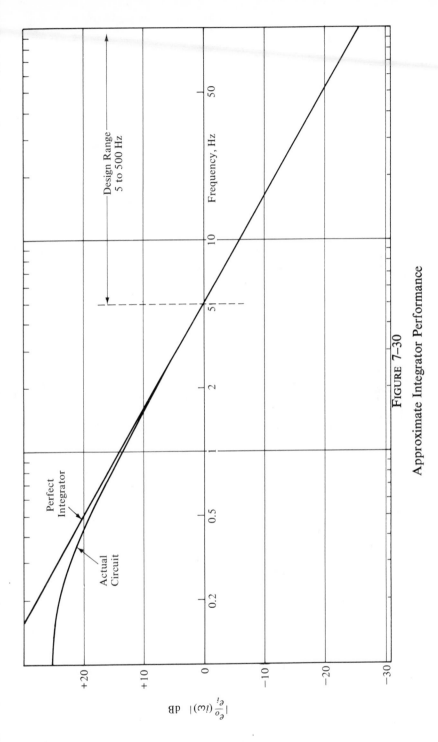

FIGURE 7-30

Approximate Integrator Performance

319

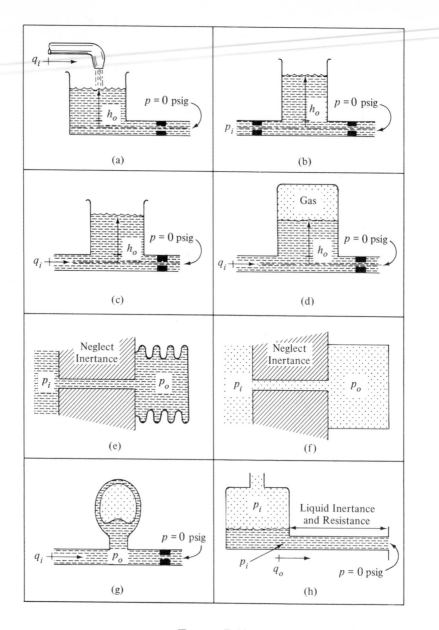

FIGURE 7–31

Some Fluid First-Order Systems

plants, etc.) where tanks are used for storage, mixing of fluids, chemical reaction vessels, etc. Figures e and f are useful models for pressure-measuring systems in which a length of tubing connects the pressure to be measured (p_i) to the chamber of a pressure transducer. In e the fluid is incompressible and the bellows represents the compliance of the transducer elastic element, while in f the compressible fluid is itself the dominant compliance. Without the accumulator in Fig. 7–31g, sudden changes in flow rate q_i would cause sudden (possibly damaging) rises in pressure p_o. The accumulator helps to reduce this effect. The model of Fig. 7–31h is intended mainly for the study of the flow response of a liquid-filled pipe to driving pressure.

Basic Laws Useful for Equation Setup. For systems involving compliances (tanks, accumulators) the conservation of mass is generally useful in deriving system equations. Over a time interval dt, one can always write

$$\text{Mass in} - \text{Mass out} = \text{Additional Mass Stored}$$

When the fluid is considered incompressible, one can substitute volume for mass. If a flow path branches, the instantaneous summation of flow rates in and out must be equal, entirely analogous to the current node law in electric circuits. This holds fundamentally for *mass* flow rates, but again the assumption of incompressibility extends it to volume flow also. For those fluid systems which are obviously in the form of "circuits" (loops) so that one can start at a chosen point, follow a loop back to the starting point and keep track of all pressure drops, the summation of pressure drops at any instant must be zero, just as in Kirchhoff's voltage loop law. Many fluid systems do not, however, "recirculate" the fluid, and attempts to manipulate them into circuit form, while possible, do not really simplify the analysis or give a better physical understanding of behavior. A direct approach which simply equates the driving pressure difference between any two points to the summation of pressure drops due to resistance and inertance is generally preferable.

Linearized and Nonlinear Analysis of a Tank/Orifice System. In Fig. 7–31a the tank discharges to atmosphere through a sharp-edge orifice. We assume the discharge pipe is short enough that its inertance and resistance are negligible compared to the orifice resistance. Inertial and frictional effects in the tank itself will almost certainly be negligible, since its large area means the velocities and accelerations ($\dot{h}_o, \ddot{h}_o$) will be small. Our idealized system is thus made up of the compliance of the tank, the orifice resistance, and a flow source q_i, ft³/sec. Conservation of mass gives us

$$q_i \, dt - K_{or}\sqrt{h_o} \, dt = A_T \, dh_o \tag{7–121}$$

where

instantaneous orifice flow rate =

$$A_{or}C_d \sqrt{\frac{2g\Delta p}{\gamma}} = \sqrt{2g}A_{or}C_d \sqrt{\frac{\gamma h_o - 0}{\gamma}} \triangleq K_{or}\sqrt{h_o} \qquad (7\text{--}122)$$

$g \triangleq$ local acceleration of gravity, ft/sec² $\qquad A_{or} \triangleq$ orifice area, ft²

$C_d \triangleq$ orifice discharge coefficient $\qquad\qquad h_o \triangleq$ tank level, ft

$A_T \triangleq$ tank cross-section area, ft²

and we are assuming an incompressible fluid. The system differential equation follows directly from Eq. (7–121)

$$A_T \frac{dh_o}{dt} + K_{or}\sqrt{h_o} = q_i \qquad (7\text{--}123)$$

Suppose that initially the inflow q_i was constant at a value q_{i0} for a long time so that the tank level had become steady at $q_{i0}^2/K_{or}^2 \triangleq h_{o0}$. Let us take this equilibrium condition as an operating point for a linearized analysis, which should be accurate as long as h_o does not vary too far from h_{o0}. We linearize $\sqrt{h_o}$ in the usual way:

$$\sqrt{h_o} \approx \sqrt{h_{o0}} + \frac{1}{2\sqrt{h_{o0}}}(h_o - h_{o0}) = \sqrt{h_{o0}} + \frac{1}{2\sqrt{h_{o0}}}h_{op} \qquad (7\text{--}124)$$

where

$$h_{op} \triangleq h_o - h_{o0} \triangleq \text{perturbation (small variation) in } h_o$$

Equation (7–123) may now be written as

$$A_T \frac{dh_{op}}{dt} + K_{or}\left[\sqrt{h_{o0}} + \frac{1}{2\sqrt{h_{o0}}}h_{op}\right] = q_{i0} + q_{ip} \qquad (7\text{--}125)$$

where

$$q_{ip} \triangleq q_i - q_{i0} \triangleq \text{perturbation in } q_i$$

$$\frac{dh_{op}}{dt} = \frac{d}{dt}[h_o - h_{o0}] = \frac{dh_o}{dt}$$

Simplification finally gives

$$\left(\frac{2\sqrt{h_{o0}}A_T}{K_{or}}\right)\frac{dh_{op}}{dt} + h_{op} = \left(\frac{2\sqrt{h_{o0}}}{K_{or}}\right)q_{ip} \qquad (7\text{--}126)$$

which we recognize to be first order with

$$K \triangleq \frac{2\sqrt{h_{o0}}}{K_{or}}, \quad \frac{\text{ft}}{\text{ft}^3/\text{sec}} \qquad \tau \triangleq \frac{2\sqrt{h_{o0}}A_T}{K_{or}}, \text{sec} \qquad (7\text{--}127)$$

and thus

$$\frac{h_{op}}{q_{ip}}(D) = \frac{K}{\tau D + 1} \qquad (7\text{--}128)$$

We should emphasize that this technique of defining perturbation variables (such as q_{ip} and h_{op}) is of very general applicability in linearized analyses of all kinds and should always be used if transfer functions are to be defined. Otherwise, when one gets the linearized system equation he will find on the right-hand side not only the input variable but also a constant term which will confuse the definition of transfer functions.

To get some feeling for the accurate range of the above linearized analysis we compare the exact nonlinear response (obtained by computer simulation) with the linearized response for perturbations of two different sizes. Let us choose the following numerical values:

$$A_T = 1.20 \text{ ft}^2 \qquad C_d = 0.60$$
$$A_{or} = 0.001056 \text{ ft}^2 \qquad h_{o0} = 2.0 \text{ ft}$$

These numbers make $q_{io} = 0.00722$ ft^3/sec. Let us take the perturbation q_{ip} to be a step change of -0.00037 ft^3/sec for the small change and -0.00212 ft^3/sec for the large change. Fig. 7-32 shows that the linearization gives almost perfect results for the small step but deviates considerably for the larger one. If we now take the same size of perturbation but make it sinusoidal at a frequency 0.002 rad/sec, i.e., $q_{ip} = 0.00037 \sin (0.002t)$ and $q_{ip} = 0.00212 \sin (0.002t)$, we get the results of Figs. 7-33a and b. Again the smaller perturbation gives excellent results; however now even the large one is quite good. This is mainly due to the frequency being high enough that the amplitude of h_o has attenuated somewhat, making the perturbation in h_o smaller than it would be for a constant q_{ip} of the same magnitude.

An Accumulator Surge-Damping System. In Fig. 7-31g we neglect inertance and compressibility of the fluid itself and consider resistance only at the flow restriction, which we assume linear. The effect of the height of liquid in the accumulator on the pressure p_o will be negligible if we deal with relatively high pressures, since it takes a column of oil about 3 feet high to produce 1 psi pressure. If we wish, a fluid circuit diagram may be drawn as in Fig. 7-34 to show clearly that the flow q_i branches into the accumulator and the resistance discharging to atmosphere, giving

$$q_i = q_c + q_R = C_f \frac{dp_o}{dt} + \frac{p_o}{R_f} \qquad (7\text{-}129)$$

where

$$C_f \triangleq \text{accumulator compliance, ft}^3/\text{psi}$$
$$R_f \triangleq \text{fluid resistance, psi}/(\text{ft}^3/\text{sec})$$

and thus

$$\tau \frac{dp_o}{dt} + p_o = Kq_i \qquad (7\text{-}130)$$

where

$$K \triangleq R_f, \text{ psi}/(\text{ft}^3/\text{sec}) \qquad \tau \triangleq R_f C_f, \text{ sec} \qquad (7\text{-}131)$$

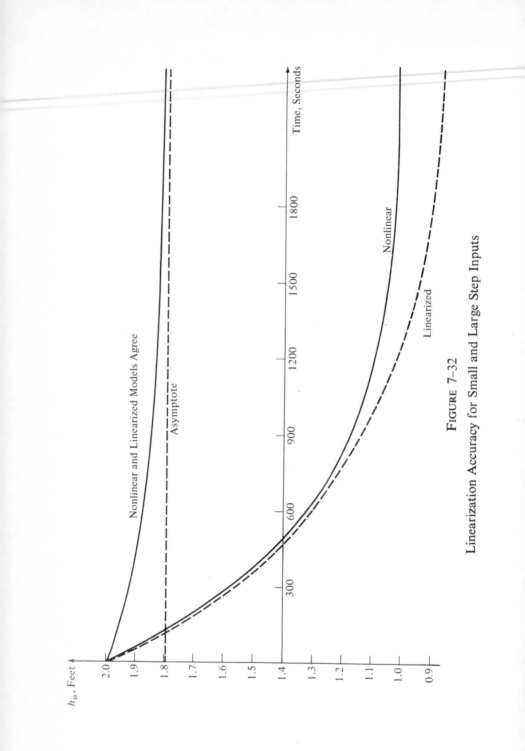

FIGURE 7-32

Linearization Accuracy for Small and Large Step Inputs

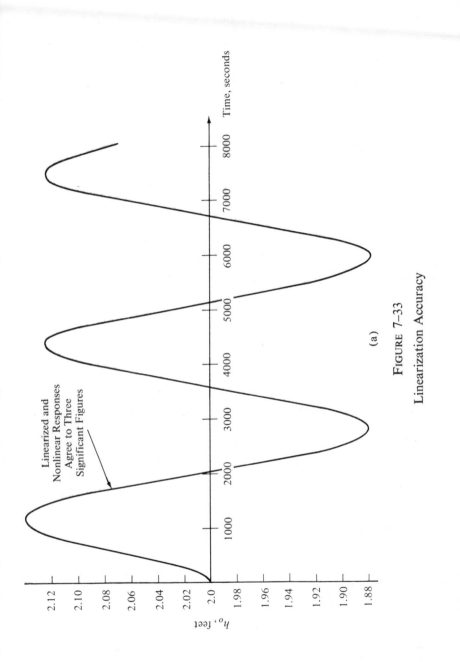

FIGURE 7-33
Linearization Accuracy

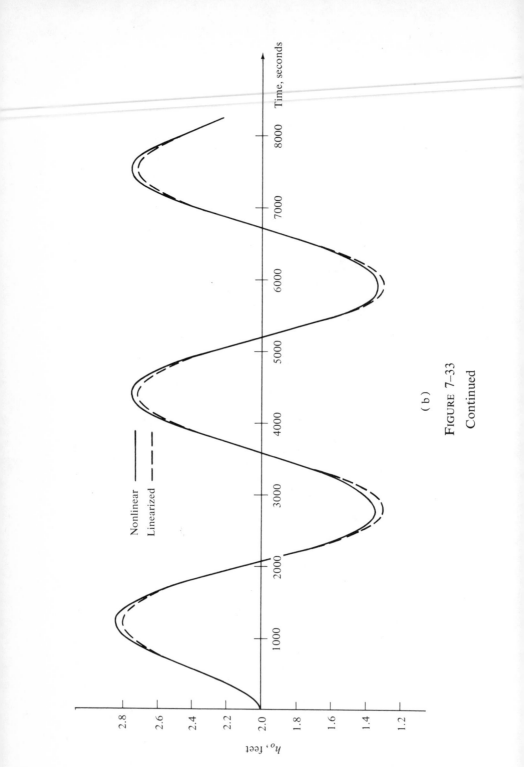

Nonlinear ———
Linearized – – –

FIGURE 7–33
Continued

(b)

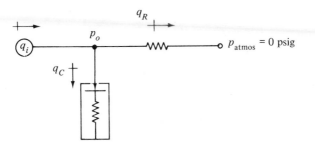

FIGURE 7–34

Fluid Circuit for Accumulator System

To show how such a system can reduce pressure surges due to flow transients, let us assume an R_f of 10,000 psi/(ft³/sec) with a steady q_i of 0.02 ft³/sec, giving a steady p_o of 200 psi. If now a flow transient of rectangular form as in Fig. 7–35 occurs, and if no accumulator were present, the peak pressure would be 800 psi. Suppose we wish to install an accumulator to reduce this pressure to 300 psi. We have already solved the necessary response problem in Fig. 7–22; thus we apply those results immediately to this case. The peak pressure, above the steady 200 psi is given by

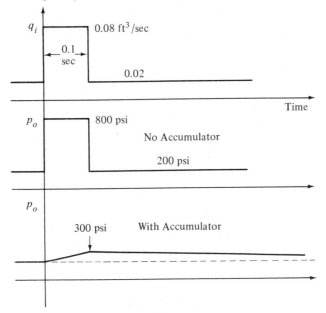

FIGURE 7–35

Pressure Transients With and Without Accumulator

$$(R_f)(0.06)(1 - e^{-0.1/\tau}) = 100 \qquad (7\text{--}132)$$

$$1 - e^{-0.1/\tau} = 0.167, \qquad \tau = 0.546 \text{ sec,}$$

$$C_f = \tau/R_f = 5.46 \times 10^{-5} \text{ ft}^3/\text{psi} \qquad (7\text{--}133)$$

We thus need an accumulator which will displace 9.43 inch³ of fluid for each 100 psi of pressure. Of course an accumulator "softer" than this will reduce the peak p_o even more; however there may be reasons, such as excessive size, weight, or cost, which would discourage use of a larger accumulator.

7-5. THERMAL FIRST-ORDER SYSTEMS.

While, in practice, thermal systems are among the most complex encountered by an engineer, the variety of *simple basic* first-order systems which one can display is more limited than in mechanical, electrical, and fluid systems, since we have only *two* basic elements rather than three, and thus the number of combinations is fundamentally less. So far as basic laws available for equation setup are concerned, conservation of energy is clearly the most generally useful. For any system, over a time interval dt, one can always write

$$\text{Energy In} - \text{Energy Out} = \text{Additional Energy Stored} \qquad (7\text{--}134)$$

In Fig. 7–36a, we show a solid body at temperature T_o immersed in a fluid at temperature T_i. This configuration is an adequate model for many practical situations, including the response of temperature-sensing instru-

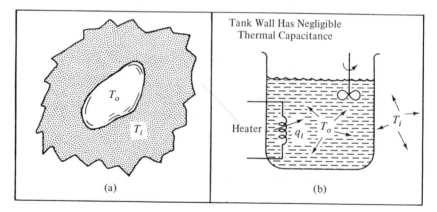

FIGURE 7–36

Some Thermal First-Order Systems

ments (thermometers, thermocouples, etc.), heat-treating of metal parts, freezing of foods, etc. We assume that the Biot number is such that the temperature of the solid body may at any instant be taken as uniform throughout, and that the surface coefficient of convective heat transfer is uniform over the surface and constant with time. The heat transferred from the fluid to the solid is all stored in the solid, so we may write over a time interval dt

$$\text{Energy in} = \text{Additional Energy Stored}$$
$$hA_s(T_i - T_o)\, dt = MC\, dT_o \qquad \textbf{(7–135)}$$

where

$h \triangleq$ surface heat transfer coefficient, Btu/(hr-ft²-°F)

$A_s \triangleq$ surface area for heat transfer, ft²

$M \triangleq$ mass of solid body, lb$_m$

$C \triangleq$ specific heat of solid body, Btu/(lb$_m$-°F)

Manipulation of Eq. (7–135) leads to

$$\frac{T_o}{T_i}(D) = \frac{1}{\tau D + 1} \qquad \textbf{(7–136)}$$

where

$$\tau \triangleq \frac{MC}{hA_s}, \text{ hr} \qquad \textbf{(7–137)}$$

As an application of this general model, consider the mercury-in-glass thermometer of Fig. 7–37. The temperature which it *indicates* is the mercury temperature T_o, which will equal the fluid temperature T_i only for

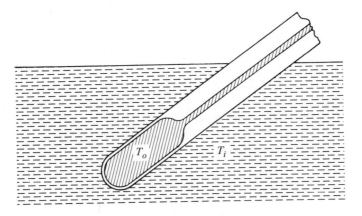

FIGURE 7–37

Mercury-in-Glass Thermometer

steady-state conditions. Its dynamic response can be found from the above type of model if we make some additional assumptions:

1. Heat loss from the mercury bulb up the thermometer stem is negligible.
2. Mass of mercury in bulb is constant.
3. Glass wall of bulb is thin enough that its energy storage is negligible.

The heat transfer between T_i and T_o is governed by a thermal resistance made up of a fluid film on the outside, the glass wall, and a fluid film on the inside; thus we should replace h in Eq. (7–135) by U, the *overall* heat transfer coefficient. In most cases U is dominated by the outside film coefficient, since the glass wall is thin and mercury is a very good conductor. A typical laboratory thermometer of this type has a cylindrical bulb about $\frac{1}{8}''$ by $\frac{1}{2}''$, giving a surface area of about 0.2 in² and a volume of about 0.006 in³. For mercury $\rho = 0.491$ lb$_m$/in³ and $C = 0.033$ Btu/(lb$_m$-°F). The film coefficient at the outside surface varies greatly with the fluid and its flow velocity; 2 Btu/(hr-ft²-°F) for still air and 500 Btu/(hr-ft²-°F) for rapidly moving water being indicative of the general range. The above numbers give time constants ranging from 0.5 to 125 seconds, which agree quite well with step-function test measurements.

When a thermometer is used in a stationary fluid, the heat transfer is by free, rather than forced, convection and the film coefficient itself depends on the temperature difference. A typical formula for h might be

$$h = K(T_i - T_o)^{1/4} \tag{7–138}$$

which would make Eq. (7–135) nonlinear as follows

$$A_s K(T_i - T_o)^{1.25} = MC \frac{dT_o}{dt} \tag{7–139}$$

Since the nonlinearity is due to h being a variable, an approach to linearization (different from those we have previously employed) would be to take h as constant at some sort of average value. For a step input, for instance, the range of $(T_i - T_o)$ would be known and one might compute an h corresponding to the mid-point of this range and set h constant at this value. This seemingly crude approach can actually give quite good results, as can be seen from Fig. 7–38 where a numerical example comparing linearized and exact response curves is given.

Systems with Several Inputs. The heated tank of Fig. 7–36b will give us some more experience with thermal systems and will also introduce some useful general ideas with regard to systems whose output is influenced by more than one input. We show a stirrer in the tank to allow us to assume a uniform temperature; the heat added by the "churning" action of the stirrer is assumed negligible. The heater (which could be electrical, a steam

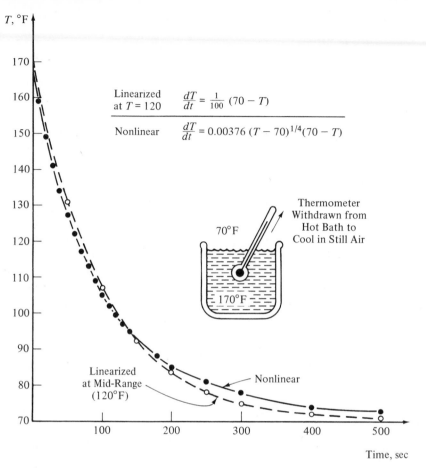

FIGURE 7–38

Linearization by Replacing Variable Parameter
with Fixed Average Value

coil, or some other type) supplies a heat input at a rate q_i Btu/sec. while the temperature T_i is that of the tank's surroundings, the so-called ambient temperature. Both q_i and T_i may vary with time in an arbitrary fashion; in a control system for the tank, q_i would be manipulated by a controller while T_i might be a random disturbing effect over which we have no control (like the outdoor temperature). Conservation of energy gives us

$$q_i \, dt - UA(T_o - T_i) \, dt = MC \, dT_o \qquad \text{(7–140)}$$

$$\tau \frac{dT_o}{dt} + T_o = (K_q)q_i + (K_T)T_i \qquad \text{(7–141)}$$

where

$$\tau \triangleq \frac{MC}{UA}, \text{ sec}$$

$M \triangleq$ mass of liquid in tank, lb_m

$C \triangleq$ specific heat of liquid in tank, $\text{Btu}/(\text{lb}_m\text{-}°\text{F})$

$U \triangleq$ overall heat transfer coefficient at tank wall, $\dfrac{\text{Btu}}{\text{sec-ft}^2\text{-}°\text{F}}$

$A \triangleq$ heat transfer area, ft^2

$K_q \triangleq 1/UA$, $°\text{F}/(\text{Btu/sec}) \triangleq$ temperature/heat flow gain

$K_T \triangleq 1.0$, $°\text{F}/°\text{F} \triangleq$ temperature/temperature gain

In Eq. (7–141) if we are given q_i and T_i as known functions of time, we can find the resulting tank temperature T_o; that is, T_o is determined by the simultaneous action of the two inputs q_i and T_i. To get transfer functions and block diagrams, we employ the principle of superposition to separate the effects of q_i and T_i. To get the transfer function $(T_o/q_i)(D)$ we momentarily consider T_i zero in Eq. (7–141) and get

$$\frac{T_o}{q_i}(D) = \frac{K_q}{\tau D + 1} \tag{7–142}$$

and then consider q_i zero to get

$$\frac{T_o}{T_i}(D) = \frac{K_T}{\tau D + 1} \tag{7–143}$$

In drawing the block diagram we must now superimpose the two effects so as to properly represent Eq. (7–141). This is done in Fig. 7–39 and can clearly be applied to any system with two inputs and obviously extended to *any* number of inputs.

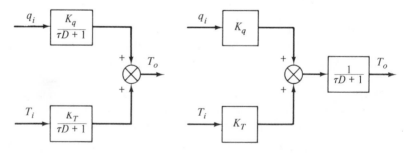

FIGURE 7–39

Alternate Forms of Block Diagram for Two-Input System

7-6. MIXED FIRST-ORDER SYSTEMS.

In Fig. 7–40 we display an electromechanical system of considerable practical importance, a DC motor driving a mechanical load. We are interested in the response of the load speed ω_o to a control voltage input e_i and a disturbing load torque T_i. We assume a permanent magnet field, so only the

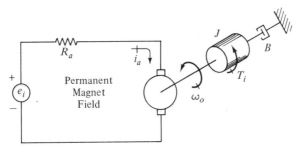

FIGURE 7–40

DC Motor and Load

armature circuit need be analyzed and we model it as resistance R_a which includes the terminal resistance of the motor and also any output resistance associated with the voltage source e_i. At any instant Kirchhoff's voltage loop law gives

$$-e_i + i_a R_a + K_E \omega_o = 0 \qquad (7\text{--}144)$$

where

$$K_E \omega_o \triangleq \text{voltage drop due to back emf of motor}$$

$$K_E \triangleq \text{motor back emf constant, volts/(rad/sec)}$$

Turning now to Newton's law at the motor shaft,

$$K_T i_a + T_i - B \omega_o = J \dot{\omega}_o \qquad (7\text{--}145)$$

where

$$K_T \triangleq \text{motor torque constant, inch-lb}_f/\text{amp}$$

$$B \triangleq \text{combined viscous damping of motor and load}$$

$$J \triangleq \text{combined inertia of motor and load}$$

The algebraic signs of the terms $K_E \omega_o$ and $K_T i_a$ may not be self-evident. We freely choose the positive sense of e_i, but once this is done it is convenient (though not necessary) to choose the positive direction of ω_o such that a positive e_i will cause a positive ω_o. To see how this was done, mentally "clamp" the motor shaft in Eq. (7–144), making $\omega_o = 0$. We see then that a positive e_i causes a positive i_a; a result of our choice for the i_a sign convention. Now in Eq. (7–145) also take $T_i = 0$. It is then clear that a positive i_a will cause a positive acceleration $\dot{\omega}_o$ and thus a positive speed ω_o

if the shaft is released; thus our sign conventions do, in fact, give a positive ω_o for a positive e_i as desired and the choice of sign on $K_T i_a$ is justified. To justify the sign of $K_E \omega_o$ we must invoke our knowledge of the *physical* nature of a motor back emf. This is that the back emf must *oppose* the voltage which caused the motion which is producing the back emf. Since a positive e_i tends to produce a positive ω_o the signs of e_i and $K_E \omega_o$ *must* be opposite in Eq. (7–144), which they are.

Equations (7–144) and (7–145) form a simultaneous set which describes our physical system. If e_i and T_i are considered known inputs, we see that there are two unknowns, ω_o and i_a. We could solve for either or both of these; since our primary interest is in the load motion we choose to eliminate i_a in favor of ω_o. From (7–144)

$$i_a = (e_i - K_E \omega_o)/R_a \qquad (7\text{–}146)$$

making (7–145)

$$(K_T/R_a)(e_i - K_E \omega_o) + T_i - B\omega_o = J\dot{\omega}_o \qquad (7\text{–}147)$$

and giving finally

$$(\tau D + 1)\omega_o = (K_{TG})T_i + (K_{EG})e_i \qquad (7\text{–}148)$$

where

$$\tau \triangleq \frac{J}{(BR_a + K_T K_E)/R_a}, \text{ sec}$$

$$K_{TG} \triangleq \text{system speed/torque gain} = \frac{R_a}{BR_a + K_T K_E}, \frac{\text{rad/sec}}{\text{inch-lb}_j}$$

$$K_{EG} \triangleq \text{system speed/voltage gain} = \frac{K_T}{BR_a + K_T K_E}, \frac{\text{rad/sec}}{\text{volt}}$$

Note that the system's basic parameters τ, K_{TG} and K_{EG} depend on *both* electrical and mechanical quantities. Note also from

$$\tau = \frac{J}{B + \dfrac{K_T K_E}{R_a}}$$

that $K_T K_E/R_a$ provides a "viscous" damping effect entirely analogous to B; even if $B = 0$, damping is still present.

Feedback Systems.   Because of the importance of automatic control in all branches of engineering, most undergraduate students will have an entire course in this area, usually somewhere after the system dynamics course. We now briefly introduce the principle of feedback, which underlies the entire automatic control area. In the absence of disturbing torque T_i the load speed ω_o is apparently completely under the control of voltage input e_i and it would seem that setting e_i at a particular value would guarantee a certain speed. This would be true if all system parameters were absolutely constant; however, in real systems quantities such as B, R_a, K_T, and K_E are all subject to drift with time, temperature, etc.; thus highly accurate control should not be expected from such an arrangement. To overcome these difficulties the feedback concept suggests that if one is interested in controlling speed he should *measure* what the speed actually is, *compare* this actual value to a desired value, and if they differ, *adjust* the voltage e_i in such a fashion as to reduce the error. Figure 7–41 shows one possible scheme for implementing this plan.

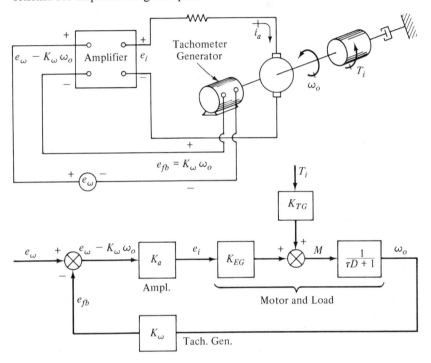

FIGURE 7–41

Speed Control by Feedback

A tachometer generator attached to the load shaft produces a voltage $K_\omega\omega_o$ which is an accurate measure of ω_o. This voltage is compared with a speed reference voltage e_ω obtained from an accurately regulated voltage source. The "error voltage" $(e_\omega - K_\omega\omega_o)$ is applied to an amplifier of gain K_a whose output supplies the armature voltage for the motor. In writing the system equation when a block diagram is already given, as in Fig. 7–41, the most efficient approach is to trace the signals through the diagram and write the equation "as we go," retaining as variables only the inputs (e_i, T_i) and outputs (ω_o) of interest. Thus, instead of writing e_{fb} for the feedback voltage we write $K_\omega\omega_o$ since our interest is in ω_o, not e_{fb}. The signal e_i is thus $K_a(e_\omega - K_\omega\omega_o)$ and the signal M is $K_{EG}K_a(e_\omega - K_\omega\omega_o) + K_{TG}T_i$, and since $\omega_o = (1/\tau D + 1) M$ we may write

$$[K_{EG}K_a(e_\omega - K_\omega\omega_o) + K_{TG}T_i]\frac{1}{(\tau D + 1)} = \omega_o \qquad (7\text{-}149)$$

"Multiplying" through by $(\tau D + 1)$ to "clear of fractions," gives

$$(\tau_{cl}D + 1)\omega_o = (K_{\omega_o})\omega_{o,\text{des}} + (K_{T_i})T_i \qquad (7\text{-}150)$$

where

$$\tau_{cl} \triangleq \text{``closed-loop'' time constant} \triangleq \frac{\tau}{1 + K_{EG}K_aK_\omega}, \text{ sec} \qquad (7\text{-}151)$$

$$K_{\omega_o} \triangleq \text{closed-loop speed gain} = \frac{1}{1 + 1/K_{EG}K_aK_\omega}, \frac{\text{rad/sec}}{\text{rad/sec}} \qquad (7\text{-}152)$$

$$K_{T_i} \triangleq \text{closed-loop torque gain} = \frac{K_{TG}}{1 + K_{EG}K_aK_\omega}, \frac{\text{rad/sec}}{\text{inch-lb}_f} \qquad (7\text{-}153)$$

$$\omega_{o,\text{des}} \triangleq \text{desired speed} \triangleq \frac{e_\omega}{K_\omega}, \text{ rad/sec} \qquad (7\text{-}154)$$

While the system with feedback (also called "closed-loop" system) is still a first-order system, some remarkable changes have occurred. These changes all depend on making the quantity $K_{EG}K_aK_\omega$, called the loop gain, a large number relative to 1.0. Fortunately this can generally be done, although great care must be used since excessive loop gain will cause instability.[1] (Our present model does not predict this instability since we have neglected several dynamic effects which, while unimportant at low and medium loop gains, become critical at very high loop gain.) Suppose we are able to make $K_{EG}K_aK_\omega = 20$. The time constant τ_{cl}, which determines the speed of response of the complete feedback system, is now $\tau/21$; the system is 21 times faster than without feedback although the motor and load have not been changed in any way! The closed-loop speed gain, which gives the relation between desired and actual speed in steady state, is

$$\left.\frac{\omega_o}{\omega_{o,\text{des}}}\right|_{\substack{\text{steady}\\\text{state}}} = K_{\omega o} = \frac{1}{1 + \dfrac{1}{[K_T/(BR_a + K_TK_E)](K_aK_\omega)}} \qquad (7\text{-}155)$$

For a simple example, suppose K_T, B, R_a, and K_E are each equal to 1.0, while $K_aK_\omega = 40$. Suppose environmental effects cause K_T to drift to the value 2.0. How much will this affect the speed? We now have

$$\left.\frac{\omega_o}{\omega_{o,\text{des}}}\right|_{\substack{\text{steady}\\\text{state}}} = \frac{1}{1 + \dfrac{1}{40}} = 0.975 \qquad (7\text{-}156)$$

---

[1] E. O. Doebelin, *Dynamic Analysis and Feedback Control* (New York: McGraw-Hill Book Company, 1962).

whereas before it was 0.952. Thus a 100% change in K_T caused only about 2% change in speed. *Without* feedback the change would have been from $K_{EG} = 0.5$ to $K_{EG} = 0.667$, a 33% change! Finally, the sensitivity K_{T_i} to disturbing torques is also 21 times less. This simple example hopefully will impress the reader with the power of the feedback concept and give some idea as to why it is used in literally thousands of different types of applications in all fields of engineering.

Hydromechanical Systems. Leaving the area of feedback, let us now consider the hydromechanical system of Fig. 7–42. A positive displacement

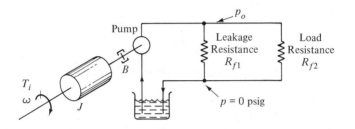

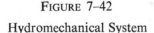

FIGURE 7–42

Hydromechanical System

pump is driven by an input torque T_i and supplies flow to a hydraulic load of pure resistance at pressure p_o. Newton's law at the pump shaft gives

$$T_i - p_o D_p - B\omega = J\dot{\omega} \qquad (7\text{-}157)$$

while conservation of mass (volume) gives

$$D_p\omega \frac{R_{f1}R_{f2}}{R_{f1} + R_{f2}} = D_p\omega R_{ft} = p_o \qquad (7\text{-}158)$$

Considering T_i as the input, p_o and ω are the unknowns; we consider p_o as the output of interest and quickly obtain

$$\frac{p_o}{T_i}(D) = \frac{K}{\tau D + 1} \qquad (7\text{-}159)$$

where

$$\tau \triangleq \frac{J}{B + D_p^2 R_{ft}}, \text{ sec} \qquad (7\text{-}160)$$

$$K \triangleq \frac{D_p R_{ft}}{B + D_p^2 R_{ft}}, \frac{\text{psi}}{\text{inch-lb}_f} \qquad (7\text{-}161)$$

Note in Eq. (7–160) that $D_p^2 R_{ft}$ plays the same role as mechanical viscous damping and thus the system is damped even if $B = 0$.

Thermomechanical Systems. In the mechanical brake of Fig. 7–43 the block is pressed against the rotating drum by the input force f_i, causing a friction torque $\mu f_i R$ and heat generation at a rate $q = 0.000107\ \mu R\omega f_i$, Btu/sec. This heat flows partially into the drum and partially into the

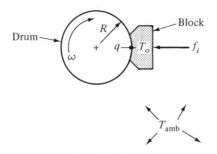

Drum

Block

FIGURE 7–43

Mechanical Brake as Thermomechanical System

block. If the block material is a much better conductor than the drum, we can assume for simplicity that all the heat goes into the block, which we assume at uniform temperature T_o. Since the block dissipates heat to the surrounding air at a rate $UA(T_o - T_{\text{amb}})$ we may write

$$[(0.000107\ \mu R\omega)f_i - UA(T_o - T_{\text{amb}})]\,dt = MC\,dT_o \qquad (7\text{–}162)$$

where

$$M \triangleq \text{mass of block}$$

$$C \triangleq \text{specific heat of block}$$

$$A \triangleq \text{heat transfer area for block}$$

$$U \triangleq \text{heat transfer coefficient for block}$$

giving finally, assuming ω constant,

$$(\tau D + 1)T_o = (K_f)f_i + T_{\text{amb}} \qquad (7\text{–}163)$$

where

$$\tau \triangleq \frac{MC}{UA},\ \text{sec}$$

$$K_f = \frac{0.000107\mu R\omega}{UA},\ \frac{°\text{F}}{\text{lb}_f}$$

7–7. FIRST-ORDER SYSTEMS WITH "NUMERATOR DYNAMICS."

In this section we briefly consider two types of systems which have first-order characteristic equations but whose dynamic behavior differs from that emphasized in this chapter because the differential equation right-hand side includes derivatives of the input signal. In transfer function form these two types are

$$\frac{q_o}{q_i}(D) = \frac{K\tau D}{\tau D + 1} \qquad (7\text{--}164)$$

and

$$\frac{q_o}{q_i}(D) = \frac{K(\tau_1 D + 1)}{\tau D + 1} \qquad (7\text{--}165)$$

The presence of the D-operator in the numerator of the transfer function leads to the "numerator dynamics" terminology.

An example of type (7–164) is afforded by the electric circuit of Fig. 7–44. Using impedance methods,

$$e_o = iR = \frac{e_i}{R + \dfrac{1}{i\omega C}} R = \frac{i\omega RC}{i\omega RC + 1} e_i \qquad (7\text{--}166)$$

$$\frac{e_o}{e_i}(i\omega) = \frac{i\omega\tau}{i\omega\tau + 1}, \qquad \tau \triangleq RC \qquad (7\text{--}167)$$

and thus

$$\frac{e_o}{e_i}(D) = \frac{\tau D}{\tau D + 1} \qquad (7\text{--}168)$$

and

$$\tau \frac{de_o}{dt} + e_o = \tau \frac{de_i}{dt} \qquad (7\text{--}169)$$

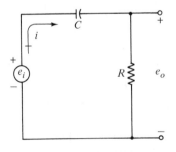

FIGURE 7–44

High-Pass Filter Circuit

Note how, in this problem, we *started* with the frequency response (AC circuit theory) and worked "backward" to the differential equation. Some people prefer this approach and find it quicker and/or easier than starting with the differential equation. The step response of such a system is found by setting e_i equal to a constant, e_{is} in Eq. (7–169). Since the derivative of a constant is zero, we get for all $t > 0$

$$\tau \frac{de_o}{dt} + e_o = 0 \qquad (7\text{-}170)$$

which has the complete solution

$$e_o = Ce^{-t/\tau} \qquad (7\text{-}171)$$

To find C we need to know $e_o(0^+)$. Whereas for the first-order systems emphasized in this chapter the output at $t = 0^-$ and $t = 0^+$ is the same, for this system there is a *sudden* change in e_o when the input is applied. Since C is assumed initially uncharged, when e_i jumps up to e_{is}, there being no voltage drop across C, Kirchhoff's voltage loop law says that e_o instantly becomes e_{is}. The complete specific solution is thus

$$e_o = e_{is}e^{-t/\tau} \qquad (7\text{-}172)$$

which is shown in Fig. 7–45a. Note that such a system will give zero steady-state output for any constant input.

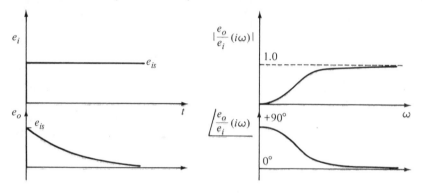

(a) Step Response   (b) Frequency Response

FIGURE 7–45

Response of $(e_o/e_i)(D) = \tau D/(\tau D + 1)$

That such systems have practical usefulness as high-pass filters becomes apparent from the frequency response

$$\frac{e_o}{e_i}(i\omega) = \frac{i\omega\tau}{i\omega\tau + 1} = \frac{\omega\tau \underline{/90°}}{\sqrt{(\omega\tau)^2 + 1} \; \underline{/\tan^{-1}\omega\tau}} \qquad (7\text{-}173)$$

which for low frequencies approaches $0 \; \underline{/90°}$, and for high frequencies approaches $1 \; \underline{/0°}$. That is, a high-pass filter rejects constant and low frequency inputs but passes high frequencies essentially unchanged. One example of the application of such a circuit is found in most oscilloscope input networks. There is generally a switch which allows connection of this type of circuit between the input terminals and the scope vertical deflection amplifiers. The time constant τ is chosen so that the range of frequencies where $(e_o/e_i)(i\omega) \approx 1 \underline{/0°}$ starts at about 2 Hz. Thus any signal with frequency content above 2 Hz is accurately measured, signal components between 0 and 2 Hz are distorted, and steady (DC) values are completely wiped out. Such action is useful, for example, when a pressure transducer and oscilloscope are used to study pressure fluctuations in the output of a reciprocating air compressor. This pressure signal has a large mean value (say 100 psi) while the fluctuations are small (say 5 psi). If one turns up the scope sensitivity to get a good look at the small oscillations, the large mean value will deflect the picture completely off the scope screen, since the scope zero-suppression control has a limited range. Now, if we switch in a high-pass filter, the large mean value is completely blocked and we can easily turn up the scope sensitivity to completely fill the screen with the oscillations. We might finally mention that systems of this general type are also useful as approximate differentiating devices; however, development of this application is left for the problems at the end of this chapter.

An example of system type (7–165) is afforded by the pneumatic phase-lag compensator used in pneumatic control systems and shown in Fig. 7–46. Such systems often operate under conditions for which the fluid

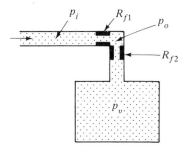

FIGURE 7–46

Pneumatic Phase-Lag Compensator

medium (air) may be considered incompressible and at constant temperature. Also the space containing p_o may be made sufficiently small that its fluid storage is negligible and thus the flow rates through the two resistances may be taken equal at any instant.

$$\frac{p_i - p_o}{R_{f1}} = \frac{p_o - p_v}{R_{f2}} \tag{7-174}$$

Conservation of mass for the tank of volume V gives

$$\rho \frac{p_o - p_v}{R_{f2}} dt = dM = \frac{V}{RT} dp_v \tag{7-175}$$

Since p_v is not of direct interest we eliminate it to get the desired relation between p_i and p_o.

$$\left(\frac{R_{f2}V}{\rho RT}D + 1\right)p_v = (\tau_1 D + 1)p_v = p_o \qquad (7\text{-}176)$$

$$p_i - p_o = \frac{R_{f1}}{R_{f2}}\left(p_o - \frac{1}{\tau_1 D + 1}p_o\right) = \frac{R_{f1}}{R_{f2}}\left(\frac{\tau_1 D}{\tau_1 D + 1}\right)p_o \qquad (7\text{-}177)$$

$$(\tau D + 1)p_o = (\tau_1 D + 1)p_i \qquad (7\text{-}178)$$

$$\frac{p_o}{p_i}(D) = \frac{\tau_1 D + 1}{\tau D + 1} \qquad \tau \triangleq \tau_1\left(1 + \frac{R_{f1}}{R_{f2}}\right) \qquad (7\text{-}179)$$

For the step response, $p_i = p_{is}$ = a constant and Eq. (7–178) becomes for $t > 0$

$$(\tau D + 1)p_o = p_{is} \qquad (7\text{-}180)$$

whose complete solution is

$$p_o = Ce^{-t/\tau} + p_{is} \qquad (7\text{-}181)$$

We again need $p_o(0^+)$ to find C and we can find it from Eq. (7–174) which holds at every instant, including $t = 0^+$. Note that p_v is still zero at 0^+, since it takes a finite time for a finite flow rate to build up a tank pressure (the tank is an "integrator"). Thus

$$\frac{p_{is} - p_o(0^+)}{R_{f1}} = \frac{p_o(0^+)}{R_{f2}} \qquad p_o(0^+) = \frac{p_{is}}{1 + \dfrac{R_{f1}}{R_{f2}}} \qquad (7\text{-}182)$$

which makes Eq. (7–181)

$$p_o = p_{is}\left[1 + \left(\frac{1}{1 + \dfrac{R_{f1}}{R_{f2}}} - 1\right)e^{-t/\tau}\right] \qquad (7\text{-}183)$$

We see (Fig. 7–47a) again that p_o experiences a sudden jump at $t = 0$. The frequency response

$$\frac{p_o}{p_i}(i\omega) = \frac{i\omega\tau_1 + 1}{i\omega\tau + 1} \qquad (7\text{-}184)$$

is shown in Fig. 7–47b.

While the system of Fig. 7–46 *must* have $\tau > \tau_1$ (see Eq. 7–179), there are other physical systems of the form

$$\frac{q_o}{q_i}(D) = \frac{\tau_1 D + 1}{\tau D + 1}$$

for which $\tau_1 > \tau$. Analysis (left for the problems) shows their response to be as shown in Fig. 7–48.

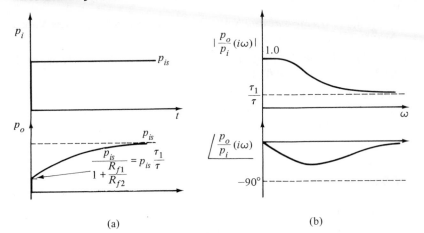

(a)   (b)

FIGURE 7–47

Response of $(p_o/p_i)(D) = (\tau_1 D + 1)/(\tau D + 1)$, $\tau > \tau_1$

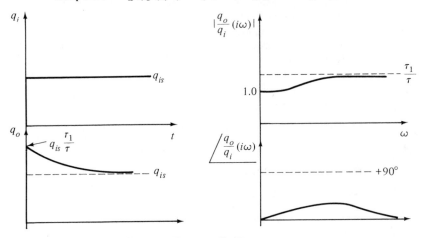

FIGURE 7–48

Response of $(q_o/q_i)(D) = (\tau_1 D + 1)/(\tau D + 1)$, $\tau_1 > \tau$

BIBLIOGRAPHY

1. Cannon, R. H., *Dynamics of Physical Systems*. New York: McGraw-Hill Book Company, 1967.

2. Doebelin, E. O., *Dynamic Analysis and Feedback Control*. New York: McGraw-Hill Book Company, 1962.

3.   Doebelin, E. O., *Measurement Systems, Application and Design*. New York: McGraw-Hill Book Company, 1966.

4.   Reswick, J. B. and C. K. Taft, *Introduction to Dynamic Systems*. Englewood Cliffs, N. J.: Prentice-Hall, Inc., 1967.

5.   Shearer, J. L., A. T. Murphy and H. H. Richardson, *Introduction to System Dynamics*. Reading, Mass.: Addison-Wesley Publishing Co., Inc., 1967.

PROBLEMS

7–1.   For the rotary system of Fig. 7–1a:

    a.   Derive the system differential equation relating the indicated output quantity to the indicated input quantity.

    b.   Put the system equation in standard form and define the standard parameters. Display both the operational and sinusoidal transfer functions and show an overall block diagram.

    c.   Obtain analog computer diagrams of both types shown in Fig. 7–10.

    d.   Write CSMP programs, both using REALPL and not using REALPL. When not using REALPL write in terms of elements (spring, damper, inertia) rather than using K and τ. Let the input be any function of time you wish and choose any numerical element values you wish.

7–2.   Repeat prob. 7–1 for the combined system of Fig. 7–1a.

7–3.   Repeat prob. 7–1 for the translational system of Fig. 7–1b.

7–4.   Repeat prob. 7–1 for the rotary system of Fig. 7–1b.

7–5.   Repeat prob. 7–1 for the translational system of Fig. 7–1c.

7–6.   Repeat prob. 7–1 for the translational system of Fig. 7–1d.

7–7.   Repeat prob. 7–1 for the rotary system of Fig. 7–1d.

7–8.   Repeat prob. 7–1 for the translational system of Fig. 7–1e.

7–9.   Repeat prob. 7–1 for the rotary system of Fig. 7–1e.

7–10.   Sketch freehand but carefully on graph paper the step response of systems with time constants of 0.5, 2.0, and 5.0 seconds (all on one sheet). What is the initial rate of change of the output in each case?

7–11.   Sketch logarithmic frequency-response curves for first-order systems with:

    a. $K = 1.0$, $\tau = 2.5$ min   b. $K = 3.4$, $\tau = 0.05$ sec

    c. $K = 0.2$, $\tau = 3.5$ hr   d. $K = 0.01$, $\tau = 0.001$ sec

7–12. In the combined system of Fig. 7–1a take:

$M = 1$ slug $\quad B = 20$ lb$_f$/(ft/sec)
$J = 0.1$ lb$_f$-ft-sec^2 $\quad T_i = 10$ ft-lb$_f$ step input
$R = 0.1$ ft

   a. What is the steady-state velocity of M and angular velocity of J?
   b. How long does it take to reach 95 percent of steady-state speed?
   c. It is suggested that M be lightened to increase response speed (decrease the time to reach steady state). What is the maximum improvement to be achieved in this way? If this is not enough, suggest other changes which would help (the steady-state speed must remain the same). Give specific numerical values which would result in a doubling of response speed.
   d. Using the numbers originally given, but with input torque as in Fig. P7–1, write a CSMP program to solve this problem.

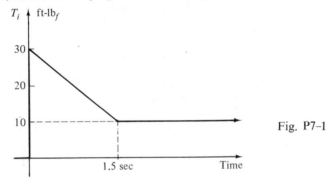

Fig. P7–1

   e. Run the program of part d and find the time to reach 95 percent of steady-state speed.

7–13. In the translational system of Fig. 7–1b the spring is nonlinear with force $= 25x_o + x_o^3$ lb$_f$, x_o in inches, while the linear damper has $B = 25$ lb$_f$/(in/sec).

   a. Find the linearized system time constant for small motions near $x_o = 0$ and also near $x_o = 5$ inches.
   b. Write a CSMP program for the nonlinear system to study the response to a force which goes from -10 lb$_f$ to $+10$ lb$_f$ in step fashion.
   c. Repeat part b for a force going from 200 to 300 lb$_f$.
   d. Run the program of part b and compare the exact response with the linearized.
   e. Run the program of part c and compare the exact response with the linearized.

7–14. Explain clearly *in words* how the systems of Fig. 7–1e respond to step inputs.

7-15. For the circuit of Fig. 7–23c:

    a. Derive the system differential equation relating the indicated output quantity to the indicated input quantity.

    b. Put the system equation in standard form and define the standard parameters. Display both the operational and sinusoidal transfer functions and show an overall block diagram.

    c. Obtain analog computer diagrams of both types shown in Fig. 7–10.

    d. Write CSMP programs, both using REALPL and not using REALPL. When not using REALPL write in terms of elements (R, L, C) rather than using K and τ. Let the input be any function of time you wish and choose any numerical element values you wish.

7-16. Repeat prob. 7–15 for the circuit of Fig. 7–23d.

7-17. Repeat prob. 7–15 for the circuit of Fig. 7–23e.

7-18. Repeat prob. 7–15 for the circuit of Fig. 7–23f.

7-19. Repeat prob. 7–15 for the circuit of Fig. 7–23h.

7-20. Get expressions for the operational impedance $Z(D)$ and sinusoidal impedance $Z(i\omega)$ for the circuits of:

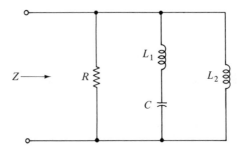

    a. Fig. P7–2

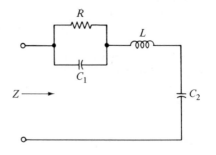

    b. Fig. P7–3

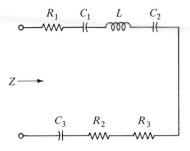

c. Fig. P7–4

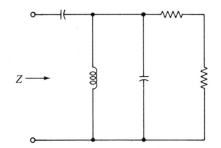

d. Fig. P7–5

7–21. Using the known behavior of L and C at very high and very low frequencies, show simplified versions of circuits:

a. Fig. P7–2 at very low frequencies.
b. Fig. P7–2 at very high frequencies.
c. Fig. P7–3 at very low frequencies.
d. Fig. P7–3 at very high frequencies.
e. Fig. P7–4 at very low frequencies.
f. Fig. P7–4 at very high frequencies.
g. Fig. P7–5 at very low frequencies.
h. Fig. P7–5 at very high frequencies.

7–22. Take all $R = 100$ ohms, all $C = 1$ μF, and all $L = 0.01$ H and apply a 60-Hz sinusoidal voltage of 110 volts *effective* value to the circuit terminals. Find the current, power factor angle, power factor, and average power for the circuits of:

a. Fig. P7–2   b. Fig. P7–3
c. Fig. P7–4   d. Fig. P7–5

7–23. The circuit of Fig. 7–23c uses the varistor of Fig. 3–2c as the resistance element; $L = 1.0$ H. Using CSMP, find the current if:

a. Voltage e_i is a step of $+10$ volts from zero.
b. Voltage e_i is steady at $+10$ volts and then drops in step fashion to zero.

    c.   Voltage e_i is steady at $+8$ volts and then jumps to $+10$ volts. Also do a linearized analytical solution and compare with the exact result.

    d.   Voltage $e_i = 9 + 1 \sin 1000t$ volts, t in seconds. Also do a linearized analytical solution and compare with the exact result.

7-24. In the circuit of Fig. 7-23f take $L = 0.01$ H, $R_1 = 1000$ ohms, $R_2 = 500$ ohms. If i_i is a rectangular pulse of duration T, what is the largest T for which the input may be approximated as an impulse?

7-25. For the fluid system of Fig. 7-31b:

    a.   Derive the system differential equation relating the indicated output quantity to the indicated input quantity, treating the flow resistances as linear. Display the transfer function.

    b.   Taking flow resistances as nonlinear (orifices), do a perturbation analysis for small changes about an equilibrium operating point to get a linearized system equation. Give the transfer function for the input and output perturbations.

    c.   Write a CSMP program for the case where the flow resistances are nonlinear (orifices).

7-26. Repeat prob. 7-25 for the system of Fig. 7-31c.

7-27. For the system of Fig. 7-31d:

    a.   Assume the gas is at constant temperature. Note that the pressure at the bottom of the tank is caused by both h_o and gas pressure. Compute the *total* compliance of this tank and then linearize it.

    b.   Assume a linear flow resistance and, using the linearized compliance from part a, derive the system differential equation relating h_o to q_i. Put in standard form and define standard parameters.

    c.   Write a CSMP program for this system using the nonlinear relations for both the compliance and the resistance (orifice).

7-28. For the system of Fig. 7-31e:

    a.   Assume linear resistance and compliance and derive the system differential equation.

    b.   If the tube is 0.08 inch in diameter and 5 inches long and the liquid has viscosity 0.005 lb_f-sec/ft^2, what is the largest tolerable compliance if the time constant cannot exceed 0.05 seconds?

    c.   Estimate how large $(p_i - p_o)$ could get before turbulent flow occurred. Fluid weighs 60 lb_f/ft^3.

7-29. For the system of Fig. 7-31f:

    a.   Using the linearized compliance for an isothermal process and assuming a linear resistance, derive the differential equation relating p_o to p_i. How do the results change if an adiabatic process is assumed?

    b.   Write a CSMP program for this system using a nonlinear isothermal compliance.

    c.   Repeat part b using a nonlinear adiabatic compliance.

7-30. For the system of Fig. 7–31h:

   a. Assuming a linear flow resistance, derive the differential equation relating q_o to p_i.

   b. After a step input pressure is applied, how long does it take to achieve steady flow? For water with viscosity 2×10^{-5} lb_f-sec/ft², specific weight 62.4 lb_f/ft³, flowing in a 0.1-ft diameter pipe 10 feet long, how long does it take? What is the largest step pressure for which laminar flow would exist? How would the analysis change for pressures larger than this?

7-31. Working directly from the block diagram of Fig. 7–39, set up a block diagram which has the heat loss $q_L = UA(T_o - T_i)$ as an output signal. From this diagram get the transfer functions $(q_L/q_i)(D)$ and $(q_L/T_i)D$. What would the response of q_L to a step of T_i look like?

7-32. Write a CSMP program for the nonlinear equation of Fig. 7–38 and run it to check the results given.

7-33. An infrared lamp projects a beam of thermal radiation of strength q_i Btu/sec on a solid body of mass M and surface area A. The surface is rough and blackened so as to absorb all the radiant flux. The body loses heat by convection (with coefficient h) to surroundings at a fixed temperature T_a.

   a. Get the transfer function relating q_i and body temperature T_o.

   b. If the body is a brass sphere of 6-inch diameter, $h = 30$ Btu/(hr-ft²-°F), $T_a = 70°F$, and $q_i = 1$ Btu/sec, what will the steady-state temperature be and how long will it take to reach 95% of this after q_i is turned on?

   c. If the sphere were hollow with 0.1 inch wall thickness, how would these results change?

7-34. In the combined system of Fig. 7–1a let a force f_i be applied directly to M (T_i is still present). Get differential equations, block diagrams, and transfer functions showing how $\dot{x}_o$ is produced by f_i and T_i.

7-35. In the system of Fig. 7–1c, if x_i is a step input, find the force which must be provided by the motion source.

7-36. In the circuit of Fig. 7–23f, if i_i is a step input, find the voltage across the current source. Get an expression for the power taken from the source.

7-37. For the circuit of Fig. 7–23b, get the transfer function relating capacitor current to i_i.

7-38. In the system of Fig. 7–1e, let a torque T_i act directly on J ($\dot{\theta}_i$ is still present). Get differential equations, transfer functions, and block diagram showing how $\dot{\theta}_o$ is produced by $\dot{\theta}_i$ and T_i.

7–39   In the system of Fig. 7–1c, if the input is sinusoidal, what is the average power drawn from the motion source in steady state? What is the torque?

7–40. In the system of Fig. 7–31b, add an inflow q_i to the top of the tank. Assume linear flow resistances and find differential equations, transfer functions, and block diagrams showing how h_o is produced by p_i and q_i.

7–41. In the system of Fig. 7–40, get differential equations, transfer functions, and block diagrams showing how i_a is produced by e_i and T_i.

7–42. For the circuit of Fig. P7–6 get the differential equation, transfer function, step response, and frequency response.

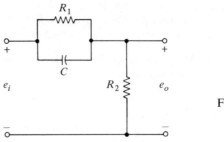

Fig. P7–6

7–43. For the system of Fig. P7–7 get the differential equation, transfer function, step response, and frequency response.

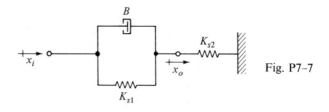

Fig. P7–7

SECOND-ORDER SYSTEMS

8–1. SECOND-ORDER SYSTEMS FORMED FROM CASCADED FIRST-ORDER SYSTEMS.

The most *general* form of second-order system encountered in practice has an equation of the form

$$a_2 \frac{d^2 q_o}{dt^2} + a_1 \frac{dq_o}{dt} + a_0 q_o = b_2 \frac{d^2 q_i}{dt^2} + b_1 \frac{dq_i}{dt} + b_0 q_i \qquad \textbf{(8–1)}$$

however, the most common and important special case, which we will emphasize, is given by

$$a_2 \frac{d^2 q_o}{dt^2} + a_1 \frac{dq_o}{dt} + a_0 q_o = b_0 q_i \qquad \textbf{(8–2)}$$

Just as in first-order systems, a widely accepted *standard form* of (8–2) has been defined and should generally be employed. We again divide by a_0 to get

$$\frac{a_2}{a_0} \frac{d^2 q_o}{dt^2} + \frac{a_1}{a_0} \frac{dq_o}{dt} + q_o = \frac{b_0}{a_0} q_i \qquad \textbf{(8–3)}$$

and define

$$\omega_n \triangleq \sqrt{\frac{a_0}{a_2}} \triangleq \text{undamped nature frequency}, \frac{\text{rad}}{\text{time}} \qquad \textbf{(8–4)}$$

$$\zeta \triangleq \frac{a_1}{2\sqrt{a_2 a_0}} \triangleq \text{damping ratio, dimensionless} \qquad (8\text{-}5)$$

$$K \triangleq \frac{b_0}{a_0} \triangleq \begin{array}{l} \text{system steady-state gain (sensitivity),} \\ \text{dimensions are those of } q_o/q_i \end{array} \qquad (8\text{-}6)$$

to get the standard form as

$$\left(\frac{D^2}{\omega_n^2} + \frac{2\zeta D}{\omega_n} + 1\right) q_o = K q_i \qquad (8\text{-}7)$$

with transfer function

$$\frac{q_o}{q_i}(D) = \frac{K}{\dfrac{D^2}{\omega_n^2} + \dfrac{2\zeta D}{\omega_n} + 1} \qquad (8\text{-}8)$$

Whenever a new physical form of second-order system is first encountered, one should *immediately* define K, ζ and ω_n, and change over to the standard form to gain all the benefits of standardization. The significance of the standard parameters K, ζ and ω_n will be developed shortly.

Second-order system models may arise naturally when a complete system is first physically analyzed as an entity. They also arise when two *components* of a system, each individually modeled as first-order, are connected in cascade to form a larger overall system. By cascade connection we mean that the output of the first component is connected as the input of the second (see Fig. 8–1). Proceeding by formal mathematics we are tempted to write

$$(\tau_2 D + 1)q_{o2} = K_2 q_{i2} = K_2 \left[\frac{K_1}{\tau_1 D + 1}\right] q_{i1} \qquad (8\text{-}9)$$

and thus

$$(\tau_2 D + 1)(\tau_1 D + 1)q_{o2} = [\tau_1 \tau_2 D^2 + (\tau_1 + \tau_2)D + 1]q_{o2} = K_1 K_2 q_{i1}$$
$$(8\text{-}10)$$

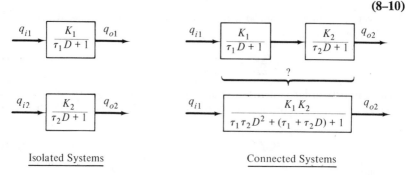

Isolated Systems   Connected Systems

FIGURE 8–1

Second-Order System from Cascaded First-Order

which we see fits the second-order form. The defect in this analysis lies in the fact that, when the two first-order models are analyzed as isolated devices, certain physical assumptions are normally made which are, at least partially, *violated* when the two systems are connected. The violation consists of the second system withdrawing from the first some *power* which was not accounted for in the model of the first system; thus the transfer function $(q_{o1}/q_{i1})(D)$ is *different* when the second system is connected and the overall transfer function $(q_{o2}/q_{i1})(D)$ is not just a simple product of the two isolated transfer functions. This phenomenon is called a *loading effect*; in some cases it is slight enough to neglect, while in others it is so large that we must re-analyze the *complete* system "from scratch" rather than trying to employ the models derived for the isolated components. Figure 8–2 shows an electrical example of this situation.

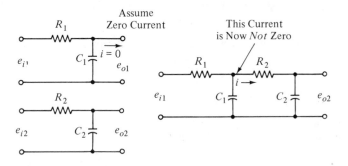

FIGURE 8–2

Electrical Example of Loading

The Significance of Loading Effects. To develop these concepts in more detail we shall analyze the system (Fig. 8–3) composed of two subsystems from Fig. 7–1 which, considered individually, would be first-order systems. Recall that the first system might represent a machine tool slide being positioned by a motor providing the force f_i; the second system could represent a velocity-measuring device which we wish to attach to the slide to measure its speed. If we simply multiply the two isolated first-order transfer functions we get for the total system

$$\frac{x_{o2}}{f_{i1}}(D) = \frac{B_2/(B_1 K_{s2})}{\dfrac{M_1 B_2}{K_{s2} B_1} D^2 + \dfrac{M_1 K_{s2} + B_1 B_2}{K_{s2} B_1} D + 1} \qquad (8\text{–}11)$$

which will be a good approximation only if the loading effect is small. To discover the nature of the loading effect we analyze the connected system "from scratch" without using the isolated first-order models at all.

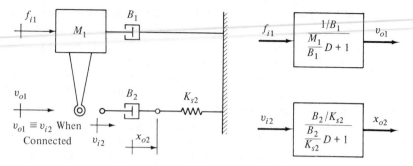

FIGURE 8–3

Cascaded Mechanical First-Order Systems

$$f_{i1} - B_1\dot{x}_{o1} - B_2(\dot{x}_{o1} - \dot{x}_{o2}) = M_1\ddot{x}_{o1} \qquad \textbf{(8-12)}$$

$$B_2(\dot{x}_{o1} - \dot{x}_{o2}) = K_{s2}x_{o2} \qquad \textbf{(8-13)}$$

Since we are interested in $(x_{o2}/f_{i1})(D)$ we eliminate x_{o1} by substituting

$$\dot{x}_{o1} = \left(\frac{B_2 D + K_{s2}}{B_2}\right) x_{o2} \qquad \textbf{(8-14)}$$

from (8–13) into (8–12) to get

$$f_{i1} - \frac{B_1}{B_2}(B_2 D + K_{s2})x_{o2} - B_2\left(\frac{B_2 D + K_{s2}}{B_2 D} - 1\right)\dot{x}_{o2}$$

$$= M_1\left(\frac{B_2 D + K_{s2}}{B_2}\right)\dot{x}_{o2} \qquad \textbf{(8-15)}$$

This leads to

$$\frac{x_{o2}}{f_{i1}}(D) = \frac{B_2/K_{s2}(B_1 + B_2)}{\dfrac{M_1 B_2}{K_{s2}(B_1 + B_2)}D^2 + \dfrac{M_1 K_{s2} + B_1 B_2}{K_{s2}(B_1 + B_2)}D + 1} \qquad \textbf{(8-16)}$$

which may be compared directly with its approximation, Eq. (8–11). Note that if B_2 is small *compared to* B_1, then (8–11) becomes a good approximation to (8–16); for example, if all parameters are 1.0 except $B_2 = 0.05$, we get

$$\frac{x_{o2}}{f_{i1}}(D) = \frac{0.05}{0.05 D^2 + 1.05 D + 1} \qquad \textbf{(8-11)}$$

$$\frac{x_{o2}}{f_{i1}}(D) = \frac{0.0476}{0.0476 D^2 + 1.00 D + 1} \qquad \textbf{(8-16)}$$

We should also point out that when the loading effect is negligible, the "internal" variable (in the above example this is $\dot{x}_{o1}$) may be calculated with good accuracy from $(x_{o1}/f_{i1})(D) = (1/B_1)/((M_1/B_1)D + 1)$. That is, the output of the first system is negligibly affected by the presence of the second system.

8-2. MECHANICAL SECOND-ORDER SYSTEMS.

We have just seen how second-order systems can arise from coupling of first-order systems. They also commonly arise in their own right when a complete physical system is not obviously "manufactured" from first-order "components." We can generate a number of practically useful systems from the first-order examples of Fig. 7–1 by deciding to develop "more accurate" models. For example, in Fig. 7–1b, the assumption of negligible inertia would be accurate only for certain operating conditions; suppose we wish to find criteria for judging rationally when we are allowed this simplification. By comparing response with and without inertia we can formulate such criteria. We will of course at the same time be solving the response problem for those systems in which inertia is *not* negligible, as in Fig. 8–4a. In Fig. 8–4b we consider a more accurate model of Fig. 7–1c in which the moving part of the damper has inertia. Similarly, 8–4c depicts 7–1d when the mass of the damper cylinder and/or the spring is included. By considering the springiness of the rod connecting the damper B_1 to the mass M in Fig. 7–1e, we get the second-order system of 8–4d. Finally, 8–4e is a version of 7–1d in which springiness in the damper and damping in the spring are included in the model. All the systems of Fig. 8–4 will lead to relations of the form of Eq. (8–2) between the indicated input and output quantities.

Step Response of Second-Order Systems. We will be using the system of Fig. 8–4a as our example for developing the general response characteristics of second-order systems. Our preference for this example rests on the fact that this configuration is widely accepted as the simplest system for introducing the basic concepts of the important field of mechanical shock and vibration. In Fig. 8–5, we re-orient the system so that motion is now in the vertical direction, so as to illustrate the method of treating gravitational forces. These forces are most conveniently treated by choosing the origin of the displacement coordinate x_o to coincide with the location of the mass when it is at rest with only its weight W and the spring force acting on it. These two forces clearly must balance each other at this point so we know that the spring force must be $-W$ when $x_o = 0$, and thus it can be written in general as $-W - K_s x_o$. Newton's law then gives

$$f_i + W - (W + K_s x_o) - B\dot{x}_o = M\ddot{x}_o \qquad (8\text{–}17)$$

where we see the gravity force W cancels out to give

$$M\ddot{x}_o + B\dot{x}_o + K_s x_o = f_i \qquad (8\text{–}18)$$

This clearly fits the standard form (8–2) so we immediately define

$$\omega_n \overset{\triangle}{=} \sqrt{\frac{K_s}{M}} \ \frac{\text{rad}}{\text{time}}, \qquad \zeta \overset{\triangle}{=} \frac{B}{2\sqrt{K_s M}}, \qquad K \overset{\triangle}{=} \frac{1}{K_s} \ \frac{\text{inch}}{\text{lb}_f} \qquad (8\text{–}19)$$

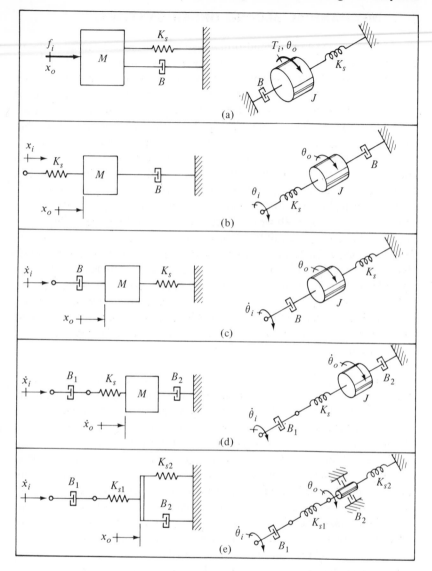

FIGURE 8–4

Some Mechanical Second-Order Systems

to get

$$\left(\frac{D^2}{\omega_n^2} + \frac{2\zeta D}{\omega_n} + 1\right) x_o = K f_i \qquad \qquad \textbf{(8–20)}$$

For a step input x_o and $\dot{x}_o$ are initially zero when f_i jumps up to f_{is}. At $t = 0^+$, x_o and $\dot{x}_o$ are both still zero, since the finite force f_{is} produces a

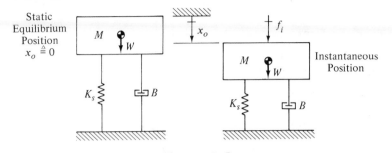

FIGURE 8–5

Basic Vibrating System

finite initial acceleration f_{is}/M, which can produce *no* finite velocity change from $t = 0$ to $t = 0^+$. The particular solution to (8–20) with $f_i = f_{is}$ is $x_{op} = Kf_{is}$. For the complementary function solution, we need the roots of the characteristic equation $D^2/\omega_n^2 + 2\zeta D/\omega_n + 1 = 0$, which are easily obtained from the quadratic formula as $-\zeta\omega_n \pm \omega_n\sqrt{\zeta^2 - 1}$. If $\zeta = 0$, which means $B = 0$ and the damper has been removed from the system, the roots $\pm i\omega_n$ are purely imaginary and the complementary solution is $x_{oc} = C \sin(\omega_n t + \phi)$, making the complete solution

$$x_o = x_{op} + x_{oc} = Kf_{is} + C \sin(\omega_n t + \phi) \qquad (8\text{–}21)$$

Initial conditions give

$$x_o(0^+) = 0 = Kf_{is} + C \sin \phi, \qquad \dot{x}_o(0^+) = 0 = C\omega_n \cos \phi \qquad (8\text{–}22)$$

and thus $\phi = 90°$ and $C = -Kf_{is}$, giving

$$x_o = Kf_{is}(1 - \sin(\omega_n t + 90°)) = Kf_{is}(1 - \cos \omega_n t) \qquad (8\text{–}23)$$

Figure 8–6 shows the motion to be a continuing oscillation at frequency ω_n, thus bringing out the significance of ω_n, and the reason for defining it as in Eq. (8–4). That is, ω_n *is the frequency of the natural oscillations of a second-order system when all damping is removed.*

Since complex roots are necessary for oscillatory complementary solutions, if we add damping to our system and gradually increase ζ from 0 toward 1.0 we will continue to get oscillations, although they now gradually die out. It is clear that $\zeta = 1.0$ is a unique point of some sort, since at this value we no longer have complex roots. For $\zeta = 1.0$ the characteristic equation is a "perfect square," giving two repeated roots $-\omega_n, -\omega_n$ with the solution

$$x_o = Kf_{is}[1 - (1 + \omega_n t)e^{-\omega_n t}] \qquad (8\text{–}24)$$

as graphed in Fig. 8–7. A second-order system with $\zeta = 1.0$ is called *critically damped*; the value of B which gives this is called the critical damping $B_{\text{crit}} \triangleq 2\sqrt{K_s M}$. Since the ratio of the actual damping to the

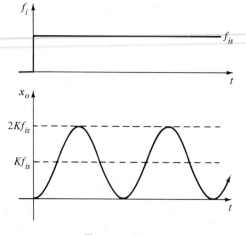

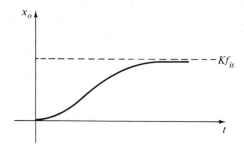

FIGURE 8–6

Step Response of Undamped Second-Order System

FIGURE 8–7

Step Response of Critically Damped Second-Order System

critical value is $B/2\sqrt{K_sM} = \zeta$ we can now appreciate why ζ is called the damping ratio and why it was defined as in (8–5). Systems with $0 < \zeta < 1.0$ are called *underdamped* and exhibit oscillations which die out; systems with $\zeta > 1.0$ are called *overdamped* and exhibit no natural oscillations.

For underdamped systems the roots are $-\zeta\omega_n \pm i\omega_n\sqrt{1 - \zeta^2}$ giving

$$x_o = Kf_{is} + Ce^{-\zeta\omega_n t} \sin(\omega_n\sqrt{1 - \zeta^2}t + \phi) \qquad (8\text{–}25)$$

From the initial conditions

$$0 = Kf_{is} + C \sin\phi, \qquad 0 = -\zeta\omega_n C \sin\phi + C\omega_n\sqrt{1 - \zeta^2} \cos\phi \qquad (8\text{–}26)$$

and thus

$$\frac{\sin\phi}{\cos\phi} = \tan\phi = \frac{\sqrt{1 - \zeta^2}}{\zeta}, \qquad \sin\phi = \sqrt{1 - \zeta^2} \qquad (8\text{–}27)$$

and $C = Kf_{is}/\sqrt{1 - \zeta^2}$ giving finally

$$x_o = Kf_{is}\left[1 - \frac{1}{\sqrt{1 - \zeta^2}} e^{-\zeta\omega_n t} \sin(\omega_n\sqrt{1 - \zeta^2}t + \phi)\right] \quad \text{(8–28)}$$

For overdamped systems we always have two real roots $(-\zeta + \sqrt{\zeta^2 - 1})\omega_n$ and $(-\zeta - \sqrt{\zeta^2 - 1})\omega_n$ which lead to

$$x_o = Kf_{is}\left[1 - \frac{\zeta + \sqrt{\zeta^2 - 1}}{2\sqrt{\zeta^2 - 1}} e^{(-\zeta + \sqrt{\zeta^2 - 1})\omega_n t}\right.$$
$$\left. + \frac{\zeta - \sqrt{\zeta^2 - 1}}{2\sqrt{\zeta^2 - 1}} e^{(-\zeta - \sqrt{\zeta^2 - 1})\omega_n t}\right] \quad \text{(8–29)}$$

When ζ is very large (B very large) the system will respond very nearly like a first-order system because the damping effect greatly overshadows the inertia effect. For example, if $\zeta = 10$,

$$x_o = Kf_{is}[1 - 1.002e^{-0.05\omega_n t} + 0.002e^{-19.95\omega_n t}] \approx Kf_{is}(1 - e^{-0.05\omega_n t}) \quad \text{(8–30)}$$

If inertia were *completely* neglected in the model, it would be first-order with $\tau = B/K_s$. For $\zeta = 10$,

$$\zeta = \frac{B}{2\sqrt{K_s M}} = \left(\frac{1}{2}\sqrt{\frac{K_s}{M}}\right)\left(\frac{B}{K_s}\right) = 10 \quad \text{(8–31)}$$

and thus

$$\tau = \frac{B}{K_s} = 20\sqrt{\frac{M}{K_s}} = \frac{20}{\omega_n} \quad \text{(8–32)}$$

which agrees with Eq. (8–30) to the number of significant figures carried there. Since no real system is free of inertia, analyses such as this are quite useful in deciding when inertia has a negligible effect and may be eliminated from the system model.

Since a second-order system has three parameters (K, ζ, ω_n), we are interested in the effect of each of these on the character of the response. The steady-state gain K has exactly the same meaning as in a first-order system; it is the amount of output produced in steady state for each unit of input, as can be seen from any of the above response equations. Since ω_n always appears in these equations as $\omega_n t$, its effect on system behavior is also simply discovered. If we keep K, ζ, and f_{is} fixed and vary ω_n, we see that increasing ω_n causes a directly proportional increase in speed of response. For example, if ω_n were doubled, equal values of the product $\omega_n t$ would be achieved in one-half the time and thus equal values of x_o occur in one-half the time. This feature is made use of in plotting a non-dimensional family of step-response curves as in Fig. 8–8. By plotting x_o/Kf_{is} versus $\omega_n t$ for various values of ζ, we get a set of curves which is

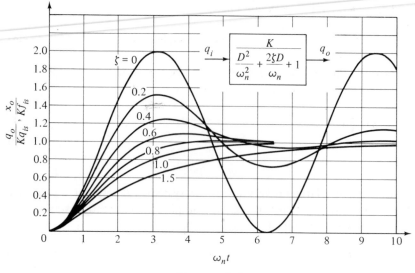

$$\frac{q_o}{Kq_{is}}, \frac{x_o}{Kf_{is}}$$

FIGURE 8–8

Second-Order Step Response Curves

useful for any second-order system, no matter what K, ζ, ω_n, or f_{is} might be. Since ζ enters the response formulas in such a complicated way, we rely on plotting of specific numerical values, as in Fig. 8–8, to reveal its effect. For underdamped systems the first overshoot (peak value) of x_o is often of practical interest and may be found by standard calculus maximizing methods to depend only on ζ, according to the formula

$$\text{percent overshoot} \triangleq 100 \frac{x_{o,\,\text{peak}} - Kf_{is}}{Kf_{is}} = 100 e^{-\zeta\pi/\sqrt{1-\zeta^2}} \quad \textbf{(8–33)}$$

which is graphed in Fig. 8–9. For a fixed value of ω_n the speed of response, as measured by the time for the output to settle within a chosen plus and minus tolerance around the final value, is determined by ζ. Using this idea of a *settling time* as a speed of response indicator we find that ζ's which are either too small or too large take longer to settle and thus an *optimum value* of ζ should exist. This depends on the size of the tolerance band chosen, but for bands of ± 2 to $\pm 10\%$ the optimum ζ is in the range 0.6 to 0.7. Many measuring instruments, which are required to respond quickly, are second-order systems and have ζ designed to be about 0.65.

Experimental Step-Function Testing. When numerical values of model parameters are to be found from experiments on an actual system, step-function testing is often useful, particularly if the system is underdamped. The simplest test often consists of giving the system some initial energy

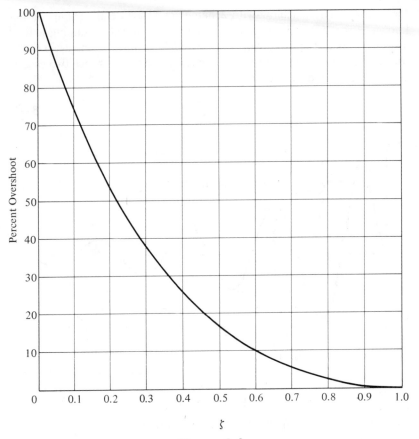

FIGURE 8–9

Effect of Damping Ratio on Overshoot

and then releasing it to perform its natural response. In Fig. 8–5 we would simply deflect the spring to a position away from zero, say $x_o = x_{o0}$, and release the mass with zero velocity and $f_i \equiv 0$. For an underdamped system the response is

$$x_o = \frac{x_{o0}}{\sqrt{1 - \zeta^2}} e^{-\zeta \omega_n t} \sin (\omega_n \sqrt{1 - \zeta^2}\, t + \phi), \qquad \sin \phi = \sqrt{1 - \zeta^2} \quad \text{(8–34)}$$

By comparison with Eq. (8–28) we see that we can use Fig. 8–8 for this response also if we replace x_o/Kf_{is} with x_o/x_{o0} and relabel the vertical axis as running from 1.0 to 0 to −1.0, rather than from 0 to 1.0 to 2.0.

Two ways of finding ζ from test data are common. If a good step input or release from initial deflection can be achieved and the first overshoot measured, Fig. 8–9 gives ζ directly. When a system is excited by a momentary pulse (such as a blow with a hammer) whose detailed shape is unknown, if the system is essentially second-order and the damping is light a response as in Fig. 8–10 will be observed. The time for one cycle of the oscillation, called the period T, and defined as the time between two successive zero-crossings, is given by

$$T = \frac{2\pi}{\omega_n \sqrt{1 - \zeta^2}} \qquad (8\text{–}35)$$

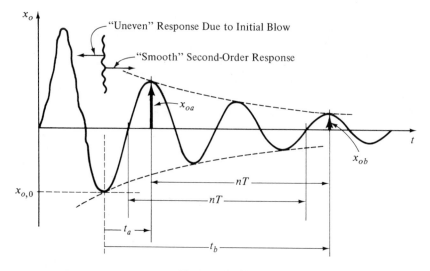

FIGURE 8–10

Response to Arbitrary Pulse Input

thus n cycles occur in a time nT. We can find ζ by measuring the decay in amplitude which occurs in n cycles. Note that once the response has become "smooth," at any peak the velocity is zero and the displacement is some value which we could call x_{o0} and treat the problem as a release from an initial deflection, using Eq. (8–34). If we assume that peaks occur in (8–34) when $\sin(\omega_n \sqrt{1 - \zeta^2}\, t + \phi) = 1.0$ (not exactly true but very close for small ζ), then the ratio of two peaks separated by a time interval nT is

$$\frac{x_{ob}}{x_{oa}} = \frac{(x_{o0}/\sqrt{1 - \zeta^2})e^{-\zeta\omega_n t_b}}{(x_{o0}/\sqrt{1 - \zeta^2})e^{-\zeta\omega_n t_a}} = \frac{e^{-\zeta\omega_n(t_a + nT)}}{e^{-\zeta\omega_n(t_a)}} = e^{-\zeta\omega_n nT} \qquad (8\text{–}36)$$

Thus if we can measure x_{ob} and x_{oa} for an n-cycle interval, ζ may be computed from

$$\frac{\zeta}{\sqrt{1 - \zeta^2}} = \frac{\log_e (x_{oa}/x_{ob})}{2\pi n} \qquad (8\text{–}37)$$

which for $\zeta < 0.2$ is closely approximated by

$$\zeta \approx \frac{\log_e (x_{oa}/x_{ob})}{2\pi n} \qquad (8\text{–}38)$$

To find ω_n we first get the *damped natural frequency* $\omega_{n,d} \triangleq \omega_n\sqrt{1 - \zeta^2}$ by measuring the transient period T as defined in (8–35). Assuming ζ has already been found, ω_n is computed from

$$\omega_n = \frac{\omega_{n,d}}{\sqrt{1 - \zeta^2}} \qquad (8\text{–}39)$$

Note that in any second-order system, any natural oscillations which might be observed always take place at frequency $\omega_n\sqrt{1 - \zeta^2}$, *not* at ω_n.

In real mechanical systems the viscous friction assumed in Eq. (8–17) rarely occurs uncontaminated by more complex and nonlinear frictional effects. When the friction is caused by bearings and/or a real damper, it is often a combination of viscous and dry or Coulomb types. The simplest model of Coulomb friction assumes a friction force which always opposes the velocity, but whose magnitude is independent of speed. If the system of Fig. 8–5 has both viscous and dry friction, its equation becomes

$$M\ddot{x}_o + B\dot{x}_o + F_{df}\frac{\dot{x}_o}{|\dot{x}_o|} + K_s x_o = f_i \qquad (8\text{–}40)$$

where

$F_{df} \triangleq$ magnitude of dry friction force, lb_f

$\dfrac{\dot{x}_o}{|\dot{x}_o|} \triangleq$ the signum function $= +1$ if $\dot{x}_o > 0$, -1 if $\dot{x}_o < 0$ (8–41)

Note that the signum function term in (8–40) gives a force which is constant in magnitude but opposes the velocity and reverses whenever velocity reverses. To show the effect of dry friction, we will solve Eq. (8–40) for three numerical examples in which the friction is all dry, all viscous, and finally partly each. We will use the CSMP digital simulation program to solve (8–40) for a case where $f_i \equiv 0$ and we consider the response to an initial displacement x_{o0}. Suppose

$$K_s = 100 \text{ lb}_f/\text{inch} \qquad F_{df} = 5 \text{ lb}_f$$
$$M = 0.01 \text{ lb}_f\text{-sec}^2/\text{inch} \qquad x_{o0} = +1.0 \text{ inch}$$
$$B = 0$$

This is clearly a system with dry friction only. The required CSMP statements would be

```
TITLE      DRY AND VISCOUS FRICTION
           XDOT2 = (−1.*B*XDOT1−1.*KS*X + FDF)/M
           XDOT1 = INTGRL (0.0, XDOT2)
```

$$X = INTGRL \ (1., XD\overline{O}T1)$$
$$FDF = FCNSW \ (XD\overline{O}T1, FFP, 0.0, FFM)$$
PARAM $M = .01, B = 0.0, KS = 100., FFP = 5., FFM = -5.$
METH$\overline{O}$D   RKSFX
TIMER $DELT = .0005, FINTIM = .05, \overline{O}UTDEL = .005$
PRTPLT $X(FDF, XD\overline{O}T1)$
END
ST$\overline{O}$P
ENDJ$\overline{O}$B

The statement defining the dry friction force FDF uses a feature of CSMP not previously discussed, the function switch FCNSW. Its general form is $Z = $ FCNSW (Y, Z1, Z2, Z3) and it enforces the relation between Z and Y shown in Fig. 8–11; that is, if $Y > 0, Z = Z3$, if $Y < 0, Z = Z1$,

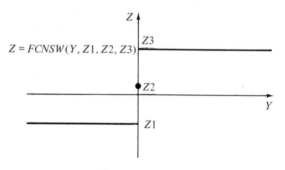

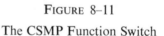

FIGURE 8–11

The CSMP Function Switch

and if $Y = 0, Z = Z2$. In our problem we use it to make the dry friction force depend on velocity XD$\overline{O}$T1 in the way desired. (Note that this feature easily handles an unsymmetrical dry friction; that is, one which has a different force for positive velocities than for negative.) When this problem was first attempted with the usual Runge-Kutta variable step-size integration algorithm it would not run to completion since the nature of the dry-friction model caused the built-in accuracy criterion to continuously cut down the step size until a built-in lower limit kicked the problem off the computer. This algorithm could probably have been altered to overcome this difficulty, but a simple change to the Runge-Kutta *fixed* step-size routine (called by the statement METH$\overline{O}$D RKSFX) solved the problem, with results as in Fig. 8–12. This figure also displays the response of this system with no dry friction and $B = 0.2$ and also when *both* dry and viscous friction are present, but at one-half their previous "strengths" so as to make the "overall" damping comparable to the first two cases.

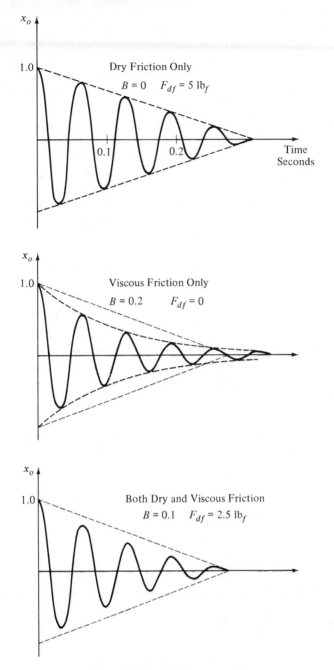

FIGURE 8–12

Effect of Dry and Viscous Friction

365

When interpreting experimental test results, the features exhibited in Fig. 8–12 are quite useful in deciding what type of friction is present. When pure dry friction exists, the decay envelope is a perfectly straight line, rather than the exponential curve $e^{-\zeta \omega_n t}$ of viscous damping. (This has been proven analytically to be true in general, not just for our example.) To check for pure viscous friction, one can compute ζ from Eq. (8–38) for $n = 1, 2, 3, \ldots$ cycles. If the value of ζ is essentially the same for all values of n, then the friction is viscous. When both types of friction are present in significant amounts, it may be possible to estimate the dry component alone by measuring the force required to "break the system loose" when it is at rest. By using this estimate of F_{df} in our CSMP computer model, we can try various combinations of B and F_{df} until our computer-calculated response curve closely matches that measured. This procedure of trial-and-error adjustment of parameters in a computer model to match experimental response curves is a powerful general procedure in modeling dynamic systems. Another feature dry friction exhibits is that the frequency of natural oscillations is exactly ω_n, the *undamped* natural frequency, not $\omega_n \sqrt{1 - \zeta^2}$ as in viscous damped systems.

Ramp Response. Let us use the rotational system of Fig. 8–4b as an example for demonstrating the response of second-order systems to ramp inputs. In this system a ramp input requires that $\theta_i = \dot{\theta}_i t$; that is, the input shaft starts rotating with a fixed angular velocity $\dot{\theta}_i$ at $t = 0$. The system equation is

$$J\ddot{\theta}_o + B\dot{\theta}_o + K_s\theta_o = K_s\theta_i = K_s\dot{\theta}_i t \qquad (8\text{–}42)$$

with initial conditions $\theta_o = \dot{\theta}_o = 0$ at $t = 0^+$. We define the standard parameters by

$$K \triangleq 1.0 \text{ rad/rad} \qquad \omega_n \triangleq \sqrt{\frac{K_s}{J}} \frac{\text{rad}}{\text{time}}$$

$$\zeta \triangleq \frac{B}{2\sqrt{JK_s}}$$

to get

$$\left(\frac{D^2}{\omega_n^2} + \frac{2\zeta D}{\omega_n} + 1 \right) \theta_o = K\dot{\theta}_i t \qquad (8\text{–}43)$$

The solutions are found to be for the overdamped case

$$\theta_o = K\dot{\theta}_i t - \frac{2\zeta\dot{\theta}_i K}{\omega_n} \left[1 + \frac{2\zeta^2 - 1 - 2\zeta\sqrt{\zeta^2 - 1}}{4\zeta\sqrt{\zeta^2 - 1}} e^{(-\zeta + \sqrt{\zeta^2 - 1})\omega_n t} \right.$$

$$\left. + \frac{-2\zeta^2 + 1 - 2\zeta\sqrt{\zeta^2 - 1}}{4\zeta\sqrt{\zeta^2 - 1}} e^{(-\zeta - \sqrt{\zeta^2 - 1})\omega_n t} \right] \qquad (8\text{–}44)$$

for the critically damped case

$$\theta_o = K\dot\theta_i t - \frac{2\dot\theta_i K}{\omega_n}\left[1 - e^{-\omega_n t}\left(1 - \frac{\omega_n t}{2}\right)\right] \qquad (8\text{–}45)$$

and for the underdamped case

$$\theta_o = K\dot\theta_i t - \frac{2\zeta\dot\theta_i K}{\omega_n}\left[1 - \frac{e^{-\zeta\omega_n t}}{2\zeta\sqrt{1 - \zeta^2}}\sin(\omega_n\sqrt{1 - \zeta^2}\,t + \phi)\right],$$

$$\tan\phi = \frac{2\zeta\sqrt{1 - \zeta^2}}{2\zeta^2 - 1} \qquad (8\text{–}46)$$

In each case the steady-state (forced) response is given by

$$\theta_{o,ss} = K\dot\theta_i t - 2\zeta\dot\theta_i K/\omega_n \qquad (8\text{–}47)$$

and the steady output velocity $\dot\theta_{o,ss} = K\dot\theta_i$. For the present example $K = 1$, so that the output shaft turns at the same speed as the input in steady state. However, it lags behind in angular position by an amount $2\zeta\dot\theta_i K/\omega_n = B\dot\theta_i/K_s$. To check the correctness of this result we note that if θ_o is rotating at speed $\dot\theta_i$, the damper B requires a torque $B\dot\theta_i$ and this torque can come only through the spring, which must deflect $B\dot\theta_i/K_s$ radians to provide this torque. The inertia J has no effect at all in steady state, since it takes no torque to drive a pure inertia at a *constant* velocity. Transient behavior is affected by ζ and ω_n in a manner similar to that for a step input; for a given ζ the speed with which the steady state is achieved is directly proportional to ω_n, while for a given ω_n the value of ζ controls the overshoot and degree of oscillation. Figure 8–13 summarizes the ramp-input behavior for a general second-order system with input q_i and output q_o.

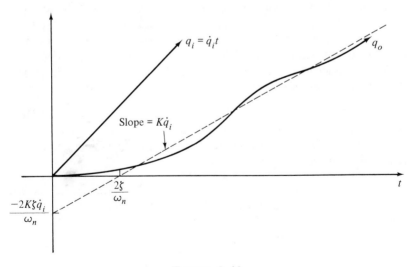

FIGURE 8–13

Ramp-Input Response of Second-Order System

Frequency Response. While the step response of the system of Fig. 8–5 might be useful in practical problems involving suddenly applied forces or "shocks," the frequency response is directly applicable to situations where oscillatory forces produce continuous vibrations which might cause mechanical damage or failure or lead to acoustic noise objectionable to human beings. A very common source of such vibration-inducing forces is unbalance in rotating or reciprocating machine parts. In a rotating part such as the rotor of an electric motor, unbalance consists of the center of mass not coinciding with the center of rotation determined by the motor bearings. To determine the force due to such unbalance the entire mass of the rotor is considered concentrated at the center of mass as in Fig. 8–14.

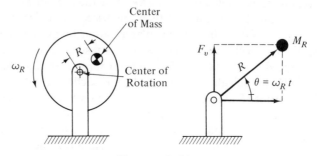

FIGURE 8–14

Rotating Unbalance as Source of Sinusoidal Force

Assuming rotation at a constant speed ω_R, the acceleration of the center of mass is directed radially inward and equal to $R\omega_R^2$. The force of the rotor *on* the bearings is thus $M_R R \omega_R^2$ directed radially outward. If the motor is attached to a structure capable of vibrating in the vertical direction, the vertical component of the unbalance force could excite these vibrations. This vertical component is clearly

$$F_v = M_R R \omega_R^2 \sin \omega_R t \qquad (8\text{–}48)$$

which we see to be a sinusoidal driving force.

To find the steady-state sinusoidal response for the system of Fig. 8–5 we use the sinusoidal transfer function

$$\frac{x_o}{f_i}(i\omega) = \frac{K}{\dfrac{(i\omega)^2}{\omega_n^2} + \dfrac{2\zeta i\omega}{\omega_n} + 1} = \frac{K}{\sqrt{\left[1 - \left(\dfrac{\omega}{\omega_n}\right)^2\right]^2 + \dfrac{4\zeta^2\omega^2}{\omega_n^2}}} \underline{/\phi} \qquad (8\text{–}49)$$

$$\phi \triangleq \tan^{-1} \frac{2\zeta}{\dfrac{\omega}{\omega_n} - \dfrac{\omega_n}{\omega}} \qquad (8\text{–}50)$$

Note that, in contrast to the step and ramp responses, a single formula suffices for all values of ζ. For the undamped ($\zeta = 0$) system, the amplitude ratio approaches infinity as the driving frequency ω approaches the system natural frequency ω_n (see Fig. 8–15). If the amplitude ratio approaches infinity, this means that a very small input force (perhaps from a small

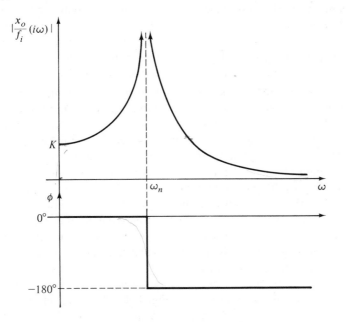

FIGURE 8–15

Frequency Response of Undamped Second-Order System

amount of rotating unbalance) can cause disastrously large vibrations. This *resonance* phenomenon is explained by noting that when $\omega = \omega_n$ the driving force is applied in exact synchronism with the natural motions and thus builds up their amplitude in the same manner as someone pushing a swing is able to add a little energy each cycle and achieve a large final motion with relatively small forces. Of course any real system must have at least a little damping, and thus the behavior for $\zeta = 0$ is somewhat academic; however, real systems with ζ as small as 0.01 are not unusual. Calculus maximization procedures applied to (8–49) show that the resonant peak occurs at a frequency ω_p given by

$$\omega_p = \omega_n \sqrt{1 - 2\zeta^2} \qquad (8\text{–}51)$$

and the amplitude ratio is

$$\frac{x_o}{f_i}(i\omega_p) = \frac{K}{2\zeta\sqrt{1 - \zeta^2}} \qquad (8\text{–}52)$$

Since a force f_i applied statically ($\omega = 0$) gives a deflection Kf_i, the *resonant magnification factor* is $1/2\zeta\sqrt{1 - \zeta^2}$. For $\zeta = 0.01$, for example, a driving force can produce 50 times the motion at resonance that it would statically.

From Eq. (8–51) we see that a resonant peak exists only if $\zeta < 1/\sqrt{2} = 0.707$. For damping greater than this the amplitude ratio decreases monotonically to zero as ω goes to infinity; thus there is no magnification. Note that the frequency of peak response shifts farther below ω_n as ζ is increased, and that this frequency is *not* the same as the frequency of damped natural oscillations $\omega_n\sqrt{1 - \zeta^2}$. Also, for a step input, overshoots occur if $\zeta < 1$, whereas for sinusoidal input a resonant peak occurs only if $\zeta < 0.707$. Figure 8–16 shows a set of universal second-order curves obtained from Eq. (8–49) by plotting x_o/Kf_i versus ω/ω_n for various values of ζ. The logarithmic plotting methods introduced with first-order systems are also applicable here. Again the low-frequency asymptote for amplitude ratio is the zero-dB line but now the high-frequency asymptote is a line of slope -40 dB/decade, with the two asymptotes intersecting at $\omega = \omega_n$. Since the corrections to the straight-line approximations depend on ζ, we now require a family of correction curves covering the usual range of ζ encountered. Figure 8–17 gives these and also the phase angle curves. Note that these are all plotted against ω/ω_n so as to be usable for any ω_n at all.

To illustrate the use of these methods, suppose we wish to establish response curves for a system as in Fig. 8–18. There an electric motor weighing 80 lb_f is mounted on a steel beam. Using the formula of Fig. 2–7c, the spring constant at the center of the beam, where the motor is located, is calculated to be $K_s = 10,000$ lb_f/inch. This would make

$$\omega_n = \sqrt{\frac{10,000}{80/386}} = 220 \frac{\text{rad}}{\text{sec}}, \qquad f_n = \frac{\omega_n}{2\pi} = 35 \text{ Hz} \qquad \textbf{(8–53)}$$

Since no intentional damper exists in this system, the damping is due to effects such as friction in bolted joints, metal hysteresis, etc., and is practically impossible to calculate from theory. To estimate the damping and also to check the theoretical frequency calculation, a vibration pickup is mounted on the beam and its output recorded as the beam is struck with a large rubber mallet. Data analysis gives $\zeta \approx 0.05$ and $f_n = 31.3$ Hz. The lower actual frequency is due to the inertia of the beam itself (neglected in our model) and the reduced stiffness of the beam spring because its ends are not perfectly fixed as the theory assumes. (The effect of beam mass has been theoretically calculated for many types of beams and may be found in vibration texts. The correction consists of adding a certain fraction of the beam's mass to the main mass.)

Having reasonably accurate values of $K = 1/K_s$, ζ and ω_n, we can now plot logarithmic frequency-response curves as in Fig. 8–19. Since $K = $

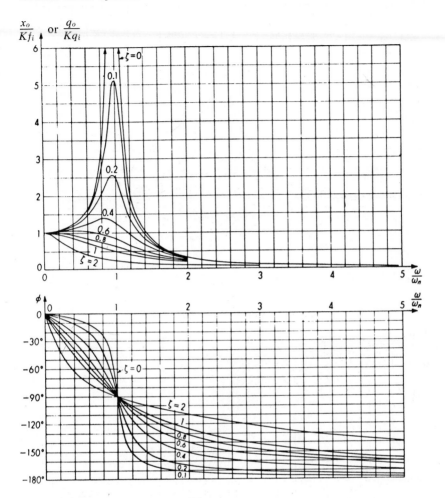

FIGURE 8–16

Second-Order Frequency-Response Curves

0.0001 inch/lb$_f$ = −80 dB, we simply "superimpose" the amplitude ratio curve for ζ = 0.05 in Fig. 8–17 on the −80-dB line. This is done by locating the breakpoint of the two straight line asymptotes at f_n = 31.3 Hz and then transferring a few points at convenient values of ω/ω_n on Fig. 8–17 to Fig. 8–19. For example, at ω/ω_n = 0.7, the correction is 6 dB above the asymptote; this point plots at f = 0.7 f_n = 21.9 Hz on Fig. 8–19. To illustrate the use of these curves, suppose the rotor of the motor weighs 40 lb$_f$, has the center of mass 0.01 inch away from the center of rotation, and turns at 1750 rpm. Due to the unbalance the exciting force is

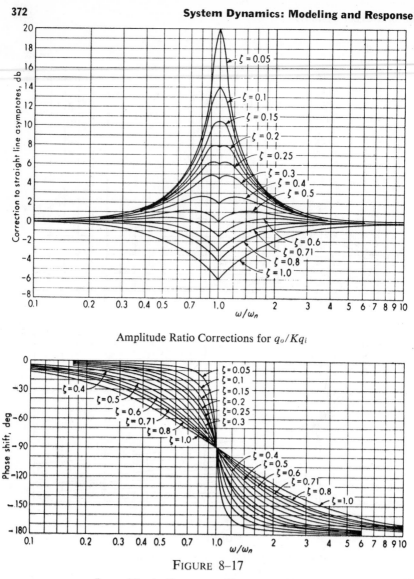

Amplitude Ratio Corrections for q_o/Kq_i

FIGURE 8–17

Logarithmic Frequency-Response Curves

$$f_i = \frac{(40)(0.01)(183)^2}{386} \sin(183t) = 34.8 \sin(183t) \quad \text{lb}_f \qquad \textbf{(8–54)}$$

At $\omega = 183$ ($f = 29.2$ Hz) the amplitude ratio x_o/f_i is -64 dB $= -60$ dB $- 4$ dB $= (0.001)(0.63) = 0.00063$ in./lb$_f$. Note that since -64 dB is not in our dB conversion table (Fig. 7–18) we break it up into -60 dB (which is obviously 10^{-3}) and -4 dB (which is in the table). The vibration amplitude x_o is thus $(34.8)(0.00063) = 0.022$ in.

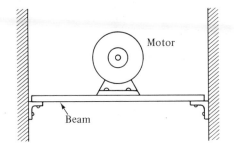

FIGURE 8–18

Practical Vibration Problem

If this vibration is judged excessive, a number of changes might be implemented to reduce it. A better balanced rotor ($R < 0.01$) would reduce f_i;
however, more careful balancing will cost money. If f_i is not reduced, we
might add a damper; Fig. 8–17 shows that for $\zeta > 0.5$, x_o/Kf_i is reduced to
0.0001 or less, but a damper also costs money and will require some system
redesign to accommodate it. Finally, since we see that the system is operating
very close to resonance ($\omega \approx \omega_n$), if we could separate the driving and
natural frequencies more, a reduction of vibration would be achieved. If
we make $\omega \ll \omega_n$, Fig. 8–19 shows that $x_o/Kf_i = 0.0001$ is the best that we
can do, while if $\omega \gg \omega_n$, the motion may be reduced to any level we wish.
In separating ω and ω_n, either or both of the frequencies might be changed;
however, if the motor speed of 1750 rpm is required by some machine which
the motor drives, we may not be allowed to vary ω.

Suppose we need to reduce the motion to 0.002 inch by changing ω_n and
that this will be done by increasing mass M (bolt a weight to the beam
below the motor) and not changing K_s or the damping. To get 0.002 inch
we need $x_o/Kf_i = 0.000057 = -84.9$ dB which Fig. 8–19 shows to occur at
frequency 52 Hz, which corresponds to $\omega/\omega_n = 52/31.3 = 1.66$. If ω is
fixed at 183 rad/sec we thus require $\omega_n = 183/1.66 = 110$ rad/sec and we
must add enough mass to achieve this. (This may or may not be practical
in an actual application in terms of space available, added stress in the
beam due to the dead weight, etc.; however, let us proceed on the assumption that it could be done.) We should point out that since $\zeta = B/2\sqrt{K_s M}$,
an increase in M will result in a decrease in ζ, giving a higher resonant
peak; however, at $\omega/\omega_n = 1.66$, even for $\zeta = 0$ the amplitude ratio is very
close to that for $\zeta = 0.05$ and thus the predicted motion should be quite
accurate.

While the above analysis shows the motion to be within the acceptable
limit for steady-state operation, since the operating frequency (183) is
above the natural frequency (110) it means that as the motor is brought up

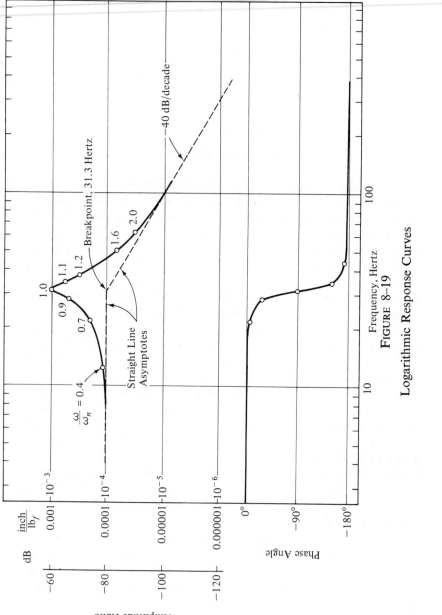

FIGURE 8-19

Logarithmic Response Curves

to speed when initially turned on, we must "pass through" the natural frequency. Intuitively one would guess that if we pass through the resonance region quickly enough there will not be time for a large transient motion to occur; however, how quickly is "quickly enough"? This may be answered analytically by writing an expression for exciting force f_i which corresponds to the acceleration of the motor from rest to operating

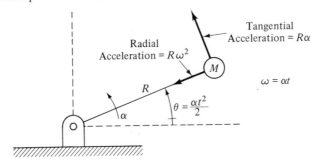

FIGURE 8–20

Accelerating Rotor

speed and then solving Eq. 8–18 using this f_i. Figure 8–20 shows that when the rotor is accelerating with constant angular acceleration α, its angular velocity is αt and its angular displacement $\theta = \alpha t^2/2$. The vertical component of the force acting on the rotor bearings is our force f_i and is given by

$$f_i = MR\omega^2 \sin\theta - MR\alpha \cos\theta = MR\left[\alpha^2 t^2 \sin\left(\frac{\alpha t^2}{2}\right) - \alpha\cos\left(\frac{\alpha t^2}{2}\right)\right]$$

$$(8\text{–}55)$$

We will assume this equation to hold from $t = 0$ until ω reaches the operating speed of 1750 rpm, at which time α becomes zero and f_i becomes $MR\omega_{ss}^2 \sin\omega_{ss}t$, where $\omega_{ss} = (1750)(2\pi/60) = 183$ rad/sec. To the author's knowledge this problem has not been solved analytically, yet it yields readily to the CSMP simulation method and makes a good example for demonstrating its utility.

```
TITLE    VIBRATION STARTUP TRANSIENTS
         FI=M*R*(ALPH*ALPH*TIME*TIME*SIN(ALPH*
             TIME*TIME/2.)−...
         ALPH*COS(ALPH*TIME*TIME/2.))
         FIA=FI*(STEP(0.0)−STEP(T1))+STEP(T1)*
             OMEGAF*OMEGAF*...
         SIN(OMEGAF*TIME)*M*R
         X=CMPXPL(XO,XDOTO,ZETA,OMEGA,FIA)
         X1=OMEGA*OMEGA*K*X
PARAM    K=.0001,ALPH=91.5,T1=2.,OMEGAF=183.
PARAM    XO=0.0,XDOTO=0.0,R=.01,M=.1037,OMEGA=110.
```

```
PARAM   ZETA=.05
TIMER   FINTIM=2.3,DELT=.0001,OUTDEL=.002
PRTPLT  X1
END
PARAM   ZETA=0.01
END
STOP
ENDJOB
```

Here the statement FI= is merely FORTRAN for Eq. (8–55). It is necessary, however, to "turn off" this force at the time (call it T1) when operating speed is reached and "turn on" the steady-state force. The statement FIA= does this, as can be seen from Fig. 8–21. [The three dots

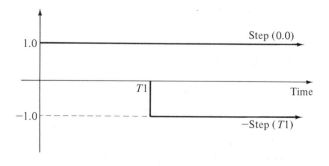

Figure 8–21

Use of STEP to Turn Functions On and Off

at the right of the first lines of the FI and FIA statements are the CSMP code for continuation cards. That is, both FI and FIA would not fit on a single card (only columns up to 72 are allowed) thus they spill over onto continuation cards signalled by the three periods.] The statement X = CMPXPL(XO,XDOTO,ZETA,OMEGA,FIA) is the standard CSMP for a second-order system with transfer function

$$\frac{X}{FIA}(D) = \frac{1}{D^2 + 2\zeta\omega_n D + \omega_n^2} \tag{8–56}$$

Note that the CSMP standard form differs from ours and thus X must be multiplied by $K\omega_n^2$ (as in the X1= statement) to get the actual motion x_o, called X1 in the program. Initial conditions are XO and XDOTO while $\zeta \triangleq$ ZETA and $\overline{\text{OMEGA}} \triangleq \omega_n$. The time to accelerate to operating speed is taken as 2 seconds and two runs are to be made, one with $\zeta = 0.05$ and one with $\zeta = 0.01$ to check our earlier conjecture about the effect of changing M on ζ. Figure 8–22 shows that with $\zeta = 0.05$ a peak motion of about 0.01 inch occurs at about $t = 1.37$, while $\zeta = 0.01$ gives 0.018 at $t = 1.42$. Both eventually go into the 0.002 inch steady state; however, the $\zeta = 0.01$

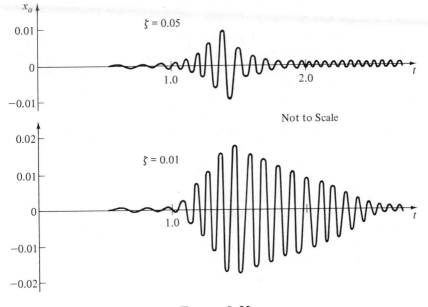

FIGURE 8–22

Vibration Startup Transients

system has a larger number of large transient oscillations. A more rapid acceleration to full speed would reduce the transients.

Impulse Response. Figure 8–23 shows an analog computer diagram for Eq. (8–18) when f_i is an arbitrary driving force. When f_i is an impulse of area A_i we cannot, of course, produce a voltage representing this; however,

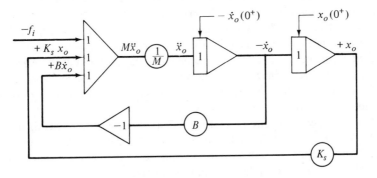

FIGURE 8–23

Second-Order Analog Computer Setup

the computer diagram shows what must be done to get the same effect. For f_i an impulse, the signal $\ddot{x}_o$ would be an impulse of area A_i/M, and thus the output of the first integrator suddenly jumps up to $-A_i/M$. We see that this is exactly the same as if we had said f_i was zero but had an initial velocity of magnitude A_i/M. Thus to find the impulse response we can set $f_i \equiv 0$ and simply set the initial condition on the first integrator at $-A_i/M$ and that on the second integrator at zero. This also tells us that in solving the equation analytically we can proceed in exactly the same way. For the underdamped case the solution is easily found to be

$$x_o = \frac{A_i}{\sqrt{K_s M(1 - \zeta^2)}}\, e^{-\zeta \omega_n t} \sin\left(\sqrt{1 - \zeta^2}\,\omega_n t\right) \tag{8–57}$$

Critically damped and overdamped cases are also readily solved, and all these results are presented in nondimensional form in Fig. 8–24. Just as in first-order systems, real force pulses will produce essentially the same response as perfect impulses of the same area if the pulse duration is "short enough." "Short enough" now means that the pulse duration T must be about $1/10$ or less of the oscillation period $2\pi/\omega_n$.

8–3. ELECTRICAL SECOND-ORDER SYSTEMS.

Low-Pass Filter. Figure 8–25 shows several electrical systems which will be found to follow the general second-order relation (Eq. 8–7) between the stated input and output quantities. The circuit of Fig. 8–25a is useful as a low-pass filter with a sharper cutoff than its first-order relative of Fig. 7–23a. Low-pass filters are intended to allow signals with frequencies below a certain range to pass through while rejecting (attenuating) signals above this range. When the pass-band and the reject-band are widely spread, for example pass 0 to 10 Hz and reject 1000 to ∞, then even a simple first-order filter can do a good job. In many practical applications, however, it is advantageous to be able to discriminate between quite closely spaced frequencies, and filters approaching the perfect cutoff characteristics are needed. Figure 8–26 shows how first- and second-order RC filters compare in this respect. (Can you guess what addition of a *third RC* section would do?)

Let us analyze the circuit of Fig. 8–25a, shown in more detail in Fig. 8–27. We have there defined two currents, i_1 and i_2. If we can find i_2 in terms of e_i, then $(e_o/e_i)(D)$ follows immediately from $e_o = (1/C_2 D)i_2$. Kirchhoff's voltage loop law around $abcda$ gives

$$i_1 R_1 + \frac{1}{C_1 D}(i_1 - i_2) - e_i = 0 \tag{8–58}$$

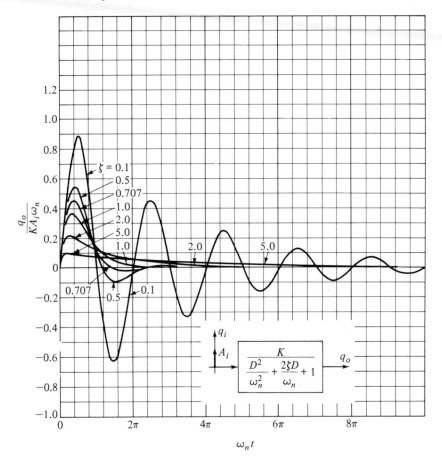

FIGURE 8–24

Second-Order Impulse Response

where we have used the current node law at b to show that the current in C_1 is $(i_1 - i_2)$. Equation (8–58) cannot be solved because there are two unknowns, i_1 and i_2; thus another equation is needed. This we get by traversing the loop *befcb*.

$$i_2 R_2 + \frac{i_2}{C_2 D} + \frac{1}{C_1 D}(i_2 - i_1) = 0 \qquad \text{(8–59)}$$

Since (8–59) has the same two unknowns as (8–58), we now have two equations in two unknowns and can solve for either or both i_1 and i_2. Our interest in e_o leads us to find i_2 and we shall use determinants to reduce our two equations in two unknowns to one equation involving only i_2. Multiply (8–58) by $C_1 D$ and gather terms to get

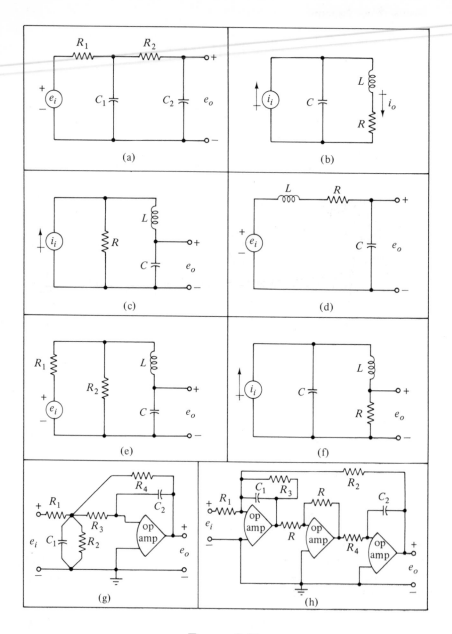

FIGURE 8–25

Some Electrical Second-Order Systems

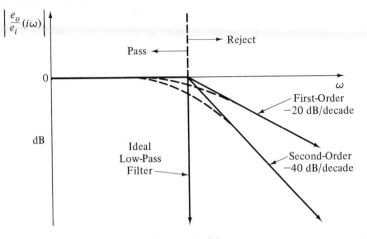

FIGURE 8–26

Comparison of Ideal and Actual Low-Pass Filters

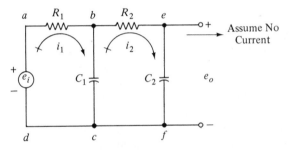

FIGURE 8–27

Low-Pass Filter Circuit

$$(R_1C_1D + 1)i_1 + (-1)i_2 = (C_1D)e_i \qquad \textbf{(8–60)}$$

and then multiply (8–59) by C_1C_2D to get

$$(-C_2)i_1 + (C_1C_2R_2D + C_1 + C_2)i_2 = 0 \qquad \textbf{(8–61)}$$

By determinants

$$i_2 = \frac{\begin{vmatrix} R_1C_1D + 1 & C_1De_i \\ -C_2 & 0 \end{vmatrix}}{\begin{vmatrix} R_1C_1D + 1 & -1 \\ -C_2 & C_1C_2R_2D + C_1 + C_2 \end{vmatrix}} \qquad \textbf{(8–62)}$$

$$= \frac{C_2De_i}{R_1R_2C_1C_2D^2 + (R_1C_1 + R_1C_2 + R_2C_2)D + 1}$$

and thus

$$\frac{i_2}{e_i}(D) = \frac{C_2 D}{R_1 R_2 C_1 C_2 D^2 + (R_1 C_1 + R_1 C_2 + R_2 C_2)D + 1} \quad (8\text{–}63)$$

and since $e_o = (1/C_2 D)i_2$

$$\frac{e_o}{e_i}(D) = \frac{1}{R_1 R_2 C_1 C_2 D^2 + (R_1 C_1 + R_1 C_2 + R_2 C_2)D + 1} \quad (8\text{–}64)$$

Since this is clearly our standard second-order form, we define

$$K \triangleq 1 \frac{\text{volt}}{\text{volt}} \qquad \omega_n \triangleq \sqrt{\frac{1}{R_1 R_2 C_1 C_2}} \frac{\text{rad}}{\text{time}}$$

$$\zeta \triangleq \frac{R_1 C_1 + R_1 C_2 + R_2 C_2}{2} \sqrt{\frac{1}{R_1 R_2 C_1 C_2}} \quad (8\text{–}65)$$

By writing

$$\zeta^2 = \frac{Z}{4} + \frac{1}{4Z} + \frac{Y}{2} + \frac{W}{2} + YW + \frac{1}{2} \quad (8\text{–}66)$$

where

$$Z \triangleq \frac{R_1 C_1}{R_2 C_2} \qquad Y \triangleq \frac{R_1}{R_2} \qquad W \triangleq \frac{C_2}{C_1} \quad (8\text{–}67)$$

we can show that ζ must always exceed 1.0 for this circuit, since the minimum value of $(Z/4 + 1/4Z)$ occurs at $Z = 1.0$ and is $1/2$. Since Y and W must both be positive, (8–66) shows that ζ^2 approaches 1.0 from above if $Z = 1.0$ and Y and W both approach zero. Note that these conditions are precisely those which would give the cascade combination of two first-order filters with identical time constants and negligible loading effect. Equation (8–66) also shows that ζ may be made arbitrarily large. While it is perfectly correct to define ω_n as in (8–65), with ζ always > 1.0 it is clear that there can never be any natural oscillations in this circuit and it is in fact generally true that *any* passive circuit (one without amplifiers) must have *both* capacitors and inductors to be capable of oscillation. Since inductors have many practical limitations, the capability of active circuits, such as those using op-amps, for producing oscillatory behavior with only R and C is of considerable importance.

Since $\zeta > 1.0$ the step response would be given by Eq. (8–29); however, we may wish to rewrite this as

$$e_o = e_{is}[1 - C_1 e^{-t/\tau_1} + C_2 e^{-t/\tau_2}] \quad (8\text{–}68)$$

where

$$C_1 \triangleq \frac{\zeta + \sqrt{\zeta^2 - 1}}{2\sqrt{\zeta^2 - 1}} \qquad C_2 \triangleq \frac{\zeta - \sqrt{\zeta^2 - 1}}{2\sqrt{\zeta^2 - 1}} \quad (8\text{–}69)$$

$$\tau_1 \triangleq \frac{1}{\omega_n(\zeta - \sqrt{\zeta^2 - 1})} \qquad \tau_2 \triangleq \frac{1}{\omega_n(\zeta + \sqrt{\zeta^2 - 1})} \quad (8\text{–}70)$$

The basic differential equation can then also be written as

$$(\tau_1 D + 1)(\tau_2 D + 1)e_o = (\tau_1\tau_2 D^2 + (\tau_1 + \tau_2)D + 1)e_o = Ke_i \quad (8\text{--}71)$$

which shows that the frequency response may be most easily calculated and graphed as a superposition of two first-order terms.

$$\frac{e_o}{e_i}(i\omega) = \frac{1}{(i\omega\tau_1 + 1)(i\omega\tau_2 + 1)} \quad (8\text{--}72)$$

Figure 8–26 showed a second-order filter with two *identical* time constants; with the present circuit this can be closely approached but not quite reached since it requires $\zeta = 1.0$. When τ_1 and τ_2 are not equal the frequency response is as in Fig. 8–28.

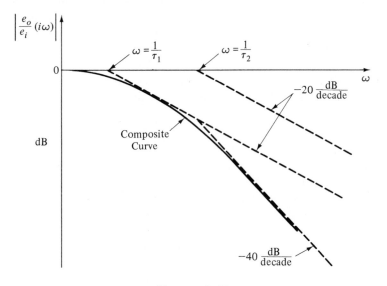

FIGURE 8–28

Overdamped Second-Order Graphed as Two First-Order

Series Resonant Circuit. Depending on the values of R, L, and C used, the system of Fig. 8–25d might serve as a model for a number of practical devices. When R is very small and L and C are chosen to give a desired natural frequency, the circuit is said to be "tuned" or series resonant. This means it will magnify a narrow band of frequencies relative to those below and above and can thus "pick out" a certain frequency, exactly what is needed, for example, in tuning a radio receiver to a station operating at a known frequency. Let us obtain $(e_o/e_i)(D)$ by impedance methods. The impedance seen by the voltage source is

$$Z(D) = LD + R + \frac{1}{CD} = \frac{LCD^2 + RCD + 1}{CD} = \frac{e_i}{i}(D) \quad (8\text{--}73)$$

and since $e_o = (1/CD)i$, we get

$$\frac{e_o}{e_i}(D) = \frac{1}{LCD^2 + RCD + 1} = \frac{K}{\dfrac{D^2}{\omega_n^2} + \dfrac{2\zeta D}{\omega_n} + 1} \qquad (8\text{-}74)$$

where

$$K \overset{\triangle}{=} 1 \frac{\text{volt}}{\text{volt}} \qquad \omega_n \overset{\triangle}{=} \sqrt{\frac{1}{LC}} \frac{\text{rad}}{\text{time}} \qquad \zeta \overset{\triangle}{=} \frac{R}{2\sqrt{LC}} \qquad (8\text{-}75)$$

From the expression for ζ we see that in this circuit the full range of behavior from undamped to overdamped is theoretically possible. Of course, just as friction B cannot really be made zero in mechanical systems, resistance R cannot be zero in a real electric circuit, and thus $\zeta = 0$ cannot be realized even if we do not intentionally "wire in" a resistor, because all inductors have parasitic resistance. When we look at op-amp circuits we will find that $\zeta \approx 0$ can be realized without any inductors at all.

If we now consider the current i produced by the input voltage e_i the impedance Z of Eq. (8–73) is of interest. For an AC system $e_i = E_i \sin \omega t$ and we want $Z(i\omega)$.

$$Z(i\omega) = i\omega L + \frac{1}{i\omega C} + R = R + i\left(\omega L - \frac{1}{\omega C}\right) \qquad (8\text{-}76)$$

Considering an input voltage of fixed amplitude but adjustable frequency, the maximum current will occur when impedance is a minimum. Since the resistive component of impedance is constant at R, we must minimize the reactive component $(\omega L - 1/\omega C)$. This will clearly be zero when

$$\omega L = \frac{1}{\omega C} \qquad \omega^2 = \frac{1}{LC} \qquad \omega = \sqrt{\frac{1}{LC}} \qquad (8\text{-}77)$$

The peak current will thus occur when $\omega = \sqrt{1/LC}$ irrespective of the value of R. Note that this "current resonance" differs from the "voltage resonance" of Eq. (8–74) in several ways. First, $(e_o/e_i)(i\omega)$ exhibits no peak at all unless $\zeta < 0.707$, while the current always has a peak. Secondly the current peak is always at the undamped natural frequency $\omega_n = \sqrt{1/LC}$ while the voltage peak (if there is one) occurs at $\omega_n\sqrt{1 - 2\zeta^2}$. Also, Eq. (8–76) shows that, at current resonance, e_i and i are precisely in phase with each other since the impedance is pure resistive at this one frequency. That is, the inductive reactance ωL and capacitive reactance $1/\omega C$ just cancel each other. Figure 8–29 shows how impedance varies with frequency.

As a numerical example, consider a circuit with $R = 22\ \Omega$, $L = 0.1$ H, negligible C and with e_i taken to be 110 volt RMS, 60 Hz from a "wall plug." We have

$$Z(i\omega)\big|_{\omega=377} = 22 + i37.7 = 43.6 \,\underline{/+59.8^\circ}\ \text{ohms} \qquad (8\text{-}78)$$

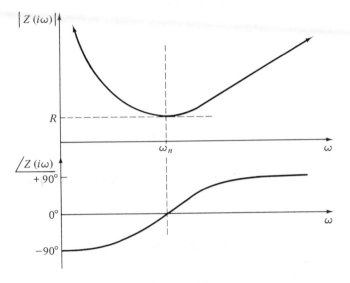

FIGURE 8-29

Impedance of Series-Resonant Circuit

The RMS current is thus $110/43.6 = 2.52$ amp and the power factor is $\cos 59.8° = 0.503$. The product of RMS voltage and current is called the *volt-amperes* and is 277 while the actual power consumed is the product of volt-amperes and power factor, in this case 139.5 watts. We can check this by computing the power I^2R in the resistor, which should be 139.5 watts also since the inductance does not actually consume any power. In AC power circuits such as those including motors, transformers, induction furnaces, etc., where the loads are inductive and the power factor lagging, one often adds capacitance to bring the power factor closer to 1.0. To get a power factor of 1.0 in our present circuit we need to add a capacitor such that

$$\frac{1}{377C} = 37.7 \qquad C = 70.4 \ \mu F \tag{8-79}$$

Then $Z = 22 \ \underline{/0°}$ ohms, the current is 5 amps RMS, and both the volt-amperes and the consumed power are 550 watts. With the power factor adjusted to 1.0, the same line voltage can now supply more power to the load.

Actually, in most practical applications the inductive "load" (series combination of L and R) must remain connected directly across the power line even after the power-factor-correcting capacitor is added; thus it must be connected in parallel as in Fig. 8-30. For this circuit

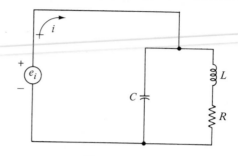

FIGURE 8–30

Use of Parallel Capacitor for Power-Factor Correction

$$Z(D) = \frac{e_i}{i}(D) = \frac{\frac{1}{CD}(R + LD)}{\frac{1}{CD} + R + LD} = \frac{LD + R}{LCD^2 + RCD + 1} \quad (8\text{–}80)$$

and the frequency response

$$Z(i\omega) = \frac{i\omega L + R}{(1 - LC\omega^2) + iRC\omega} \quad (8\text{–}81)$$

will have a zero phase angle (power factor = 1.0) if the numerator phase angle equals that of the denominator.

$$\tan^{-1}\frac{\omega L}{R} = \tan^{-1}\frac{RC\omega}{1 - LC\omega^2}, \qquad \frac{\omega L}{R} = \frac{RC\omega}{1 - LC\omega^2} \quad (8\text{–}82)$$

$$C = \frac{L}{R^2 + L^2\omega^2} = \frac{0.1}{484 + 1420} = 52.5 \ \mu F \quad (8\text{–}83)$$

If this capacitor is connected in parallel the total impedance at $\omega = 377$ is 86.5 $\underline{/0°}$ ohms, making the current drawn from the power line $110/86.5 = 1.27$ amp RMS. Since the load is connected across the power line exactly the same as if C were not there, the *load* current and power are identical with our original case; however, the current drawn from the power line has been reduced from 2.52 amp to 1.27. This reduction in line current is really the main reason for correcting the power factor to near 1.0 and explains why the power company encourages users to do this by granting better rates if they do. Basically, generating and transmission equipment is sized to handle a required current; thus if currents can be kept down while still providing the required consumed power, the equipment is more efficiently utilized. When the power factor is far from 1.0, large amounts of power are pumped into the load in one part of the cycle but most of it is returned to the line in the next part. Thus the current is large without really providing much useful power to the load.

Op-Amp Circuits. Figures 8–25g and h show two active circuits using operational amplifiers which will exhibit second-order behavior. As mentioned earlier, such active circuits can give oscillatory response even though no inductors whatever are involved. The circuit of Fig. 8–25h uses three op-amps connected as summing integrator, inverter, and integrator, with proper feedback to the summer much as it would be done in a general-purpose analog computer. While this circuit is flexible and easily adjusted to a wide range of parameters, it is rather wasteful of components and expensive, and thus might not be preferred for some special-purpose applications where cost was a factor and flexibility was less significant. Since 8–25g also uses a configuration different from that (Fig. 3–26a) which we have assumed up to now, let us analyze it to show the techniques involved.

Figure 8–31 shows this circuit with parameter values as recommended[1] by one of the large op-amp manufacturers. While the circuit appears

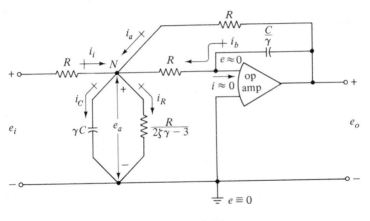

FIGURE 8–31

Single-Amplifier Second-Order Op-Amp Circuit

complex at first glance it yields readily to application of the current node law at the location N.

$$i_b + i_i + i_a = i_C + i_R \qquad (8\text{–}84)$$

The voltage e_a at node N is found from

$$e_a = -i_b R = \left(-\frac{C}{\gamma} De_o\right) R \qquad (8\text{–}85)$$

since the capacitor of value C/γ has one end at e_o and the other at the op-

---

[1] *The Lightning Empiricist*, G. A. Philbrick Researches, Vol. 13, Nos. 1, 2, 3, 4, Dedham, Mass., 1965.

amp input terminal, which we recall to be a virtual ground; that is, nearly at zero potential. Since the op-amp input current is also nearly zero, the current in C/γ must also go through R to reach N. Since the right end of R is at zero potential, N must be at potential $-i_b R$. We can now express all the currents in Eq. (8–84) in terms of e_i and e_o and thereby get the desired relation between e_i and e_o.

$$i_a = \frac{e_o - e_a}{R} = \frac{\left(1 + \dfrac{RC}{\gamma} D\right) e_o}{R}, \qquad i_i = \frac{e_i - e_a}{R} = \frac{e_i + \dfrac{RC}{\gamma} De_o}{R}$$

$$\text{(8–86)}$$

$$i_C = \gamma C De_a = -RC^2 D^2 e_o, \qquad i_R = \frac{e_a}{R/(2\varsigma\gamma - 3)} = -\frac{C(2\varsigma\gamma - 3)}{\gamma} De_o$$

$$\text{(8–87)}$$

Substitution in (8–84) leads to

$$(R^2 C^2 D^2 + 2\varsigma RCD + 1)e_o = -e_i \qquad \text{(8–88)}$$

and thus

$$\frac{e_o}{e_i}(D) = \frac{K}{\dfrac{D^2}{\omega_n^2} + \dfrac{2\varsigma D}{\omega_n} + 1} \qquad \text{(8–89)}$$

where

$$K \triangleq -1\,\frac{\text{volt}}{\text{volt}} \qquad \omega_n \triangleq \frac{1}{RC}\frac{\text{rad}}{\text{sec}} \qquad \varsigma \triangleq \varsigma \qquad \text{(8–90)}$$

In designing a circuit of this type, recommendations given in the reference of footnote one are helpful. These include

$$\gamma \geq \frac{3}{2\varsigma} \qquad \varsigma > 0.87\sqrt{\frac{f_n}{P_{GB}}} \qquad \text{(8–91)}$$

where $f_n \triangleq \omega_n/2\pi$ and $P_{GB} \triangleq$ gain-bandwidth product of the op-amp in Hertz, a performance specification commonly quoted for op-amps and typically about 10^6 Hz for good quality amplifiers. To judge whether the circuit will load the device supplying the input voltage e_i we must know the input impedance at these terminals; it is approximately R. Using these guidelines let us design a circuit with $f_n = 100$ Hz, $\varsigma = 0.1$, and an input impedance of 50,000 Ω. Since R must be 50,000 to get the desired input impedance, Eq. (8–90) gives

$$C = \frac{1}{(50,000)(628)} = 0.0319 \times 10^{-6}\text{farad} = 0.0319\ \mu\text{F} \quad \text{(8–92)}$$

while (8–91) gives $\gamma \geq 15$, let us take $\gamma = 30$. Then the two required capacitors are $\gamma C = 0.955\ \mu$F and $C/\gamma = 0.0016\ \mu$F while the resistor $R/(2\varsigma\gamma - 3) = 16,600\ \Omega$. From (8–91) the gain-bandwidth product of our op-amp must be at least 7500, which is easily met.

The circuit type discussed above is useful for natural frequencies in the range of about 0.01 to 100,000 Hz and a wide range of ζ values from underdamped to overdamped. For very small or zero (undamped) values of ζ the more conventional analog computer type circuits as in Fig. 8–23 or 8–25h may be more appropriate. In these circuits $\zeta = 0$ is achieved by severing a feedback path completely, rather than by setting a component at a particular numerical value. They are thus able to realize $\zeta = 0$ with sufficient accuracy that they are regularly used as oscillators to generate sine and cosine waves in analog computer setups. In Fig. 8–23, for example, if we set $f_i = 0$, disconnect the feedback voltage $B\dot{x}_o$, set $-\dot{x}_o(0) = 0$ and $x_o(0) = 10$ volts, when the computer is turned on we will get $x_o = 10 \cos \omega_n t$, where $\omega_n = \sqrt{K_s/M}$.

8–4. FLUID SECOND-ORDER SYSTEMS.

Figure 8–32 displays some examples of fluid systems whose behavior, with respect to the labelled input and output quantities, will be found to follow the basic second-order equation (8–3). In Fig. 8–32a we see an example of cascaded first-order systems in which no loading effect whatsoever occurs. That is, the addition of the second tank has no influence at all on the response of the first. When connected as in 8–32b, the usual loading effect occurs. Figures 8–32c and d again might represent pressure-measuring systems as did their first-order counterparts in Fig. 7–31; however, we now no longer neglect fluid inertance in the tube.

Let us analyze the system of Fig. 8–32b. During a time interval dt a volume inventory for the left tank gives

$$q_i \, dt - \left[\frac{\gamma(h_1 - h_o)}{R_{f1}} \right] dt = A_{T1} dh_1 \tag{8–93}$$

where

$$\gamma \triangleq \text{specific weight of fluid}$$

$$h_1 \triangleq \text{level in left tank}$$

$$A_{T1} \triangleq \text{cross-section area of left tank}$$

$$R_{f1} \triangleq \text{fluid resistance between tanks}$$

Similarly, for the left tank,

$$\left[\frac{\gamma(h_1 - h_o)}{R_{f1}} - \frac{\gamma h_o}{R_{f2}} \right] dt = A_{T2} dh_o \tag{8–94}$$

where

$$R_{f2} \triangleq \text{outlet resistance}$$

$$A_{T2} \triangleq \text{cross-section area of right tank}$$

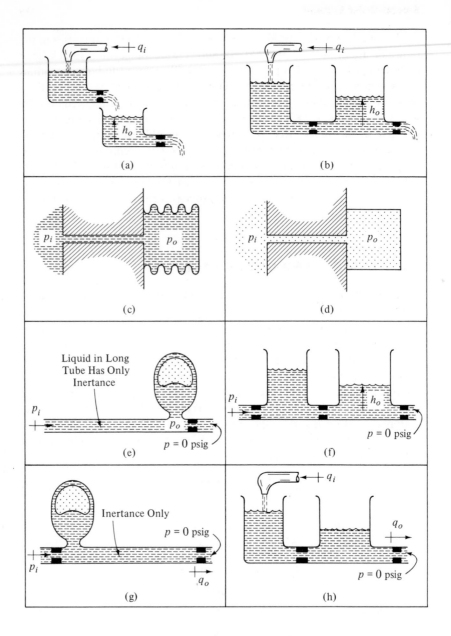

FIGURE 8–32

Some Fluid Second-Order Systems

These two equations may be written as

$$\left(\frac{R_{f1}A_{T1}}{\gamma}D + 1\right)h_1 + (-1)h_o = \left(\frac{R_{f1}}{\gamma}\right)q_i \qquad (8\text{–}95)$$

$$\left(-\frac{R_{f2}}{R_{f1} + R_{f2}}\right)h_1 + \left(\frac{A_{T2}R_{f1}R_{f2}}{\gamma(R_{f1} + R_{f2})}D + 1\right)h_o = 0$$

We can now easily solve for either or both h_1 and h_o; our interest here is in h_o so we write

$$h_o = \frac{\begin{vmatrix} \dfrac{R_{f1}A_{T1}}{\gamma}D + 1 & \dfrac{R_{f1}}{\gamma}q_i \\[2ex] -\dfrac{R_{f2}}{R_{f1} + R_{f2}} & 0 \end{vmatrix}}{\begin{vmatrix} \dfrac{R_{f1}A_{T1}}{\gamma}D + 1 & -1 \\[2ex] -\dfrac{R_{f2}}{R_{f1} + R_{f2}} & \dfrac{A_{T2}R_{f1}R_{f2}}{\gamma(R_{f1} + R_{f2})}D + 1 \end{vmatrix}}$$

$$= \frac{\dfrac{R_{f1}R_{f2}}{\gamma(R_{f1} + R_{f2})}q_i}{\dfrac{R_{f1}^2R_{f2}A_{T1}A_{T2}}{\gamma^2(R_{f1} + R_{f2})}D^2 + \left(\dfrac{R_{f1}A_{T1}}{\gamma} + \dfrac{A_{T2}R_{f1}R_{f2}}{\gamma(R_{f1} + R_{f2})}\right)D + \dfrac{R_{f1}}{R_{f1} + R_{f2}}} \qquad (8\text{–}96)$$

Cross-multiplying and simplifying gives

$$\left(\frac{R_{f1}R_{f2}A_{T1}A_{T2}}{\gamma^2}D^2 + \frac{A_{T1}R_{f1} + A_{T1}R_{f2} + A_{T2}R_{f2}}{\gamma}D + 1\right)h_o = \left(\frac{R_{f2}}{\gamma}\right)q_i$$
$$(8\text{–}97)$$

When a fairly complicated result such as Eq. (8–97) is obtained it is always desirable to check it for possible mistakes. Several types of checks may be applied to most situations; let us carry some of these out for the present problem to illustrate their general applicability. When an expression contains a sum of terms, all terms must have the same dimensions. Thus, in the Dh_o coefficient above, if the first term were $A_{T1}R_{f1}^2$ we would suspect a mistake. Furthermore, the dimensions of the four terms in (8–97) must also be consistent; for example, $R_{f2}q_i/\gamma$ must have the same dimensions as h_o. Does it? Check the Dh_o and D^2h_o terms also, recalling that D has dimensions $1/\text{time}$ and D^2 has $1/(\text{time})^2$. Another type of checking involves evaluation of special cases for which the answer is obvious or easily found. Often the steady-state response to a step input provides such a case. For q_i equal to a constant, say q_{is}, Eq. (8–97) predicts h_o will become steady at $h_o = q_{is}R_{f2}/\gamma$. Does this make sense? If h_o is steady at $q_{is}R_{f2}/\gamma$, the outflow of the right tank would be q_{is}, exactly the same as the inflow

q_i. Since inflow = outflow corresponds to an equilibrium condition, h_o could indeed remain steady at $q_{is}R_{f2}/\gamma$, verifying at least the steady-state aspect of Eq. (8–97). Another type of checking lets selected parameters become zero or infinity and then determines whether the general equation collapses into a simpler form corresponding to a known result. This *limiting case* check may be applied to the flow resistance R_{f1} by letting it become zero in Eq. (8–97), giving

$$\left(\frac{R_{f2}(A_{T1} + A_{T2})}{\gamma} D + 1 \right) h_o = \left(\frac{R_{f2}}{\gamma} \right) q_i \qquad (8\text{–}98)$$

A little reflection shows this to be the equation of a *single* tank of area $(A_{T1} + A_{T2})$ discharging through a resistance R_{f2}. When R_{f1} is zero, the levels in the two tanks are always identical; thus the two tanks really are equivalent to one with an area equal to the sum of the individual areas, and thus Eq. (8–97) appears to handle this limiting case correctly. Finally let us set $R_{f1} = 0$ (same as one big tank) and $R_{f2} = \infty$ (the outflow from the right tank is completely shut off). The relation between q_i and h_o should now be that of a pure integrator, since all the inflow is captured by the tank. To check Eq. (8–97), first divide through by R_{f2} and then set $R_{f1} = 0$, $R_{f2} = \infty$ to get the result

$$\frac{A_{T1} + A_{T2}}{\gamma} Dh_o = \frac{1}{\gamma} q_i \qquad (8\text{–}99)$$

$$h_o = \left(\frac{1}{A_{T1} + A_{T2}} \right) \frac{q_i}{D} = \frac{1}{A_{T1} + A_{T2}} \int q_i \, dt \qquad (8\text{–}100)$$

While these various checking schemes do not *guarantee* the validity of a result, they are most helpful in discovering mistakes and establishing confidence in computed results. The reader should thus develop the habit of using such checks; we all make mistakes, the good engineer discovers and corrects his before they become disastrous.

Returning to Eq. (8–97) we see that it fits our second-order pattern and thus define

$$\frac{h_o}{q_i} (D) = \frac{K}{\dfrac{D^2}{\omega_n^2} + \dfrac{2\zeta D}{\omega_n} + 1} \qquad (8\text{–}101)$$

$$\left. \begin{array}{c} K \triangleq \dfrac{R_{f2}}{\gamma} \dfrac{\text{ft}}{\text{ft}^3/\text{sec}} \qquad \omega_n \triangleq \dfrac{\gamma}{\sqrt{R_{f1}R_{f2}A_{T1}A_{T2}}} \dfrac{\text{rad}}{\text{sec}} \\[3ex] \zeta \triangleq \dfrac{A_{T1}R_{f1} + A_{T1}R_{f2} + A_{T2}R_{f2}}{2\sqrt{R_{f1}R_{f2}A_{T1}A_{T2}}} \end{array} \right\} \qquad (8\text{–}102)$$

By analogy with Eq. (8–65) we can show that here also ζ cannot be less than 1.0; oscillatory behavior is impossible. This is basically due to our neglecting all inertial effects of the fluid, which will generally be a good assumption for the fluid in the tanks themselves since $\ddot{h}_o$ will normally be small. Inertia of fluid in the pipe connecting the two tanks could conceivably lead to oscillatory behavior, but only if fluid resistance were very small, such as might be associated with pipe wall friction for a low-viscosity liquid. For concentrated flow resistances such as orifices or partially closed valves, the pressure drops due to inertia will normally be negligible compared to the resistive drops and our overdamped model will be valid. If desired, Eq. (8–101) can be cast into the two-time-constant form as was done in (8–68)–(8–71).

For our final example of fluid second-order systems, let us consider the tank/tubing system of Fig. 8–32d in which the fluid medium is a gas. The tank and tube walls are considered rigid, and tube volume is small compared to the tank so that fluid compliance effects are predominantly in the tank. Fluid inertance and resistance characterize the gas in the tube. In addition to its usefulness as a model for the dynamic response of pressure-measuring systems, this configuration also corresponds to the *Helmholtz resonator* of classical acoustics, and is thus applicable to the study of certain problems in acoustic noise reduction. Due to the low viscosity of air, such systems will generally be oscillatory. If the oscillation frequency predicted by our lumped-parameter model becomes too high, we approach the regime where distributed-parameter models (or else lumped models with many lumps) become necessary for accuracy.

A criterion for judging the validity of our simple model consists of comparing the characteristic lengths of the system (the length of the tube and the greatest transverse dimension of the tank) with the wavelength associated with the propagation of pressure waves in the fluid medium. If system dimensions are small compared to the wavelength then our simple model should be reasonably accurate. The wavelength varies with frequency according to

$$\text{wavelength} \triangleq \lambda = \frac{\text{velocity of wave propagation}}{\text{frequency}} \triangleq \frac{c}{f} \quad \textbf{(8–103)}$$

where

$$c = \sqrt{k\,g\,R\,T} \quad \frac{\text{ft}}{\text{sec}} \quad \textbf{(8–104)}$$

$k \triangleq$ ratio of specific heats for the gas $\qquad g \triangleq 32.2\,\dfrac{\text{ft}}{\text{sec}^2}$

$R \triangleq$ gas constant $\qquad\qquad\qquad\qquad T \triangleq$ absolute temperature, °R

For air at 70°F, for example,

$$c = \sqrt{(1.4)(32.2)(53.3)(530)} = 1120\,\frac{\text{ft}}{\text{sec}} \quad \textbf{(8–105)}$$

Thus if the maximum dimension of a system were, say, two feet, and if we assume that good accuracy will be obtained if the wavelength is 10 times the characteristic length, we are limited to frequencies f lower than 56 cycles/sec.

To proceed with our analysis we assume that p_i and p_o are both initially steady at some value p_m when p_i changes in an arbitrary fashion but with magnitude small compared to p_m. For instance if $p_m = 100$ psia we might restrict p_i to the range 90 to 110 psia, that is, a $\pm 10\%$ range about the mean value. This restriction is necessary since for gases the fluid compliance, inertance, and resistance are all nonlinear and a linearized model will be accurate only for small changes around some operating point. (These sorts of assumptions are excellent for acoustic systems, since the pressure oscillations [sound waves] are a *very* small fraction of the mean [atmospheric] pressure.) Since small tube pressure drop $p_i - p_o$ tends to encourage laminar flow, we take the inertance and resistance for this condition. The compliance of the tank depends on the type of compression process assumed; for rapid oscillations there is not enough time for heat transfer to occur and the adiabatic (no heat flow) process is a good model. For such a process the compliance dV/dp is given by V/kp. We have thus

$$I_f = \frac{16\rho L}{3\pi D_t^2} \qquad R_f = \frac{128\mu L}{\pi D_t^4} \qquad C_f = \frac{V}{kp_m} \qquad (8\text{–}106)$$

Note that:

a. C_f really varies with p_o but is assumed constant at a value corresponding to p_m.
b. The density ρ actually varies from one end of the tube to the other at any instant of time and also with time due to changing p_i and p_o. We assume it constant at a value corresponding to p_m and system temperature T (also assumed constant in the tube).

If we now consider p_{ip} and p_{op} to be the perturbations ($p_i - p_m$) and ($p_o - p_m$) we may write the instantaneous flow rate q as

$$q = \frac{(p_{ip} - p_{op})}{R_f + DI_f} \frac{\text{ft}^3}{\text{sec}} \qquad (8\text{–}107)$$

For the tank compliance,

$$\int q \, dt = \frac{q}{D} = C_f p_{op} \text{ ft}^3 \qquad (8\text{–}108)$$

thus

$$\frac{p_{ip} - p_{op}}{R_f + DI_f} = C_f D p_{op} \qquad (8\text{–}109)$$

and finally

$$(C_f I_f D^2 + C_f R_f D + 1)p_{op} = p_{ip} \qquad (8\text{–}110)$$

This is clearly second-order with

$$K \triangleq 1.0 \qquad \omega_n \triangleq \sqrt{\frac{1}{C_f I_f}} = \frac{D_t}{4}\sqrt{\frac{3\pi k p_m}{L\rho V}}$$

$$\zeta \triangleq \frac{16\mu}{D_t^3}\sqrt{\frac{3VL}{\pi\rho k p_m}} \tag{8-111}$$

As a numerical example, consider an air system with a tank of volume 2 inch³, a tube with $D_t = 0.2$ inch, and $L = 3$ inch, with $p_m = 14.7$ psia and temperature 75°F. The viscosity μ of air at 14.7 psia and 75°F is found from a table of fluid properties to be 3.8×10^{-7} lb$_f$-sec/ft² while the density ρ is calculated from the perfect gas law, $\rho = p_m/gRT = (14.7)(144)/[(32.2)(53.3)(535)] = 0.00231$ lb$_f$-sec²/ft⁴. We have then

$$\omega_n = \frac{0.2}{48}\sqrt{\frac{(3)(3.14)(1.4)(14.7)(144)}{(3/12)(0.00231)(2/1728)}} = 852.\frac{\text{rad}}{\text{sec}}, \qquad f_n = 136 \text{ Hz} \tag{8-112}$$

$$\zeta = \frac{(16)(3.8 \times 10^{-7})}{(0.2/12)^3}\sqrt{\frac{(3)(2/1728)(3/12)}{(3.14)(0.00231)(1.4)(14.7)(144)}} = 0.00836 \tag{8-113}$$

The calculated frequency value would be quite close to measured values for pressure perturbations up to a few psi, however the ζ value would probably be found to be much higher than 0.00836 due to turbulent flow and energy losses at the tube entrance and exit. In fact, the calculation of damping ratio for any oscillatory fluid system is subject to considerable uncertainty, and experimental testing is often necessary if accurate values are important.

8-5. THERMAL SECOND-ORDER SYSTEMS.

In contrast with mechanical, electrical, and fluid systems, which have two types of energy storage and thus may exhibit natural oscillations, thermal systems have only one type of storage element and oscillation is theoretically impossible. By taking enough "lumps" of resistance and capacitance we can get a system equation of arbitrarily high order; however the roots of the characteristic equation will always be found to be real, and thus no oscillations can occur.

In Fig. 8–33 we reconsider the system of Fig. 7–36b for those applications in which the energy storage of the heater itself is comparable to that of the liquid in the tank; that is, $(MC)_{\text{heater}}$ is the same order of magnitude as $(MC)_{\text{fluid}}$, and thus cannot be neglected. To get the overall system equation we write the conservation of energy for both the heater and the fluid. For the heater

$$q_i \, dt - U_h A_h(T_h - T_o) \, dt = M_h C_h \, dT_h \tag{8-114}$$

where

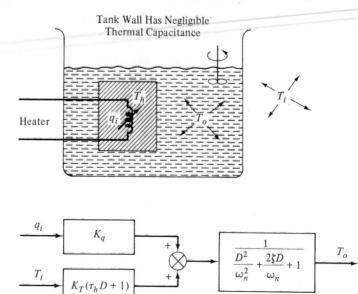

FIGURE 8-33

Thermal Second-Order System

$U_h \triangleq$ overall heat transfer coefficient between heater and fluid

$A_h \triangleq$ surface area of heater, $T_h \triangleq$ heater temperature

$M_h \triangleq$ heater mass, $C_h \triangleq$ specific heat of heater material

Similarly for the fluid in the tank,

$$U_h A_h (T_h - T_o)\, dt - U_w A_w (T_o - T_i)\, dt = M_f C_f\, dT_o \qquad \textbf{(8–115)}$$

where

$U_w \triangleq$ overall heat transfer coefficient between fluid and environment

$A_w \triangleq$ surface area of vessel, $T_o \triangleq$ fluid temperature

$M_f \triangleq$ fluid mass, $C_f \triangleq$ fluid specific heat

Our two simultaneous equations may be written as

$$(\tau_h D + 1)T_h + (-1)T_o = (1/U_h A_h)q_i \qquad \textbf{(8–116)}$$
$$(-U_h A_h / U_w A_w)T_h + (\tau_f D + 1 + U_h A_h / U_w A_w)T_o = T_i$$

where

$$\tau_h \triangleq \frac{M_h C_h}{U_h A_h} \qquad \tau_f \triangleq \frac{M_f C_f}{U_w A_w} \qquad \textbf{(8–117)}$$

Solving for T_o by determinants,

$$T_o = \frac{\begin{vmatrix} \tau_h D + 1 & \dfrac{q_i}{U_h A_h} \\[2ex] \dfrac{-U_h A_h}{U_w A_w} & T_i \end{vmatrix}}{\begin{vmatrix} \tau_h D + 1 & -1 \\[2ex] \dfrac{-U_h A_h}{U_w A_w} & \tau_f D + 1 + \dfrac{U_h A_h}{U_w A_w} \end{vmatrix}} = \frac{(\tau_h D + 1)T_i + \left(\dfrac{1}{U_w A_w}\right) q_i}{\tau_f \tau_h D^2 + \left(\tau_f + \tau_h + \dfrac{M_h C_h}{U_w A_w}\right) D + 1}$$

$$(8\text{–}118)$$

Cross-multiplying gives

$$\left[\tau_f \tau_h D^2 + \left(\tau_f + \tau_h + \frac{M_h C_h}{U_w A_w}\right) D + 1\right] T_o = (\tau_h D + 1)T_i + \left(\frac{1}{U_w A_w}\right) q_i$$

$$(8\text{–}119)$$

which is of the form

$$\left(\frac{D^2}{\omega_n^2} + \frac{2\zeta D}{\omega_n} + 1\right) T_o = K_T(\tau_h D + 1)T_i + K_q q_i \qquad (8\text{–}120)$$

where

$$\omega_n \triangleq \sqrt{\frac{1}{\tau_f \tau_h}} \qquad \zeta \triangleq \frac{\tau_f + \tau_h + \dfrac{M_h C_h}{U_w A_w}}{\sqrt{\tau_f \tau_h}} \qquad K_T \triangleq 1.0 \qquad K_q \triangleq \frac{1}{U_w A_w}$$

$$(8\text{–}121)$$

Since we know that the system will be overdamped we may wish to rewrite Eq. (8–120) in the two-time-constant form as in Eqs. (8–68)–(8–71). Note that $(T_o/q_i)(D)$ will be our standard second-order form, while $(T_o/T_i)(D)$ has "numerator dynamics" and thus has a different response than that discussed so far in this chapter.

8-6. MIXED SECOND-ORDER SYSTEMS.

In Fig. 8–34 we show an electromechanical system similar to that of Fig. 7–40 except now the motor armature is supplied with a constant current i_a and thus control of load speed ω_o is accomplished by manipulation of the field voltage e_i. Note that the assumption of constant armature current makes the armature resistance R_a and the motor back emf irrelevant since i_a is fixed by the source, rather than being determined by R_a and motor speed, as would be the case with an armature supplied with constant voltage. For the field we have

$$i_f R_f + L \frac{di_f}{dt} - e_i = 0 \qquad (8\text{–}122)$$

$$(\tau_f D + 1)i_f = \frac{1}{R_f} e_i \qquad \tau_f \triangleq \frac{L_f}{R_f} \qquad (8\text{–}123)$$

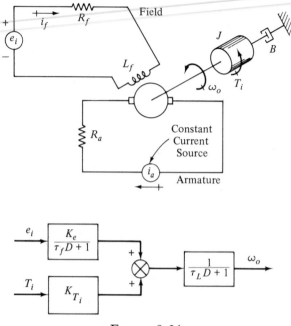

FIGURE 8–34

Field-Controlled DC Motor and Load

We see that (8–122) is completely independent of events in the armature circuit or the mechanical load motion, and thus we are able to immediately find i_f from e_i only. Since i_a is predetermined by the current source, there is no need to write an armature circuit equation and we go to Newton's law for the load motion.

$$T_i + K_T i_f - B\omega_o = J\dot{\omega}_o \qquad (\tau_L D + 1)\omega_o = K_{T_i}T_i + \frac{K_T}{B}i_f \quad \text{(8–124)}$$

where

$$\tau_L \triangleq \frac{J}{B} \qquad K_{T_i} \triangleq \frac{1}{B} \qquad K_T \triangleq \text{motor torque constant} = \frac{\text{inch-lb}_f}{\text{amp of } i_f}$$

$$\text{(8–125)}$$

Since i_f is known from (8–123),

$$(\tau_L D + 1)\omega_o = K_{T_i}T_i + \frac{K_T}{B}\left(\frac{1/R_f}{\tau_f D + 1}\right)e_i \qquad \text{(8–126)}$$

$$(\tau_L D + 1)(\tau_f D + 1)\omega_o = K_{T_i}(\tau_f D + 1)T_i + K_e e_i \qquad K_e \triangleq K_T/BR_f$$

$$\text{(8–127)}$$

To draw a block diagram we need the transfer functions

$$\frac{\omega_o}{e_i}(D) = \frac{K_e}{\tau_f \tau_L D^2 + (\tau_f + \tau_L)D + 1}$$

$$\frac{\omega_o}{T_i}(D) = \frac{K_{T_i}(\tau_f D + 1)}{(\tau_L D + 1)(\tau_f D + 1)} = \frac{K_{T_i}}{\tau_L D + 1} \qquad \text{(8-128)}$$

Note that the "cancellation" of $(\tau_f D + 1)$ makes the response of speed to the external torque T_i a first-order type, while the response to field voltage is second-order. This somewhat unusual result may be attributed to the "one-way" nature of the coupling between the field circuit and the load motion. Application of field voltage directly produces torque which obviously influences load motion; however, application of torque T_i, while influencing load motion, has *no* effect on what is happening in the field circuit, as Eq. (8-122) shows. Thus, we should not expect the speed/torque transfer function to involve τ_f and, of course, Eq. (8-128) shows that it does not. Mathematically, Eqs. (8-122) and (8-124) are not truly a simultaneous set, since each contains only *one* unknown (i_f in one case, ω_o in the other). That is, the two equations are not "bilaterally" coupled as has been the case in the other systems requiring more than one equation which we have previously encountered.

If, in Fig. 8-34, the constant-current armature supply is replaced with a constant-voltage source e_a, the overall system equation becomes linear with a time-varying coefficient. This class of equation allows neither a ready analytical solution nor the convenience of the transfer function methods. In the field circuit, Eqs. (8-122) and (8-123) still hold; however, in the armature we now have

$$i_a R_a + K_{be} i_f \omega_o - e_a = 0 \qquad i_a = (e_a - K_{be} i_f \omega_o)/R_a \qquad \text{(8-129)}$$

Newton's law yields

$$K_{T1} i_a i_f + T_i - B\omega_o = J\dot{\omega}_o \qquad \text{(8-130)}$$

where $K_{T1} \triangleq$ motor torque constant, inch-lb$_f$/amp^2. Combining (8-129) and (8-130) gives

$$J\dot{\omega}_o + B\omega_o + \frac{K_{T1}K_{be}}{R_a}\omega_o i_f^2 = T_i + \frac{K_{T1}e_a}{R_a}i_f \qquad \text{(8-131)}$$

Since i_f can be found immediately when e_i is given, the term $(K_{T1}K_{be}i_f^2/R_a)\omega_o$ gives the equation a time-varying coefficient. Such terms may be approximated by the same technique used for nonlinearities, if the analysis is limited to small perturbations around an equilibrium operating point. We have

$$i_f^2\omega_o \approx i_{fo}^2\omega_{o0} + \frac{\partial(i_f^2\omega_o)}{\partial\omega_o}\bigg|_{\omega_{o0},i_{fo}}\omega_{op} + \frac{\partial(i_f^2\omega_o)}{\partial i_f}\bigg|_{\omega_{o0},i_{fo}}i_{fp} \qquad \text{(8-132)}$$

$$i_f^2 \omega_o \approx i_{fo}^2 \omega_{o0} + i_{fo}^2 \omega_{op} + 2 i_{fo} \omega_{o0} i_{fp} \qquad \omega_{op} \overset{\triangle}{=} \omega_o - \omega_{o0} \qquad i_{fp} \overset{\triangle}{=} i_f - i_{fo}$$

$$(8\text{-}133)$$

we now rewrite (8–131) as an approximation

$$J\dot{\omega}_{op} + B(\omega_{o0} + \omega_{op}) + \frac{K_{T1}K_{be}}{R_a}(i_{fo}^2\omega_{o0} + i_{fo}^2\omega_{op} + 2 i_{fo}\omega_{o0}i_{fp})$$

$$= T_{io} + T_{ip} + \frac{K_{T1}e_a}{R_a}(i_{fo} + i_{fp}) \quad (8\text{-}134)$$

The equilibrium operating point refers to a situation where i_{fo} and T_{io} have been applied and ω_o has become steady at ω_{o0}. For this initial steady state the torques are balanced and thus

$$B\omega_{o0} + \frac{K_{T1}K_{be}}{R_a}i_{fo}^2\omega_{o0} = T_{io} + \frac{K_{T1}e_a}{R_a}i_{fo} \qquad (8\text{-}135)$$

Subtracting this equation from (8–134) gives

$$J\dot{\omega}_{op} + \left(B + \frac{K_{T1}K_{be}i_{fo}^2}{R_a}\right)\omega_{op} = T_{ip} + \left(\frac{K_{T1}e_a}{R_a} - \frac{2K_{T1}K_{be}\omega_{o0}i_{fo}}{R_a}\right)i_{fp} \quad (8\text{-}136)$$

which is now linear with *constant* coefficients and may be treated in our usual way. That is,

$$(\tau D + 1)\omega_{op} = K_1 T_{ip} + K_2 i_{fp} \qquad (8\text{-}137)$$

where

$$\tau \overset{\triangle}{=} \frac{JR_a}{BR_a + K_{T1}K_{be}i_{fo}^2} \quad K_1 \overset{\triangle}{=} \frac{R_a}{BR_a + K_{T1}K_{be}i_{fo}^2} \quad K_2 \overset{\triangle}{=} \frac{K_{T1}(e_a - 2K_{be}\omega_{o0}i_{fo})}{BR_a + K_{T1}K_{be}i_{fo}^2}$$

$$(8\text{-}138)$$

To resolve the question of how large the "small perturbations" i_{fp} and T_{ip} may get before accuracy suffers we may again appeal to digital simulation of the "exact" equations. Using the response of ω_o to a step change in e_i only ($T_i \equiv 0$), a CSMP program might go as follows.

```
TITLE     MOTŌR RESPŌNSE TŌ FIELD VŌLTAGE STEP
          EI = EIO+ES*STEP (0.0)
          IF1 = EI/RF
          IF = REALPL(IFO,TAUF,IF1)
          WŌDŌT = (KT1*IA*IF−B*WŌ)/J
          WŌ = INTGRL(WŌO,WŌDŌT)
          IA = (EA−KBE*WŌ*IF)/RA
INCŌN     IFO = .5, WŌO = 40.
PARAM     EIO = .5, ES = .05, TAUF = .05, RF = 1.
PARAM     KT1 = 1., EA = 20., RA = 2., KBE = .2
PARAM     B = .1, J = .01
TIMER     FINTIM = .4, DELT = .001, ŌUTDEL = .004
PRTPLT    WŌ, WŌDŌT, IF, EI
END
STŌP
ENDJŌB
```

Note that the field voltage EI jumps from 0.5 to 0.55 volts at $t = 0$, a 10% step change. The approximate and exact response should be quite close for such a small perturbation. By increasing the step size ES in subsequent runs until the approximate and exact responses diverge significantly, the valid range of the approximate analysis may be found. Figure 8–35 shows the CSMP result for the exact response of ω_o with numerical values as quoted above. The reader may wish to obtain the approximate result analytically for comparison.

8–7. SYSTEMS WITH NUMERATOR DYNAMICS.

While our emphasis has been on second-order systems of the type described by Eq. (8–2), systems of form (8–1) with b_1 and/or b_2 not zero do occur in practice and we here wish to briefly treat them. Our first example is from the field of vehicle dynamics; specifically the lateral (steering) dynamics of the automobile. Vehicle dynamics is of course a vast and complex subject; however, useful information can be obtained by taking a simplified view of a restricted area. The reader interested in pursuing the subject further will find numerous references.[1]

We consider an automobile proceeding down a straight and flat road at constant speed V ft/sec when the driver initiates a steering maneuver by turning the front wheels a small angle δ. This will cause the car to deviate from its original straight path and we are interested in this motion. Experiments and more comprehensive analysis have shown that for maneuvers which are not too violent (sidewise acceleration less than 0.3 to 0.4 g's) a linearized study gives results which compare well with actual behavior. Also, the rolling motion of the car on its suspension may be neglected as may aerodynamic forces. Under these assumptions the automobile's motion is a combination of translation and rotation in a horizontal plane and the forces causing this motion are those developed by the tires at the tire/road interface.

A key to the rational analysis of automobile steering dynamics is an understanding of how a pneumatic tire develops a sidewise force (called "cornering force"). The slip angle concept provides this understanding and its discovery, years ago, was a significant breakthrough. In Fig. 8–36, when the center plane of the tire and the velocity of its axle (viewed from above) are aligned, no sidewise force is developed at the road/tire interface. To develop a cornering force, a slip angle β_t must exist and experiments show that if β_t is less than 5° to 7° the cornering force is proportional to β_t. The

---

[1] W. F. Millikan et al., *Research in Automobile Stability and Control and in Tyre Performance* (London: The Inst. of Mech. Eng., 1956).

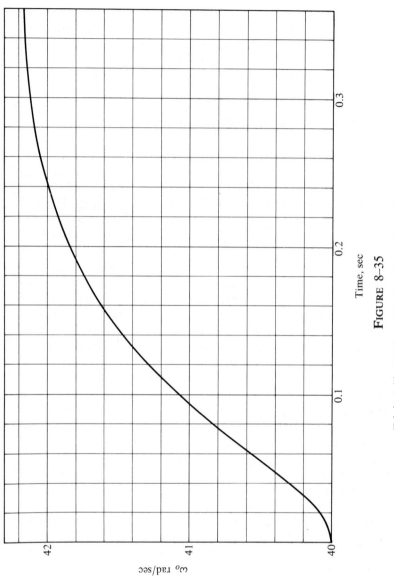

FIGURE 8–35

Digital Simulation Solution for Motor Response

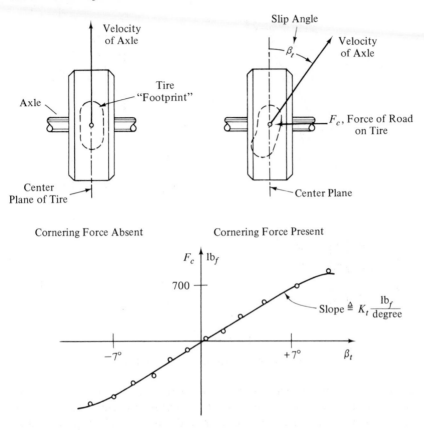

FIGURE 8–36

Tire Cornering Force and Slip Angle

proportionality factor K_t is called the cornering stiffness and is the order of 50 to 200 lb_f/degree for automobile tires.

Since the car's motion is a combination of rotation about a vertical axis and sidewise translation superimposed on the constant-velocity forward motion, the pertinent physical laws are

$$\sum F_Y = MA_Y \qquad (8\text{–}139)$$

$$\sum T_Z = J_Z \ddot{\psi} \qquad (8\text{–}140)$$

Figure 8–37 shows that A_Y is the sidewise acceleration of the center of mass and ψ (called the yaw angle) denotes the angular displacement of the car body centerline about the vertical Z axis. The moment of inertia about the Z axis is called J_Z. Since the direction of the velocity V need not coincide with that of the car centerline, the angle β (called sideslip angle) is defined

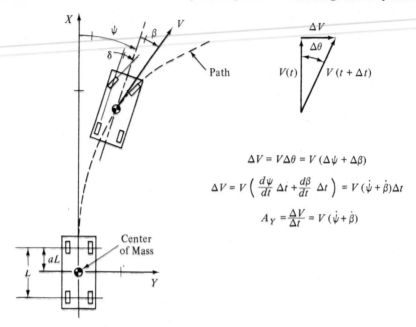

$$\Delta V = V\Delta\theta = V\left(\Delta\psi + \Delta\beta\right)$$

$$\Delta V = V\left(\frac{d\psi}{dt}\,\Delta t + \frac{d\beta}{dt}\,\Delta t\right) = V(\dot\psi + \dot\beta)\Delta t$$

$$A_Y = \frac{\Delta V}{\Delta t} = V(\dot\psi + \dot\beta)$$

FIGURE 8–37

Automobile Motion Coordinates

as the angle between them. Assumption of small angles allows us to approximate the sidewise acceleration A_Y in terms of $\dot\psi$ and $\dot\beta$ as shown in Fig. 8–37. It is next necessary to express the forces and torques in Eqs. (8–139) and (8–140) using the cornering stiffness of the tires and the slip angles. Note that the four tire/road interfaces are the *only* places where forces act on the car since we neglect aerodynamic forces. At the front wheels the sidewise velocity is $V\sin\beta + \dot\psi aL \approx V\beta + \dot\psi aL$, since $\sin\beta \approx \beta$ for small angles. If the steer angle δ were zero, the front wheel slip angle β_{tf} would be given by

$$\beta_{tf} \approx \tan\beta_{tf} = \frac{V\beta + \dot\psi aL}{V} = \beta + \frac{\dot\psi aL}{V} \qquad (8\text{–}141)$$

Since δ is not zero, the actual slip angle will be

$$\beta_{tf} \approx \beta + \frac{\dot\psi aL}{V} - \delta \qquad (8\text{–}142)$$

The two front wheels behave nearly identically so we may write the total front wheel force as

$$F_{cf} = 2K_{tf}\left(\beta + \frac{\dot\psi aL}{V} - \delta\right) \qquad (8\text{–}143)$$

where $K_{tf} \triangleq$ cornering stiffness of each front tire. Similarly for the rear wheels,

$$F_{cr} = 2K_{tr}\left(\beta - \frac{(1-a)L\dot{\psi}}{V}\right) \tag{8-144}$$

where we note that the effect of rotation $\dot{\psi}$ now *subtracts* from the side velocity $V\beta$.

We are now ready to substitute into (8–139) and (8–140) to get

$$\sum F_Y = -2K_{tf}\left(\beta + \frac{\dot{\psi}aL}{V} - \delta\right) - 2K_{tr}\left(\beta - \frac{\dot{\psi}(1-a)L}{V}\right)$$

$$= MV(\dot{\psi} + \dot{\beta}) \tag{8-145}$$

$$\sum T_Z = -2K_{tf}\left(\beta + \frac{\dot{\psi}aL}{V} - \delta\right)aL$$

$$+ 2K_{tr}\left(\beta - \frac{\dot{\psi}(1-a)L}{V}\right)(1-a)L = J_Z\ddot{\psi} \tag{8-146}$$

It is conventional to treat the yaw rate $\dot{\psi}$ and the sideslip angle β as the unknowns in this set of equations with steer angle δ of course being the input.

$$[MVD + 2(K_{tf} + K_{tr})]\beta + \left[MV + \frac{2L}{V}(a(K_{tf} + K_{tr}) - K_{tr})\right]\dot{\psi} = 2K_{tf}\delta$$
$$\tag{8-147}$$

$$[2L[a(K_{tf} + K_{tr}) - K_{tr}]]\beta$$

$$+ \left[J_Z D + \frac{2L^2}{V}[a^2(K_{tf} + K_{tr}) + (1 - 2a)K_{tr}]\right]\dot{\psi} = 2K_{tf}aL\delta \tag{8-148}$$

If we choose to solve for $\dot{\psi}$ we ultimately find

$$\frac{\dot{\psi}}{\delta}(D) = \frac{K_{\dot{\psi}}(\tau_{\dot{\psi}}D + 1)}{\dfrac{D^2}{\omega_n^2} + \dfrac{2\zeta D}{\omega_n} + 1} \tag{8-149}$$

Often $K_{tf} \approx K_{tr} \triangleq K_t$ and then

$$K_{\dot{\psi}} \triangleq \frac{2K_t}{MV(1 - 2a) + \dfrac{2K_tL}{V}} \frac{\text{rad/sec}}{\text{rad}} \qquad \tau_{\dot{\psi}} \triangleq \frac{aVM}{2K_t} \text{ sec} \tag{8-150}$$

$$\omega_n \triangleq \sqrt{\frac{2K_tL\left[MV(1 - 2a) + \dfrac{2K_tL}{V}\right]}{MVJ_Z}} \frac{\text{rad}}{\text{sec}} \tag{8-151}$$

$$\zeta \triangleq \frac{2K_tJ_Z + MK_tL^2(2a^2 - 2a + 1)}{\sqrt{2VJ_ZMK_tL\left[MV(1 - 2a) + \dfrac{2K_tL}{V}\right]}} \tag{8-152}$$

Rewriting (8–149) as a differential equation

$$\left[\frac{D^2}{\omega_n^2} + \frac{2\zeta D}{\omega_n} + 1\right]\psi = K_\psi(\tau_\psi D + 1)\delta \tag{8-153}$$

let us consider the response of ψ to a step input of δ of size 0.01 radian. For $t > 0$, the right-hand side of (8–153) is 0.01 K_ψ since the derivative of a constant is zero. Thus the form of the *general* solution will be identical with that for a system with only $K_\psi\delta$ on the right-hand side; the presence of $\tau_\psi D$ has no effect here. The *specific* solution will, however, be different from that graphed in Fig. 8–8 because the initial condition $D\psi(0^+)$ is now *not* zero. This may be seen by integrating Eq. (8–153) between $t = 0$ and $t = 0^+$.

$$\frac{D\psi}{\omega_n^2}\bigg|_0^{0^+} + \frac{2\zeta\psi}{\omega_n}\bigg|_0^{0^+} + \psi\bigg|_0^{0^+} = K_\psi\tau_\psi\delta\bigg|_0^{0^+} + K_\psi\int_0^{0^+}\delta\, dt \tag{8-154}$$

The last two terms on the left must both be zero, since the sudden turning of the front wheels through a finite angle δ produces only a finite force which cannot suddenly change $\dot\psi$ and ψ from the zero values they had just *before* the wheels were turned. On the right, δ is finite and thus $\int \delta\, dt$ is zero over the infinitesimal interval $0 \to 0^+$. We thus have

$$D\psi(0^+) = \omega_n^2 K_\psi\tau_\psi\delta = \frac{2K_t aL}{J_Z}\delta = \begin{matrix}\text{yaw angular acceleration}\\\text{at } t = 0^+\end{matrix} \tag{8-155}$$

This same result can also be obtained from Eq. (8–146) at $t = 0^+$. We see then that the "numerator dynamics" $(\tau_\psi D + 1)$ affects the step response by making the initial rate of change of the output quantity not zero; thus the curves of Fig. 8–8 cannot be used for such systems.

It should be clear that for inputs which are not constant, the presence of numerator dynamics will also cause changes in the particular solution, not just the initial conditions. The frequency response is also obviously affected since the numerator term $(\tau_\psi i\omega + 1)$ contributes a *leading* phase angle and an increasing amplitude ratio to the total response. Analog simulation of these systems requires some discussion also, since the term $K_\psi\tau_\psi(d\delta/dt)$ implies that the input signal must be *differentiated*, a process generally to be avoided on any computer due to its noise-accentuating properties. One method of treating this difficulty is to integrate Eq. (8–153) and then set up this new equation on the computer.

$$\left[\frac{D}{\omega_n^2} + \frac{2\zeta}{\omega_n} + \frac{1}{D}\right]\psi = K_\psi\tau_\psi\delta + K_\psi\frac{\delta}{D} \tag{8-156}$$

$$\frac{D\psi}{\omega_n^2} = K_\psi\tau_\psi\delta + K_\psi\int\delta\, dt - \frac{2\zeta}{\omega_n}\psi - \int\psi\, dt \tag{8-157}$$

Figure 8–38 shows the computer diagram, which also gives another way of looking at the initial condition question treated in Eq. (8–155).

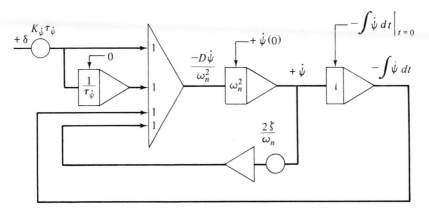

FIGURE 8–38

Analog Setup to Avoid Differentiation

Another analog technique useful for systems with numerator dynamics breaks the transfer function into two parts and defines a new variable E as follows:

$$\frac{E_o}{E_i}(D) = \frac{P(D)}{Q(D)} \qquad \frac{E}{E_i}(D) = \frac{1}{Q(D)} \qquad E_o = P(D)E \quad \text{(8–158)}$$

In our example

$$\frac{\psi}{\delta}(D) = \frac{K_{\dot\psi}(\tau_{\dot\psi} D + 1)}{\dfrac{D^2}{\omega_n^2} + \dfrac{2\zeta D}{\omega_n} + 1} \qquad \frac{E}{\delta}(D) = \frac{1}{\dfrac{D^2}{\omega_n^2} + \dfrac{2\zeta D}{\omega_n} + 1} \quad \text{(8–159)}$$

$$\psi = K_{\dot\psi}(\tau_{\dot\psi} D + 1)E \quad \text{(8–160)}$$

Figure 8–39 shows how this technique is applied. Finally, we should not fail to point out that if we directly set up the two simultaneous Eqs. (8–147) and (8–148) then no differentiations of δ appear and the above two "tricks" are not needed. With regard to digital simulation, CSMP does provide a differentiator; however, the use of integrators and the schemes of Figs. 8–38 or 8–39 or else direct simulation from the simultaneous equations would generally be preferred.

As our final example let us consider the circuit of Fig. 8–40, which will be found to have an equation of form (8–1) with *all* terms present. This circuit has considerable practical importance as a lead-lag controller in feedback control systems. An impedance approach is probably the quickest route to the transfer function $(e_o/e_i)(D)$ which we desire.

$$e_o(D) = i(D)Z_1(D) = i(D)\left[R_2 + \frac{1}{C_2 D}\right] = i(D)\left(\frac{R_2 C_2 D + 1}{C_2 D}\right)$$

$$\text{(8–161)}$$

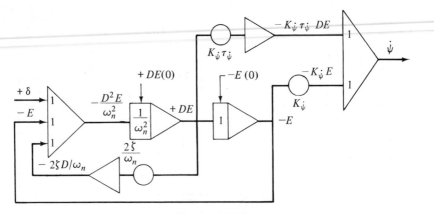

FIGURE 8–39

Alternative Scheme to Avoid Differentiation

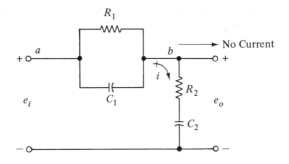

FIGURE 8–40

Lead-Lag Controller Circuit

$$i(D) = \frac{e_i}{Z_{\text{total}}} = \frac{e_i}{\dfrac{(R_1)\left(\dfrac{1}{C_1 D}\right)}{R_1 + \dfrac{1}{C_1 D}} + R_2 + \dfrac{1}{C_2 D}}$$

$$= \frac{C_2 D(R_1 C_1 D + 1)e_i}{R_1 C_1 R_2 C_2 D^2 + (R_1 C_1 + R_2 C_2 + R_1 C_2)D + 1} \qquad (8\text{–}162)$$

$$\frac{e_o}{e_1}(D) = \frac{(R_1 C_1 D + 1)(R_2 C_2 D + 1)}{R_1 C_1 R_2 C_2 D^2 + (R_1 C_1 + R_2 C_2 + R_1 C_2)D + 1}$$

$$= \frac{(\tau_1 D + 1)(\tau_2 D + 1)}{(\tau_3 D + 1)(\tau_4 D + 1)} \qquad (8\text{–}163)$$

where

$$\tau_1 \triangleq R_1C_1 \qquad \tau_2 \triangleq R_2C_2 \qquad \tau_3\tau_4 \triangleq R_1C_1R_2C_2$$

$$\tau_3 + \tau_4 \triangleq R_1C_1 + R_2C_2 + R_1C_2 \qquad \textbf{(8–164)}$$

In (8–163) the numerator is clearly overdamped second-order and since the denominator has the same ω_n and a larger D term ($\tau_3 + \tau_4 > \tau_1 + \tau_2$) the denominator is also overdamped, as we would expect from a passive R-C circuit without inductors.

In practical applications τ_3 is often chosen to be some fraction of τ_1, which then makes τ_4 a multiple of τ_2. For example, suppose we want $\tau_1 = 0.01$ sec, $\tau_2 = 0.05$ sec, and $\tau_3 = (1/10)\tau_1 = 0.001$ sec. Then τ_4 must be 0.5 sec and from (8–164) we find that $R_1C_2 = 0.501 - 0.060 = 0.441$. Since $R_1C_1 = 0.01$ and $R_2C_2 = 0.05$ we now have 3 equations in 4 unknowns, giving an infinite number of possible solutions. Since the impedance presented to the source e_i has a minimum value of R_2 at high frequencies we might wish to choose R_2 sufficiently large so as to not load the source excessively. Having chosen R_2, the design then becomes unique; for example $R_2 = 10,000\ \Omega$ gives $C_2 = 5\ \mu F$, $R_1 = 88,200\ \Omega$, $C_1 = 0.113\ \mu F$.

The step response of such a system is found from the differential equation

$$[\tau_3\tau_4D^2 + (\tau_3 + \tau_4)D + 1]e_o = [\tau_1\tau_2D^2 + (\tau_1 + \tau_2)D + 1]e_i \qquad \textbf{(8–165)}$$

For e_i a step of, say, 1 volt, the right-hand side becomes the number 1.0 for any $t > 0$ and thus the particular solution is clearly $e_{op} = 1.0$. The complementary solution will have the standard form for overdamped second order; however the initial conditions will *not* conform to those of a system without numerator dynamics and again the curves of Fig. 8–8 are not applicable. Any of the several methods for finding $e_o(0^+)$ and $De_o(0^+)$ which we have used previously may be employed here. Using physical reasoning from the circuit diagram, at $t = 0^+$ the capacitor C_1 is still uncharged since it takes a finite current a finite time to charge a capacitor. Since there is no voltage *drop* across C_1, the potential at point b (which is the output voltage e_o) must be the same as at a; thus e_o instantly jumps up to e_i. The current i must be finite since it must go through the resistor R_2; in fact $i(0^+)$ is e_i/R_2. Thus, initially, all the current i goes through C_1 (*none* of it through R_1) and through R_2 and C_2. To find $De_o(0^+)$ we note that $e_o = e_i + e_{ab}$ and thus $De_o = De_i + De_{ab}$. At $t = 0^+$, e_i has become constant; thus $De_i = 0$ and $De_o = De_{ab}$. The voltage rise e_{ab} is the voltage across C_1, and since C_1 carries the current $i = e_i/R_2$, we have $De_{ab} = -i/C_1 = -e_i/R_2C_1$ and thus $De_o(0^+) = -e_i/R_2C_1$. These same results for $e_o(0^+)$ and $De_o(0^+)$ may be found by integrating (8–165) twice between

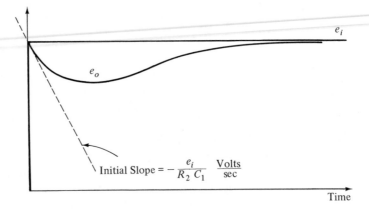

FIGURE 8–41

Step Response for Circuit of Fig. 8–40

$t = 0$ and $t = 0^+$ or by studying an analog computer diagram for the equation. Figure 8–41 shows the general shape of the step response while 8–42 displays the logarithmic frequency-response curves for a system with the numerical values quoted earlier.

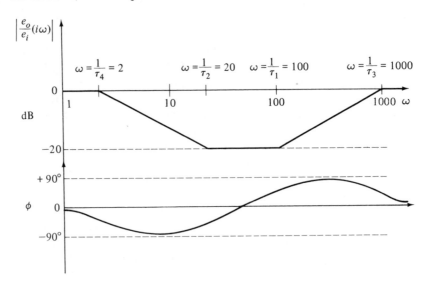

FIGURE 8–42

Frequency Response of Circuit of Fig. 8–40

BIBLIOGRAPHY

1. Cannon, R. H., *Dynamics of Physical Systems*. New York: McGraw-Hill Book Company, 1967.

2. Doebelin, E. O., *Dynamic Analysis and Feedback Control*. New York: McGraw-Hill Book Company, 1962.

3. Doebelin, E. O., *Measurement Systems, Application and Design*. New York: McGraw-Hill Book Company, 1966.

4. Reswick, J. B. and C. K. Taft, *Introduction to Dynamic Systems*. Englewood Cliffs, N. J.: Prentice-Hall, Inc., 1967.

5. Shearer, J. L., A. T. Murphy, and H. H. Richardson, *Introduction to System Dynamics*. Reading, Mass.: Addison-Wesley Publishing Co., Inc., 1967.

PROBLEMS

8-1. For the rotary system of Fig. 8–4a:

    a. Write the differential equation, put it in standard form and define the standard parameters, give the operational and sinusoidal transfer functions and show a block diagram.

    b. Draw an analog computer diagram in terms of elements (springs, inertias, dampers).

    c. Draw an analog computer diagram in terms of standard parameters.

    d. Write a CSMP program using CMPXPL and standard parameters.

    e. Write a CSMP program not using CMPXPL and using element values.

8-2. Repeat prob. 8–1 for the translational system of Fig. 8–4b.

8-3. Repeat prob. 8–1 for the rotary system of Fig. 8–4b.

8-4. Repeat prob. 8–1 for the translational system of Fig. 8–4c.

8-5. Repeat prob. 8–1 for the rotary system of Fig. 8–4c.

8-6. Repeat prob. 8–1 for the translational system of Fig. 8–4d.

8-7. Repeat prob. 8–1 for the rotary system of Fig. 8–4d.

8-8. Repeat prob. 8–1 for the translational system of Fig. 8–4e.

8-9. Repeat prob. 8–1 for the rotary system of Fig. 8–4e.

8-10. The rotary system of Fig. 8–5b may be used as a model for the shaft (K_s) and cutting blade (J) of a rotary lawnmower. The shaft is 0.5 inch diameter and 3 inches long; the blade is a rectangular flat 0.1 by 2 by 18 inches, both are steel. If the engine providing the driving motion runs at 1000 rpm would torsional vibration problems be expected? The motion θ_i may contain fluctuations at frequencies up to three times the engine speed.

8–11. A more accurate model of the lawnmower of prob. 8–10 places another inertia J_1 at the other end of the spring to represent the motor inertia. The input now is a driving torque T_i applied to J_1. Find the natural frequency of this model. If J_1 is 0.05 in-lb$_f$-sec^2 and other conditions are as in prob. 8–10, are vibration problems predicted?

8–12. For the accelerometer of Fig. 2–24a, get the differential equation relating input acceleration $\ddot{x}_i$ to output displacement x_o. If the proof mass weighs 0.05 lb$_f$, what spring constant is needed to give 0.1 inch x_o for a steady acceleration of 10 g's? Estimate the range of frequencies for which this instrument would measure properly if the input acceleration is a sinusoidal vibration.

8–13. The automobile of Fig. 8–37 becomes unstable when the numerator of Eq. (8–152) becomes negative. Why? Derive an expression for the critical speed V above which instability occurs. What is this speed for a car with $a = 0.6$, $M = 109$ slugs, $K_t = 11{,}460$ lb$_f$/rad, and $L = 9$ ft?

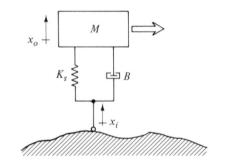

Fig. P8–1

8–14. Figure P8–1 shows a simplified model of a car riding over a rough road.

a. Derive the differential equation relating road profile x_i to car body displacement x_o.

b. Get operational and sinusoidal transfer functions in standard form and sketch the general shape of the frequency-response curves.

c. If $M = 100$ slugs and $K_s = 6000$ lb$_f$/ft, what value of B is needed to get critical damping?

d. For the numerical values of part c, plot the logarithmic frequency-response curves.

e. Write a CSMP program to find x_o if the car is driven over the bump of Fig. P8–2 at 40 mph.

f. Repeat part e for speeds of 10 and 80 mph.

g. Using the numerical values of part c, run the programs of parts e and f.

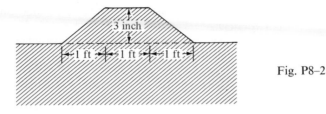

Fig. P8–2

8–15. Plot logarithmic frequency-response curves for second-order systems with:

    a. $K = 10$, $\zeta = 0.5$, $\omega_n = 25$ rad/sec
    b. $K = 0.04$, $\zeta = 0.2$, $\omega_n = 15{,}000$ rad/sec

8–16. For the circuit of Fig. 8–25b:

    a. Derive the differential equation relating the indicated input and output quantities and define the standard parameters.
    b. Give the operational and sinusoidal transfer functions and show a block diagram.
    c. Draw an analog computer diagram in terms of elements (R, L, C).
    d. Draw an analog computer diagram in terms of standard parameters.
    e. Write a CSMP program using CMPXPL and standard parameters.
    f. Write a CSMP program not using CMPXPL and using element values.

8–17. Repeat prob. 8–16 for the system of Fig. 8–25c.

8–18. Repeat prob. 8–16 for the system of Fig. 8–25e.

8–19. Repeat prob. 8–16 for the system of Fig. 8–25f.

8–20. Repeat prob. 8–16 for the system of Fig. 8–25h.

8–21. In Fig. 8–25e, what is the impedance seen by the voltage source e_i? Find the steady-state sinusoidal current drawn from the source if $R_1 = 1000$ ohms, $R_2 = 800$ ohms, $L = 0.2$ H, $C = 0.5$ μF, and e_i is 100 sin 377t volts, t is seconds.

8–22. The system shown in Fig. P8–3 is proposed as a means of differentiating e_i signals with frequency content up to about 1 Hz. The noise filter is necessary since e_i also contains small amplitude but high-frequency (50 Hz and above) noise components.

Fig. P8–3

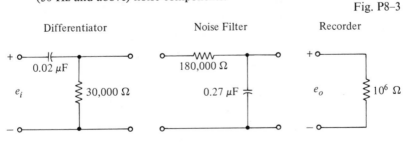

Differentiator   Noise Filter   Recorder

    a. Get transfer functions for the differentiator and noise filter separately.

    b. Assuming the e_o terminals open circuit (recorder not connected), plot logarithmic frequency-response curves for $(e_o/De_i)(i\omega)$, assuming no loading effect between noise filter and differentiator.

    c. To check the validity of the no-loading assumption get $(e_o/De_i)(i\omega)$ *exactly*, by analyzing the whole circuit (including recorder) as an entity.

    d. Does the circuit perform its intended functions?

8-23. For the system of Fig. 8–32a:

    a. Derive the differential equation relating the indicated input and output quantities, define the standard parameters and give the transfer function and block diagram. Assume complete linearity.

    b. Draw an analog computer diagram using element values.

    c. Draw an analog computer diagram using standard parameters.

    d. Write a CSMP program using CMPXPL and standard parameters.

    e. Write a CSMP program using element values and not using CMPXPL.

8-24. Repeat prob. 8–23 for Fig. 8–32c.

8-25. Repeat prob. 8–23 for Fig. 8–32e.

8-26. Repeat prob. 8–23 for Fig. 8–32f.

8-27. Repeat prob. 8–23 for Fig. 8–32g.

8-28. Repeat prob. 8–23 for Fig. 8–32h.

8-29. In Fig. 8–32b, treat the level in the left tank as the output quantity. Get differential equation, transfer function, and block diagram.

8-30. In Fig. 8–32b, add an inflow $q_{i1}(t)$ to the right tank, giving a system with two inputs. Get differential equations, transfer functions, and block diagram showing the response of *both* tank levels to *both* inflows.

8-31. In Fig. P8–4 a thermometer is inserted into a thin-wall metal well containing fluid with mass M_w and specific heat c_w. The well wall is

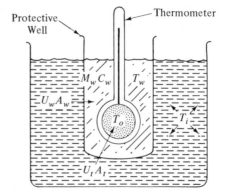

Fig. P8–4

considered a pure resistance and the thermometer bulb contains fluid with mass M_t and specific heat c_t. Get the differential equation relating T_o to T_i.

8–32. In the system of Fig. 5–2 let e_f be replaced by a constant-current source and let the electrical load be pure resistance. Get differential equations, transfer functions, and block diagrams relating input torque T to output speed ω and voltage e_G.

8–33. In prob. 8–32 let the field current be a variable input quantity $i_f(t)$. Write a CSMP program which would solve for ω and e_G if T and i_f were known inputs.

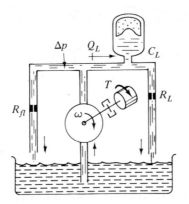

Fig. P8–5

8–34. The system of Fig. 5–8 is made specific in Fig. P8–5. Treating the system as strictly linear, get differential equations, transfer functions, and block diagram showing how outputs Δp, Q_L, and ω are produced by input torque T.

8–35. For the linearized model of Eq. (8–137), get the step response corresponding to the exact result obtained in the text using CSMP and compare the two.

8–36. In the automobile example of section 8–7, get the transfer function $(\beta/\delta)(D)$.

9

DYNAMICS OF GENERAL LINEAR SYSTEMS

9-1. TRANSIENT BEHAVIOR AND STABILITY.

It is probably clear to the reader at this stage that the modeling of systems comprised of large numbers of elements, whether mechanical, electrical, fluid, or thermal, will lead to high-order differential equations. As long as linearized, constant coefficient models are adequate, the analytical methods of Chapter 6 will yield information on the nature of system behavior, even though computers may be desirable for evaluating specific numerical cases. These analytical methods tell us that the system response will always be made up of a natural part (the complementary solution) and a forced part (the particular solution). Furthermore, the natural part must always be comprised of terms of the form Ce^{at}, corresponding to real roots and $Ce^{at} \sin(bt + \phi)$, corresponding to complex root-pairs (plus possibly some terms of form $Ct^n e^{at}$ or $Ct^n e^{at} \sin(bt + \phi)$ due to repeated roots).

We should at this point carefully note that our detailed study of first-order systems has made us familiar with the nature of the response associated with *any* real root while our second-order system studies have done the same for *any* complex root-pair. That is, the more complex higher order systems do not hold any fundamental surprises, they merely allow the simultaneous presence of several (or many) first-order and/or second-order modes of behavior in the total response. In mechanical vibrating systems, for example, models with many masses, springs, and dampers will exhibit *many* natural frequencies, but the behavior of each of these is not essentially different from the basic second-order system.

With regard to procedures for *deriving* the differential equations for complex systems, it is basically a matter of applying the same physical laws we have been using but applying them *more times*. In the mechanical system of Fig. 9–1a the application of Newton's law to each mass in turn will readily lead us to a set of three simultaneous second-order equations describing the system. Similarly, for the electrical and fluid systems of Fig. 9–1, a straightforward application of familiar laws leads to the system equations. When systems became *very* complex, "bookkeeping" matters such as knowing when a complete set of equations has been obtained and just how many unknowns there really are may complicate matters. Systematic methods such as *Lagrange's energy approach* for mechanical systems and *network topology* for electrical circuits may be helpful in such situations, but are beyond the scope of this text.

Let us analyze the system of Fig. 9–1a to give a concrete example of the procedure. Application of Newton's law leads directly to

$$(M_1D^2 + B_1D + K_{s1})x_1 + (-B_1D - K_{s1})x_2 + (0)x_3 = f_{i1} \quad \text{(9–1)}$$

$$(-B_1D - K_{s1})x_1 + [M_2D^2 + B_1D + (K_{s1} + K_{s2})]x_2 + (-K_{s2})x_3 = 0 \quad \text{(9–2)}$$

$$(0)x_1 + (-K_{s2})x_2 + [M_3D^2 + B_2D + (K_{s2} + K_{s3})]x_3 = f_{i3} \quad \text{(9–3)}$$

Since the system characteristic equation will turn out to be sixth degree, it will be impossible to proceed with an analytical solution while system parameters are given only as letters, since the roots of algebraic equations of degree greater than four can only be found when the coefficients are known numbers. While there is nothing to prevent us from working with literal coefficients until we reach the root-solving stage, let us assume some numerical values now, strictly for convenience. Let us take

$$M_1 = M_2 = M_3 = 1.0 \quad B_1 = B_2 = 0.1 \quad K_{s1} = K_{s2} = K_{s3} = 1.0 \quad \text{(9–4)}$$

and choose to solve for x_1 by determinants.

$$x_1 = \frac{\begin{vmatrix} f_{i1} & (-0.1D - 1) & 0 \\ 0 & (D^2 + 0.1D + 2) & -1 \\ f_{i3} & -1 & (D^2 + 0.1D + 2) \end{vmatrix}}{\begin{vmatrix} (D^2 + 0.1D + 1) & (-0.1D - 1) & 0 \\ (-0.1D - 1) & (D^2 + 0.1D + 2) & -1 \\ 0 & -1 & (D^2 + 0.1D + 2) \end{vmatrix}} \quad \text{(9–5)}$$

Expansion of the determinants gives

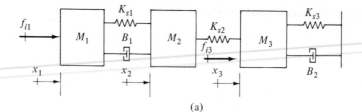

(a)

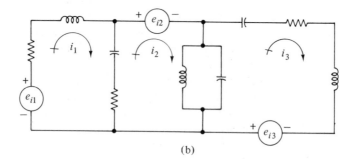

(b)

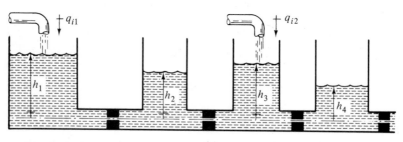

(c)

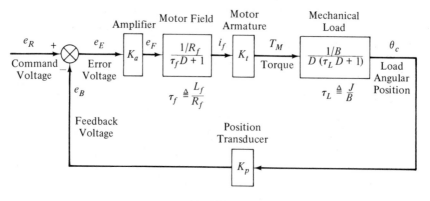

(d)

FIGURE 9–1

Examples of Systems Higher Than Second-Order

$$(D^6 + 0.3D^5 + 5.02D^4 + 0.8D^3 + 6.01D^2 + 0.2D + 1)x_1$$
$$= (D^4 + 0.2D^3 + 4.01D^2 + 0.4D + 3)f_{i1} + (0.1D + 1)f_{i3} \quad \text{(9–6)}$$

The system characteristic equation is

$$D^6 + 0.3D^5 + 5.02D^4 + 0.8D^3 + 6.01D^2 + 0.2D + 1 = 0 \quad \text{(9–7)}$$

and, using a digital computer root-finding program (available in most computer-system libraries) we find the roots to be (using about 0.5 sec of computer time)

$$-0.00644 \pm i0.445 \qquad -0.0741 \pm i1.79 \qquad -0.0694 \pm i1.24 \quad \text{(9–8)}$$

Suppose we wish to find the response of x_1 to a step input of f_{i1} of size 1.0. With the roots available an analytical solution is possible, though quite tedious since we need to find six initial conditions on x_1 and then calculate six constants of integration. At this point the availability of a computer is most appreciated. Using analog or digital simulation we can work with the original set of equations (9–1)–(9–3) which will give us x_1, x_2, and x_3 simultaneously without ever having to go through the determinant expansion procedure needed in the analytical approach. Using CSMP the program might take the form:

```
TITLE     STEP RESPONSE OF 3−MASS SYSTEM
          X1DOT2 = (FI1+KS1*X2+B1*X2DOT−KS1*X1−B1*X1DOT)/M1
          X1DOT = INTGRL (0.0, X1DOT2)
          X1 = INTGRL (0.0, X1DOT)
          X2DOT2 = (KS2*X3+B1*X1DOT+KS1*X1−B1*X2DOT−
                   (KS1+KS2)*X2)/M2
          X2DOT = INTGRL (0.0, X2DOT2)
          X2 = INTGRL (0.0, X2DOT)
          X3DOT2 = (FI3+KS2*X2−B2*X3DOT−(KS2+KS3)*X3)/M3
          X3DOT = INTGRL (0.0, X3DOT2)
          X3 = INTGRL (0.0, X3DOT)
PARAM     FI1 = 1.0, FI3 = 0.0
PARAM     KS1 = 1., KS2 = 1., KS3 = 1.
PARAM     M1 = 1., M2 = 1., M3 = 1.
PARAM     B1 = .1, B2 = .1
TIMER     FINTIM = 5., DELT = .01, OUTDEL = .05
PRTPLT    X1, X2, X3
END
STOP
ENDJOB
```

Figure 9–2 shows the response curves for x_1, x_2, and x_3. The roots (9–8) show the existence of three natural frequencies, 0.445, 1.24, and 1.79 rad/sec, each very lightly damped; however, the response to a step of f_{i1} appears to be almost entirely dominated by the lowest frequency mode of vibration, that is, the 0.445 rad/sec mode. Only in x_3, and there for just the first cycle or so, can we see some of the higher frequency modes. The relative strengths of each of the three modes of vibration are quite sensitive

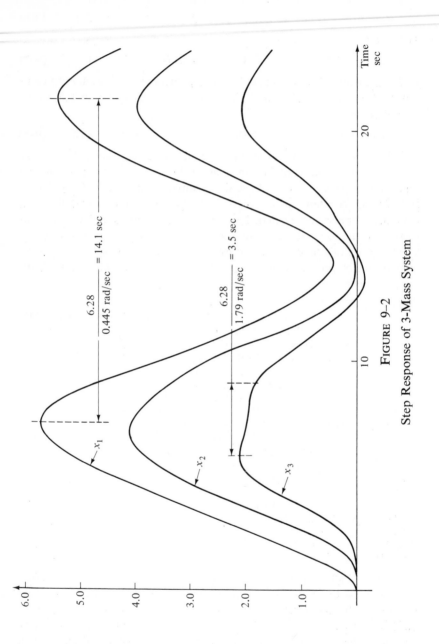

FIGURE 9–2

Step Response of 3-Mass System

to the nature and point of application of the external forces and/or the initial conditions. Figure 9–3 shows the response for $f_{i1} = f_{i3} = 0$ with initial conditions $x_1(0) = -2$, $x_2(0) = 0$, and $x_3(0) = 1$. There the mode with frequency 1.24 rad/sec is quite obvious in all three coordinates during the early phase of the response; however, after about fifty seconds the two higher frequency modes have damped out and the terminal phase is dominated by the lowest frequency mode in all three coordinates.

From Eq. (9–6) we can define operational and sinusoidal transfer functions in the usual ways.

$$\frac{x_1}{f_{i1}}(D) = \frac{D^4 + 0.2D^3 + 4.01D^2 + 0.4D + 3}{D^6 + 0.3D^5 + 5.02D^4 + 0.8D^3 + 6.01D^2 + 0.2D + 1} \qquad (9\text{–}9)$$

$$\frac{x_1}{f_{i1}}(i\omega) = \frac{(i\omega)^4 + 0.2(i\omega)^3 + 4.01(i\omega)^2 + 0.4(i\omega) + 3}{(i\omega)^6 + 0.3(i\omega)^5 + 5.02(i\omega)^4 + 0.8(i\omega)^3 + 6.01(i\omega)^2 + 0.2(i\omega) + 1}$$
$$(9\text{–}10)$$

$$\frac{x_1}{f_{i3}}(D) = \frac{0.1D + 1}{D^6 + 0.3D^5 + 5.02D^4 + 0.8D^3 + 6.01D^2 + 0.2D + 1} \qquad (9\text{–}11)$$

$$\frac{x_1}{f_{i3}}(i\omega) = \frac{0.1i\omega + 1}{(i\omega)^6 + 0.3(i\omega)^5 + 5.02(i\omega)^4 + 0.8(i\omega)^3 + 6.01(i\omega)^2 + 0.2(i\omega) + 1}$$
$$(9\text{–}12)$$

Furthermore, if we choose to, we can obtain equations similar to (9–6) for both x_2 and x_3 and thereby define other transfer functions:

$$\frac{x_2}{f_{i1}}(D) \qquad \frac{x_2}{f_{i3}}(D) \qquad \frac{x_3}{f_{i1}}(D) \qquad \frac{x_3}{f_{i3}}(D) \qquad (9\text{–}13)$$

We could also add a force f_{i2} on mass M_2 and analyze to define

$$\frac{x_1}{f_{i2}}(D) \qquad \frac{x_2}{f_{i2}}(D) \qquad \frac{x_3}{f_{i2}}(D) \qquad (9\text{–}14)$$

If this is done the following interesting fact will be observed:

$$\frac{x_1}{f_{i2}}(D) = \frac{x_2}{f_{i1}}(D) \qquad \frac{x_1}{f_{i3}}(D) = \frac{x_3}{f_{i1}}(D) \qquad \frac{x_2}{f_{i3}}(D) = \frac{x_3}{f_{i2}}(D) \qquad (9\text{–}15)$$

This is not a coincidence peculiar to this example but rather a manifestation of a general result for linear systems, the so-called *reciprocity theorem*. Stated informally this theorem says that, in a multi-degree-of-freedom system, if we apply force at location one and measure the response motion at location two, if we then apply the same force at location two it will produce the identical motion at location one.

While this section is not devoted specifically to frequency response, it may be instructive to display the amplitude ratio and phase angle curves for $(x_1/f_{i1})(i\omega)$. These curves can be obtained from Eq. (9–10) in the usual way by substituting numerical values of ω and computing the magnitude and angle of the resulting complex number. It is clear that equations of

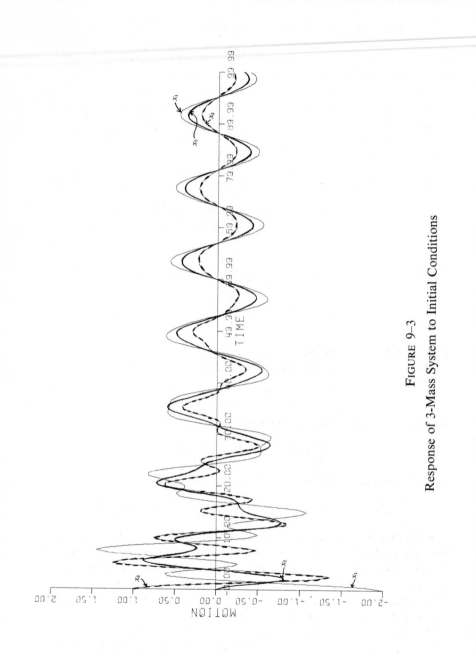

Figure 9–3

Response of 3-Mass System to Initial Conditions

this form may be programmed for evaluation on a digital computer, and in fact such programs may be made *general* so that they need be worked out only once and then easily applied to any numerical example which might arise. The reason this can be done for linear systems is that the transfer functions are *always* expressible in the form

$$\frac{q_o}{q_i}(D) = \frac{b_m D^m + b_{m-1} D^{m-1} + \ldots + b_1 D + b_0}{a_n D^n + a_{n-1} D^{n-1} + \ldots + a_1 D + a_0} \qquad (9\text{-}16)$$

so that one need merely tell the program the numerical values of the a's and b's and the values of ω which are of interest. Such programs are widely available in computer libraries and one such was used to compute the curves of Fig. 9–4, 500 frequency points at 0.01 rad/sec intervals between $\omega = 0.0$ and 5.00 being calculated in about 4 seconds.

Examination of Fig. 9–4 reveals that at the lowest natural frequency the amplitude ratio is about 75. Since the response to a static force ($\omega \to 0$) has a ratio of 3, the resonant magnification is 25. Even though the roots for the two higher natural frequencies exhibit very light damping (real parts small compared to imaginary), the curves of Fig. 9–4 indicate that sinusoidal forcing at these frequencies does *not* cause a large response; in fact the amplitude ratio is *less* than 3, the static value. Thus it appears that the higher frequency modes are more difficult to excite than the lowest, which may explain why the step input of Fig. 9–2 produced motion mainly at the lowest frequency. While this "preference" for the lowest mode of vibration cannot be stated as a hard-and-fast rule, it is very commonly observed in multi-degree-of-freedom systems. This behavior is of course sensitive to the amount and location of damping within the system since, for example, for no damping at all, all three resonant peaks would be of *infinite* height.

Stability. In the area of automatic feedback control systems the question of system stability assumes a role of central importance. Even some systems which are not obviously recognizable as feedback systems may be unstable. The automobile of section 8–7, for example, will exhibit unstable behavior for certain values of system parameters. By instability we mean behavior wherein the slightest disturbances to the system, even if applied and then removed, will cause the response to tend toward infinity. For linear systems with constant coefficients the determination of stability is straightforward and unequivocal; *all the roots of the system characteristic equation must lie in the left half of the complex plane;* that is, they must have negative real parts. For other types of systems no *general* stability criteria are available.

The truth of the above-stated stability criterion is easily demonstrated with numerical examples. For any positive real root the system response

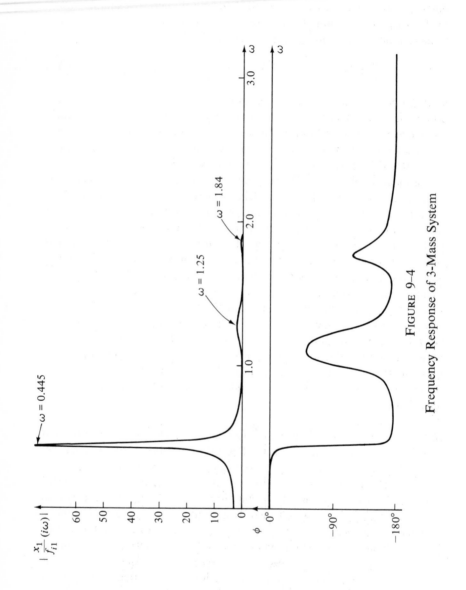

FIGURE 9-4

Frequency Response of 3-Mass System

424

will contain a term of form Ce^{at}, and with a positive this must go to infinity even if there is no sustained external forcing function. Similarly a complex root-pair $+2 \pm i7$, for example, will give a term $Ce^{2t} \sin (7t + \phi)$, which clearly is an oscillation of ever increasing amplitude. For equations of order higher than four, the characteristic equation's roots can be found only if system parameters are known as numbers, which is undesirable from a designer's point of view, since results in *letter* form may be studied to reveal useful design relations. Fortunately, E. J. Routh in the 1870s discovered a method of detecting the *presence* of unstable roots without actually getting their numerical values. We now give a brief treatment of this Routh Stability Criterion; more complete discussions are available in the literature.[1]

To apply Routh's criterion arrange the coefficients of the system characteristic equation in the form

$$a_n \qquad a_{n-2} \quad a_{n-4} \quad a_{n-6}$$
$$a_{n-1} \quad a_{n-3} \quad a_{n-5} \quad \cdots$$

carrying these first two rows until you "run out" of coefficients. Then form a third row

$$b_1 \qquad b_2 \qquad b_3 \cdots$$

from the first two using the rule

$$b_1 \triangleq \frac{a_{n-1}a_{n-2} - a_n a_{n-3}}{a_{n-1}} \qquad b_2 \triangleq \frac{a_{n-1}a_{n-4} - a_n a_{n-5}}{a_{n-1}}$$

and continuing in this established pattern until the b's become zero. A *fourth* row is then constructed from the second and third rows in exactly the same way as the third was formed from the first two. By continuing this row-forming process until all zeros are obtained, a triangular array is obtained. The procedure is actually much quicker and simpler than the general instructions imply. Take Eq. (9–6) as an example.

1.0	5.02	6.01	1.0
0.3	0.8	0.2	0
2.35	5.34	1.0	0
0.0757	0.0724	0	0
3.09	1.0	0	0
0.0479	0	0	0
1.0	0	0	0
0	0	0	0

While it is necessary to form the entire array, *stability is determined by the first column only. The number of changes of algebraic sign in this column*

[1] E. O. Doebelin, *Dynamic Analysis and Feedback Control* (New York: McGraw-Hill Book Company, 1962).

is equal to the number of unstable roots. Since our example exhibits no sign changes, the system is stable, which of course agrees with the roots (9–8) which were actually found earlier.

To show how useful results are obtained when the system parameters are known only as letters, let us consider the position control system of Fig. 9–1d, which might be used in an automatic machine tool. The system equation relating output angle θ_c to input voltage e_R is written directly from the block diagram as

$$(e_R - K_p\theta_c) \frac{K_aK_t/R_fB}{D(\tau_fD + 1)(\tau_LD + 1)} = \theta_c \qquad (9\text{–}17)$$

$$(\tau_f\tau_LD^3 + (\tau_f + \tau_L)D^2 + D + K)\theta_c = (K/K_p)e_R \qquad (9\text{–}18)$$

where

$$K \triangleq \text{loop gain} \triangleq \frac{K_aK_tK_p}{R_fB} \quad \frac{\text{volt/sec}}{\text{volt}} \qquad (9\text{–}19)$$

The Routh array is

$$
\begin{array}{ccc}
\tau_f\tau_L & 1 & 0 \\[2mm]
\tau_f + \tau_L & K & 0 \\[2mm]
\dfrac{\tau_f + \tau_L - K\tau_f\tau_L}{\tau_f + \tau_L} & 0 & 0 \\[2mm]
K & 0 & 0
\end{array}
$$

Since τ_f, τ_L, and K are normally positive numbers, it is clear that a sign change (and thus instability) will result if

$$K > \frac{\tau_f + \tau_L}{\tau_f\tau_L} = \frac{1}{\tau_f} + \frac{1}{\tau_L} \qquad (9\text{–}20)$$

Since, in feedback systems, we normally desire the highest possible loop gain to achieve accuracy and speed, relation (9–20) is of great practical importance since it puts an upper limit on K and also shows what must be done if higher K must be obtained. For example, if $\tau_f = 0.01$ sec and $\tau_L = 0.1$ sec, instability is predicted for $K > 110$. If we need $K = 200$, we clearly must change the motor and/or the load to reduce τ_f and/or τ_L. Note that one would actually use a K considerably *less* than the critical value, since for K at this value the roots are *right on the imaginary axis* (neither in the left or the right half plane) and the system, while not going to infinity, would be so oscillatory as to be unacceptable.

9–2. DEAD-TIME ELEMENTS.

We here introduce an element useful in modeling the dynamic behavior of certain physical systems which does not fit the pattern of linear, constant-coefficient differential equations but still allows a fairly simple analytical approach. When this *dead-time element* is combined with our more familiar first-order and second-order modes of response we considerably expand our capability in modeling real systems. The definition of the dead-time element whose input is q_i and output is q_o is given by

$$q_i = f(t) = \text{any time function}$$

$$q_o = f(t - \tau_{DT}), \qquad t \geq \tau_{DT} \qquad \tau_{DT} \triangleq \text{dead time, sec} \qquad \text{(9–21)}$$

That is, the output is identical in form to the input, but is *delayed* in occurrence by τ_{DT} seconds (see Fig. 9–5).

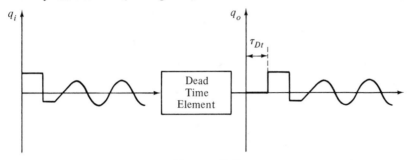

FIGURE 9–5

Behavior of Dead-Time Element

A long pneumatic transmission line as used in some remote-control systems gives a good example of dead-time behavior. Since the propagation velocity of a pressure disturbance in air is about 1120 ft/sec, a step pressure p_i in Fig. 9–6 produces *no response whatever* at the receiver location p_0 until 1 sec has gone by. Then the response follows approximately a first-order curve. A model for such a system might thus be a dead time

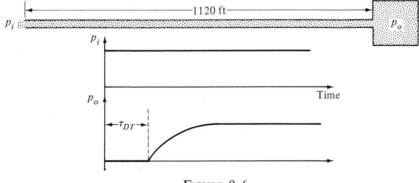

FIGURE 9–6

Measured Response of Long Pneumatic Transmission Line

connected in cascade with a first-order system. For unmanned lunar vehicles steered by radio remote control from the earth, a delay (dead time)
of about 1.3 seconds (one-way) or 2.6 seconds (round trip) will be encountered since the distance is 240,000 miles and the speed of radio wave
propagation is nearly 186,000 miles/sec. If the robot has a television "eye"
which sends a picture back to the human controller on earth to guide his
steering inputs, the 2.6 sec delay between command and observed response makes steering quite "unnatural" and difficult. Other practical
examples of dead times are found in the many industrial processes involving continuous webs or sheets of material (sheet steel, plastic films, paper
machines, etc.) where changes in thickness, composition, or temperature at
one station on the moving sheet are not felt at the "downstream" locations
until a definite time has passed.

When a dead time appears embedded in a cascade of "ordinary" linear
elements, as in Fig. 9–7a, it presents no analytical difficulties since it does
nothing except to shift the origin of the t axis by τ_{DT} seconds for every
response variable "downstream" from it. If the overall response (q_i to q_o)
is of interest, the dead time may be, for analysis purposes, shifted as in

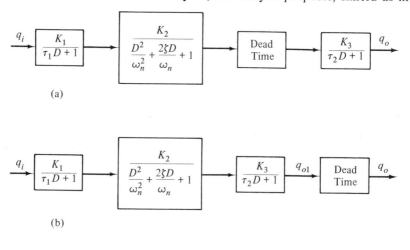

(a)

(b)

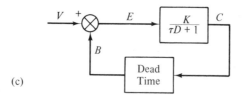

(c)

FIGURE 9–7

Dead Times Embedded in Conventional Systems

9–7b so that all the ordinary dynamics precede it. Then one solves for q_{o1}, and delays this response by τ_{DT} seconds to get q_o. When a dead time appears in a feedback system (Fig. 9–7c), then real analytical problems arise. Trying to get an equation relating C to V will give

$$[V - C(t - \tau_{DT})]\,\frac{K}{\tau D + 1} = C \qquad (9\text{–}22)$$

$$\tau\frac{dC}{dt} + C + KC(t - \tau_{DT}) = KV \qquad (9\text{–}23)$$

The term $KC(t - \tau_{DT})$ makes this a differential/difference equation; a class *not* readily solvable by analytical methods. Analog computers also do not have the capability of accurately simulating dead times unless a tape recorder/reproducer is employed. (Can you guess how the tape machine would be used?) Digital simulations such as CSMP handle dead times very easily and accurately since it is a simple matter to tell the computer to "remember" what was happening τ_{DT} seconds ago and make it "happen again now." The standard CSMP statement for a dead-time element with input X and output Y is

$$Y = DELAY(N,DT,X) \qquad (9\text{–}24)$$

where N is the number of computing increments contained in the dead time DT. Thus if we have specified a computing increment DELT = 0.1 sec on a TIMER card, and if the dead time were 5 seconds we would write Y = DELAY(50,5.0,X). The number N *must* be an integer. A complete program for the feedback system of Fig. 9–7c would be

```
TITLE    FEEDBACK SYSTEM WITH DEAD TIME
         E = V−B
         Cl = REALPL (0.0, 10., E)
         V = STEP (0.0)
         B = DELAY (50, 5.0, C)
         C = 3.7*Cl
TIME     FINTIM = 200., DELT = 0.1, O̅UTDEL = .5
PRTPLT   C, E, V
END
STO̅P
ENDJO̅B
```

Since a dead time is *not* a nonlinearity, a precise transfer function can be found for it; however, it will *not* be a ratio of polynomials in the D operator as is the case for "ordinary" lumped linear systems. Actually, frequency response and the sinusoidal transfer function are perhaps the most direct way to the desired result. If in (9–21) we let $q_i = A_i \sin \omega t$, then $q_o = A_i \sin [\omega(t - \tau_{DT})] = A_i \sin (\omega t - \phi)$ where $\phi \triangleq \omega\tau_{DT}$. Then

$$\frac{q_o}{q_i}(i\omega) = 1 \ \underline{/-\omega\tau_{DT}} \tag{9-25}$$

and we see that a dead time has an amplitude ratio of one at all frequencies and a lagging phase angle directly proportional to frequency (Fig. 9-8).

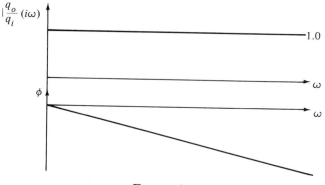

FIGURE 9-8

Dead-Time Frequency Response

In searching for a function of $i\omega$ which gives these results it is found that

$$\frac{q_o}{q_i}(i\omega) = e^{-i\omega\tau_{DT}} \tag{9-26}$$

gives the desired result, since

$$e^{-i\omega\tau_{DT}} = \cos\omega\tau_{DT} - i\sin\omega\tau_{DT}$$

$$= \sqrt{\cos^2\omega\tau_{DT} + \sin^2\omega\tau_{DT}} \ \underline{/-\omega\tau_{DT}} \tag{9-27}$$

Since we can always substitute D for $i\omega$ we get

$$\frac{q_o}{q_i}(D) = e^{-\tau_{DT}D} \tag{9-28}$$

for the operational transfer function. While this is a correct expression, it is of little use except as a means of generating *approximations* by the Taylor series expansion

$$e^{-\tau_{DT}D} = 1 - \tau_{DT}D + \tau_{DT}^2\frac{D^2}{2} - \tau_{DT}^3\frac{D^3}{6} + \ldots \tag{9-29}$$

The simplest usable approximation is

$$e^{-\tau_{DT}D} \approx 1 - \tau_{DT}D \tag{9-30}$$

which we see is an "ordinary" type of transfer function and thus will *not* lead to differential/difference equations and their associated difficulties, but *will* suffer from the uncertainties and inaccuracies always associated with approximations.

9-3. GENERALIZED FREQUENCY RESPONSE.

Throughout this text we have been using the concept of sinusoidal transfer function for finding frequency response but have only *proven* its validity in a few simple cases, elsewhere simply accepting it on faith. We now wish to establish its validity in general terms and then also extend the logarithmic plotting methods to arbitrarily complex systems. For a general linear system with constant coefficients and no dead times the differential equation is of form

$$(a_n D^n + a_{n-1}D^{n-1} + \ldots + a_1 D + a_0)q_o$$
$$= (b_m D^m + b_{m-1}D^{m-1} + \ldots + b_1 D + b_0)q_i \quad \text{(9-31)}$$

If $q_i = A_i \sin \omega t$ the method of undetermined coefficients tells us that the forced (steady-state) solution must have the form $q_o = A \sin \omega t + B \cos \omega t$, which can always be written as $q_o = A_o \sin (\omega t + \phi)$.

Our analysis uses the rotating-vector or "phasor" method of representing sinusoidal quantities. From a basic trigonometric identity

$$Ae^{i\theta} = A(\cos \theta + i \sin \theta) = A \cos \theta + iA \sin \theta \quad \text{(9-32)}$$

The complex number given by the right-hand side is shown in Fig. 9–9. We wish to represent both q_i and q_o in this way so we let $A = A_i$ and $\theta = \omega t$ for q_i and $A = A_o$ with $\theta = \omega t + \phi$ for q_o. For a given frequency ω the two phasors rotate at a constant angular velocity, always maintaining the fixed angle ϕ between them. We will need to be able to differentiate phasors with respect to time, so we note that

$$\frac{d}{dt}(A_i e^{i\omega t}) = (i\omega)A_i e^{i\omega t}, \qquad \frac{d}{dt}(A_o e^{i(\omega t+\phi)}) = (i\omega)A_o e^{i(\omega t+\phi)} \quad \text{(9-33)}$$

and furthermore,

$$\frac{d^n}{dt^n}(A_i e^{i\omega t}) = (i\omega)^n A_i e^{i\omega t} \quad \text{(9-34)}$$

$$\frac{d^n}{dt^n}(A_o e^{i(\omega t+\phi)}) = (i\omega)^n A_o e^{i(\omega t+\phi)} \quad \text{(9-35)}$$

so that differentiating n times is the same as multiplying by $(i\omega)^n$.

We now are ready to examine Eq. (9–31) when both q_i and q_o are in sinusoidal steady state. Then *every term* in that equation will be sinusoidal and representable by an appropriate phasor. In fact we may write

$$a_n(i\omega)^n A_o e^{i(\omega t+\phi)} + \ldots + a_1(i\omega)A_o e^{i(\omega t+\phi)} + a_0 A_o e^{i(\omega t+\phi)}$$
$$= b_m(i\omega)^m A_i e^{i\omega t} + \ldots + b_1(i\omega)A_i e^{i\omega t} + b_0 A_i e^{i\omega t} \quad \text{(9-36)}$$

Equation (9–36) will lead directly to the sinusoidal transfer function but we must first show that when Eq. (9–36) holds, the basic differential Eq.

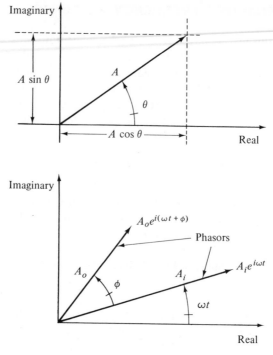

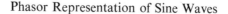

FIGURE 9–9

Phasor Representation of Sine Waves

(9–31) also holds, since they are *not* the same equation. Equation (9–36) is a complex algebraic (not differential) equation and may be split up into real and imaginary parts. Since a complex equation is satisfied only if *both* the real and imaginary parts are separately satisfied, we will show that satisfaction of the imaginary parts guarantees that the basic differential equation is also satisfied. This is done by examining each term; however, two on each side will be sufficient to establish the pattern.

$$Im[a_1(i\omega)A_o e^{i(\omega t+\phi)}] = a_1\omega A_o \cos(\omega t + \phi) = a_1 Dq_o \qquad (9\text{–}37)$$

$$Im[a_0 A_o e^{i(\omega t+\phi)}] = a_0 A_o \sin(\omega t + \phi) = a_0 q_o \qquad (9\text{–}38)$$

$$Im[b_1(i\omega)A_i e^{i\omega t}] = b_1\omega A_i \cos \omega t = b_1 Dq_i \qquad (9\text{–}39)$$

$$Im[b_0 A_i e^{i\omega t}] = b_0 A_i \sin \omega t = b_0 q_i \qquad (9\text{–}40)$$

Clearly, all the other terms will reduce in the same way, thus we can use Eq. (9–36) and be assured that the system differential equation will be satisfied.

$$[a_n(i\omega)^n + \ldots + a_1(i\omega) + a_0]A_o e^{i(\omega t+\phi)} \qquad (9\text{–}41)$$
$$= [b_m(i\omega)^m + \ldots + b_1(i\omega) + b_0]A_i e^{i\omega t}$$

$$\frac{A_o e^{i(\omega t + \phi)}}{A_i e^{i\omega t}} = \frac{A_o}{A_i} e^{i\phi} = \frac{A_o}{A_i}(\cos \phi + i \sin \phi) = \frac{A_o}{A_i} \underline{/\phi}$$

$$= \frac{a_n(i\omega)^n + \ldots + a_1(i\omega) + a_0}{b_m(i\omega)^m + \ldots + b_1(i\omega) + b_0} \triangleq \frac{q_o}{q_i}(i\omega) \qquad (9\text{-}42)$$

This last result is of course the sinusoidal transfer function, whose validity we have now established.

Our next objective is to extend the logarithmic plotting methods to linear systems of arbitrary complexity, as given by Eq. (9–16). We first note that one can always factor the polynomials in the numerator and denominator of (9–16) and express the transfer function as a *product* of first-order and/or second-order terms.

$$\frac{q_o}{q_i}(D) = \frac{K(\tau_1 D + 1)\ldots\left[\dfrac{D^2}{\omega_{n1}^2} + \dfrac{2\zeta_1 D}{\omega_{n1}} + 1\right]\ldots}{(\tau_a D + 1)\ldots\left[\dfrac{D^2}{\omega_{na}^2} + \dfrac{2\zeta_a D}{\omega_{na}} + 1\right]\ldots} \qquad (9\text{-}43)$$

For example, Eq. (9–11) can be so expressed using the known roots (9–8).

$$\frac{x_1}{f_{i3}}(D) = \frac{0.1D + 1}{(D + .00644 + i.445)(D + .00644 - i.445)}$$

$$\times \frac{1}{(D + .0741 + i1.79)(D + .0741 - i1.79)} \qquad (9\text{-}44)$$

$$\times \frac{1}{(D + .0694 + i1.24)(D + .0694 - i1.24)}$$

$$\frac{x_1}{f_{i3}}(D) = \frac{0.1D + 1}{\left[\dfrac{D^2}{(.445)^2} + \dfrac{2 \times .0144 D}{.445} + 1\right]\left[\dfrac{D^2}{(1.79)^2} + \dfrac{2 \times .0415 D}{1.79} + 1\right]}$$

$$\times \frac{1}{\left[\dfrac{D^2}{(1.24)^2} + \dfrac{2 \times .056 D}{1.24} + 1\right]} \qquad (9\text{-}45)$$

While Eq. (9–43) has an arbitrary number of factors, if we show the validity of our method for a single pair of factors it obviously holds for any number. Let

$$\frac{q_o}{q_i}(D) = G_1(D)G_2(D) \qquad (9\text{-}46)$$

and thus

$$\frac{q_o}{q_i}(i\omega) = G_1(i\omega)G_2(i\omega) = (M_1 \underline{/\phi_1})(M_2 \underline{/\phi_2}) = M_1 M_2 \underline{/\phi_1 + \phi_2} \qquad (9\text{-}47)$$

Now, applying the definition of the decibel to the amplitude ratio we get

$$\text{dB value of overall amplitude ratio} = 20 \log_{10}(M_1 M_2) \qquad (9\text{-}48)$$

$$= 20 \log_{10} M_1 + 20 \log_{10} M_2 = \text{sum of individual dB values} \qquad (9\text{-}49)$$

Thus if we graph dB amplitude ratio curves for any number of individual factors, the overall amplitude ratio is obtained by simple graphical addition. Equation (9–47) shows that the phase angle curves are similarly combined.

As an example, consider the transfer function

$$\frac{q_o}{q_i}(D) = \frac{1.0(0.05D + 1)}{\left[\dfrac{D^2}{1^2} + \dfrac{2 \times .05D}{1} + 1\right]\left[\dfrac{D^2}{5^2} + \dfrac{2 \times .05D}{5} + 1\right]} \qquad (9\text{–}50)$$

whose frequency response is graphed in Fig. 9–10. The curves shown there were obtained in just a few minutes by the following stepwise procedure.

1. Note that (9–50) contains two second-order terms with ω_n's of 1 and 5 rad/sec respectively and one first-order numerator term with $\tau = 0.05$ sec and thus a breakpoint frequency of 20 rad/sec.
2. Procure 3-cycle semi-log graph paper and lay out a frequency scale which will include the breakpoints 1, 5, and 20 rad/sec with "a little to spare" on both the low- and high-frequency ends.
3. Consult Fig. 8–17 and note that for $\zeta = 0.05$ a peak amplitude ratio of 20 dB may be expected. Since the two second-order terms will, at high frequency, ultimately produce $-360°$ phase shift while the first-order produces $+90°$, a phase angle scale from $+90°$ to $-360°$ will surely suffice. With these guides in mind lay out dB and ϕ scales.
4. To get the amplitude-ratio curve, first put in its straight line asymptote. Since the overall gain is 1, the low-frequency amplitude ratio is 0 dB. The first breakpoint is encountered at $\omega = 1$; draw the asymptote sloping downward to the right at -40 dB/decade. This line continues unbroken until $\omega = 5$, then the slope changes to -80 dB/decade since another -40 dB/decade is now added on. At $\omega = 20$ an upward slope of $+20$ dB/decade is added, so the composite slope is now -60 dB/decade from here out to infinity.
5. Using Figs. 7–17 and 8–17, correct the straight line amplitude ratio in the neighborhood of the breakpoints and also plot the phase angle curve.

We might note from Fig. 9–10 that when a system has several natural frequencies, even though the damping associated with each might be equally light ($\zeta = 0.05$ in this case), the higher natural frequency is not nearly as "prominent." (In Fig. 9–10 the first peak has amplitude ratio 10 while the second is 0.63.) The separation of the overall transfer function into its "component parts" and the logarithmic plotting scheme make the reason for this behavior quite obvious. Once the frequency "passes" the first peak, all subsequent peaks will be superimposed on the -40 dB/ decade line caused by the first resonance. Thus if the higher ω_n's are not very close to the first one it is difficult for them to contribute peaks with

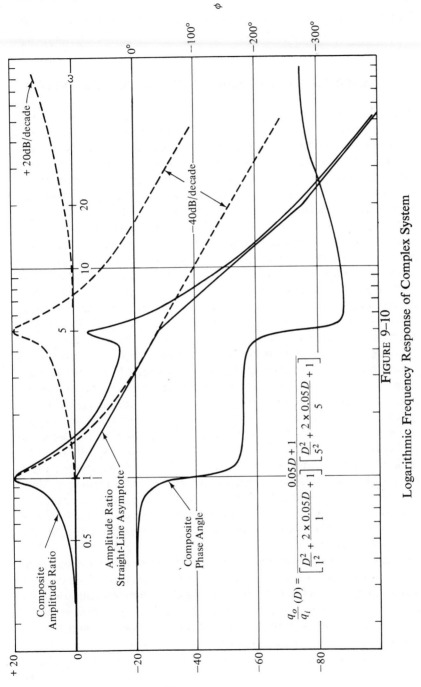

$$\frac{q_o}{q_i}(D) = \frac{0.05D+1}{\left[\dfrac{D^2}{1^2} + 2 \times 0.05D + 1\right]\left[\dfrac{D^2}{5^2} + 2 \times 0.05D + 1\right]}$$

FIGURE 9-10

Logarithmic Frequency Response of Complex System

435

high magnification even if their damping is slight. This helps to explain the behavior exhibited in Fig. 9–4.

9–4. RESPONSE TO PERIODIC, NONSINUSOIDAL INPUTS.

While a computer technique such as CSMP allows us to find the response of a very complex system to a very complex form of driving input, it is still desirable to have available generalized analytical methods for such problems. These methods are necessary, not so much to grind out numerical solutions to specific problems (the computer is generally unbeatable at this) as to provide *insight* into the qualitative aspects of system behavior. Such insight is essential in the conceptual phases of the design of new systems and is also necessary in deciding just which computer analyses are really needed to demonstrate system performance. For linear systems with constant coefficients such generalized analytical tools are available. They consist of a system description in terms of its frequency response and an input signal description in terms of its frequency spectrum. When both these descriptions are given one can always calculate the system response.

We have just shown in the previous section how to get the frequency-response curves of any linear system. The problem of finding the frequency spectrum of a given input signal is best treated by separating signals into classes for which a certain approach is applicable. Most (but not all) signals of practical importance can be classified in three broad classes: periodic, transient, and random (see Fig. 9–11). A periodic signal is one which exhibits a definite cycle and repeats itself over and over unendingly. A transient signal has a beginning and an end. It is zero for all times before its beginning and after its end, but may have any shape in between. Random signals continue unendingly but exhibit no predictable pattern or cycle. Frequency spectrum methods are available[2] for treating all three types of signals analytically. Furthermore, electronic instruments called *spectrum analyzers* allow direct calculation of frequency spectra from real physical signals, the so-called *real-time* analyzers often doing the job in a fraction of a second.

The intended scope of this text limits our treatment of spectrum analysis to the simplest case, which is the class of periodic functions. The necessary mathematical tool is the Fourier series, which we now introduce without proof. It can be shown that any periodic function $q_i(t)$ which is single-valued, finite, and has a finite number of discontinuities and maxima and

---

[2] E. O. Doebelin, *Measurement Systems* (New York: McGraw-Hill Book Company, 1967), pp. 143–188.

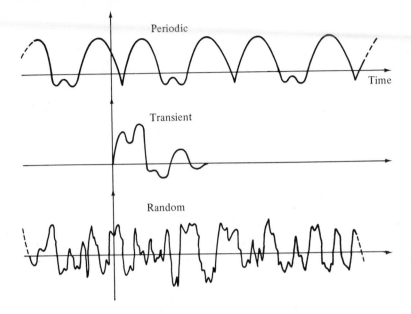

FIGURE 9–11

Common Classes of Signals

minima in one cycle (conditions easily met by any real physical signal) may be represented by the Fourier series:

$$q_i(t) = q_{i,av} + \frac{1}{L}\left(\sum_{n=1}^{\infty} a_n \cos\frac{n\pi t}{L} + \sum_{n=1}^{\infty} b_n \sin\frac{n\pi t}{L}\right) \quad (9\text{-}51)$$

where

$$q_{i,av} \triangleq \text{average value of } q_i = \frac{1}{2L}\int_{-L}^{L} q_i(t)\, dt \quad (9\text{-}52)$$

$$a_n \triangleq \int_{-L}^{L} q_i(t) \cos\frac{n\pi t}{L}\, dt \quad (9\text{-}53)$$

$$b_n \triangleq \int_{-L}^{L} q_i(t) \sin\frac{n\pi t}{L}\, dt \quad (9\text{-}54)$$

In Fig. 9–12 the origin of the t coordinate may be chosen wherever convenient. As an example, the square wave of Fig. 9–13 gives

$$q_{i,av} = 0, \text{ by inspection}$$

$$a_n = \int_{-.01}^{0} \cos\frac{n\pi t}{.01}\, dt + \int_{0}^{.01} -\cos\frac{n\pi t}{.01}\, dt = \frac{\sin n\pi - \sin n\pi}{100 n\pi} = 0 \quad (9\text{-}55)$$

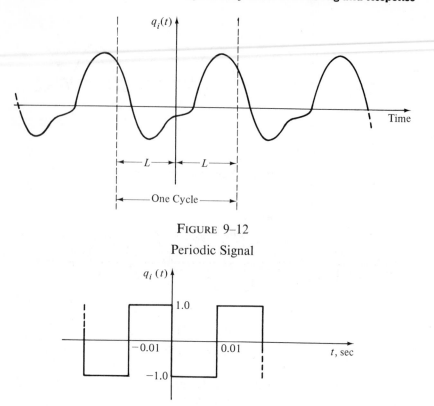

FIGURE 9–12

Periodic Signal

FIGURE 9–13

Square Wave

$$b_n = \int_{-.01}^{0} \sin \frac{n\pi t}{.01} \, dt + \int_{0}^{.01} -\sin \frac{n\pi t}{.01} \, dt = \frac{\cos(n\pi) - 1}{50n\pi} \quad \textbf{(9–56)}$$

Equation 9–51 then gives us

$$q_i(t) = 100 \sum_{n=1}^{\infty} \frac{\cos n\pi - 1}{50n\pi} \sin \frac{n\pi t}{.01} \quad \textbf{(9–57)}$$

$$= -\frac{4}{\pi} \left(\frac{\sin 100\pi t}{1} + \frac{\sin 300\pi t}{3} + \frac{\sin 500\pi t}{5} + \cdots \right)$$

To get the series (9–57) to *perfectly* fit the square wave, we must take an infinite number of terms. If we will accept an approximate representation the series may be truncated at a finite number of terms.

While Eq. (9–51) indicates that, in general, both sine and cosine terms are obtained, it is often most convenient to combine sines and cosines of like frequency by using the identity

$$A \cos \omega t + B \sin \omega t \equiv C \sin (\omega t + \alpha)$$

$$C \triangleq \sqrt{A^2 + B^2} \qquad (9\text{-}58)$$

$$\alpha \triangleq \tan^{-1} B/A$$

We thus get a series of only sine terms with phase angles. The *frequency spectrum* of a periodic function is given by its Fourier series since the series tells us how the function is "built up" from sine waves of various frequencies, amplitudes, and phase angles. This information is best displayed graphically, as in Fig. 9–14 for our example square wave. The

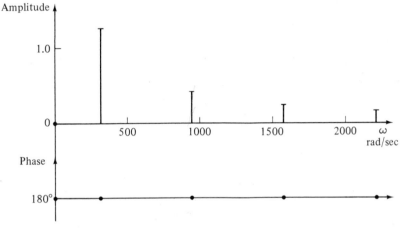

FIGURE 9–14

Frequency Spectrum of Square Wave

lowest frequency component is called the *fundamental* or *first harmonic*. Components associated with $n = 2, 3, 4$, etc., are called, respectively, the second, third, and fourth harmonics.

The response of a linear system to a periodic input is obtained by use of the superposition theorem and the system's frequency response, together with the Fourier series for the input. The Fourier series gives the driving function as a sum of sine waves, the superposition theorem allows us to get the total response as a sum of the responses to the individual sine waves, and the frequency-response curves let us get the response to any one sine wave quickly and easily. If the Fourier series for the input $q_i(t)$ is given by

$$q_i(t) = C_0 + C_1 \sin (\omega_1 t + \alpha_1) + C_2 \sin (2\omega_1 t + \alpha_2) + \ldots \quad (9\text{-}59)$$

We may then compute the response to each term separately if we know the amplitude ratio and phase shift of the system at $\omega = 0, \omega_1, 2\omega_1$, etc. These individual response terms are then simply added up to get the total system response, which will also be a periodic function. We should emphasize that the response found in this way is the *steady-state* response; that is, the output which exists when the input has been applied for a long enough time that system transients (natural response) have died out. If the periodic input is applied to a system at rest at $t = 0$, the Fourier series approach gives *no* information on how the response builds up to its final steady state. Figure 9–15 shows graphically how the steady-state response is calculated. Note that two factors are operating which allow us to neglect the higher frequency components of the Fourier series and thereby deal with a *finite*, rather than infinite, number of terms. First, the amplitudes of

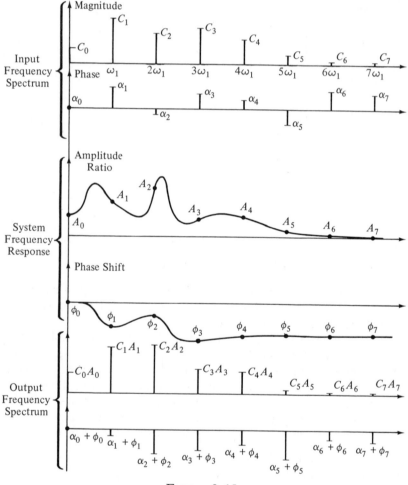

FIGURE 9–15

System Response Calculation for Periodic Input

the higher harmonics in a Fourier series always ultimately tend toward zero, since the frequency spectrum of real physical signals cannot extend to infinity. Second, the frequency response of any real system also cannot extend to infinity but must gradually reduce to zero amplitude ratio. Since the output (response) is the *product* of input spectrum and system amplitude ratio, when both factors approach zero for high values of ω the response must clearly also drop off. Thus, beyond some range of frequencies, the response amplitude is a small fraction of what it is at low frequencies and may legitimately be neglected relative to the large low-frequency values.

While simple periodic functions describable by mathematical formulas (such as the square wave) may be Fourier analyzed quickly and easily using Eqs. (9–51)–(9–53a), in many practical problems the periodic function is complicated and also may be given by an *experimental graph* for which no formula is known. Since Eqs. (9–51)–(9–53a) are all susceptible to numerical calculation, one can write a general-purpose digital computer program to compute the Fourier series. The input data to such a program are generally a set of points obtained by sampling the graph of $q_i(t)$, usually at equally spaced intervals. The number of sample points needed to accurately describe the function depends on its frequency content. If there are many "wiggles" within one cycle of the function, it will take many sample points to document them. A program available to the author allows up to 200 points and will compute any number of harmonics asked for, up to one-half the number of data points. A 50-point sample can thus provide up to 25 harmonics.

Figure 9–16 shows one cycle of a periodic input which is to be applied to the linear system of Fig. 9–10. Since no formula is available for this experimentally measured $q_i(t)$ the above-mentioned computer program was used to obtain the first 10 harmonics from a 20-point sample. The program also "reconstructs" the original function from the truncated Fourier series so that one can see whether it is a good fit of the given curve. Calculation of the ten harmonics and reconstruction of both a 5-harmonic and a 10-harmonic curve fit are accomplished in 2 seconds. We see that a 5-harmonic approximation gives a quite respectable curve fit while the 10-harmonic is almost perfect. This is substantiated in Fig. 9–17 where we see that the 9th and 10th harmonics have become very small.

Figures 9–10 and 9–17 now allow us to calculate the steady-state response $q_{oss}(t)$ for the $q_i(t)$ of Fig. 9–16. We have

$$q_{oss}(t) = (0.35)(1.0) + (0.56)(1.66) \sin (0.628t + 118°)$$
$$+ (0.65)(2.8) \sin (1.26t - 235°) + (0.23)(0.46) \sin (1.88t - 210°)$$
$$+ (0.27)(0.27) \sin (2.51t - 198°) + \ldots \qquad \textbf{(9–60)}$$

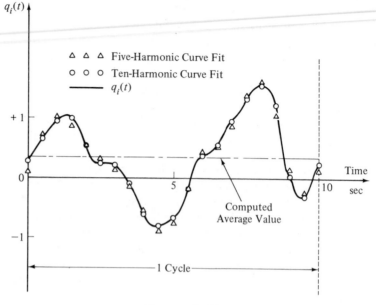

FIGURE 9-16

Fourier Series for Periodic Function

Note that the largest component of $q_{oss}(t)$ occurs at the second harmonic frequency since there we have *both* a large input signal and also a system amplitude ratio greater than 1.0. Any harmonic frequency at which this occurs will give a large response component. In vibrating systems where large response motions are *not* desired, this method of analysis gives a clear picture of just which features of the input force and system response are causing the trouble and the system designer can try to change either or both to reduce overall vibration. In Eq. (9–60) the fourth harmonic amplitude in q_{oss} is only 4% of the second harmonic, and thus it and the higher harmonics contribute little to the total q_{oss} and might be safely neglected. No matter how many harmonics are retained in $q_{oss}(t)$ it will be a periodic function when plotted against time.

If it is necessary to know how the system behaves during the time the steady state is being built up in a system which starts from rest, a complete solution including both complementary and particular solutions must be obtained. This is analytically rather difficult in our example, since the input $q_i(t)$ is given by a graph rather than a formula. A computer approach using CSMP, however, again saves the day. The function $q_i(t)$ is easily produced using the arbitrary function generator AFGEN and made periodic by manufacturing a time-sweep using ramps and steps. Let us give the program and then explain it.

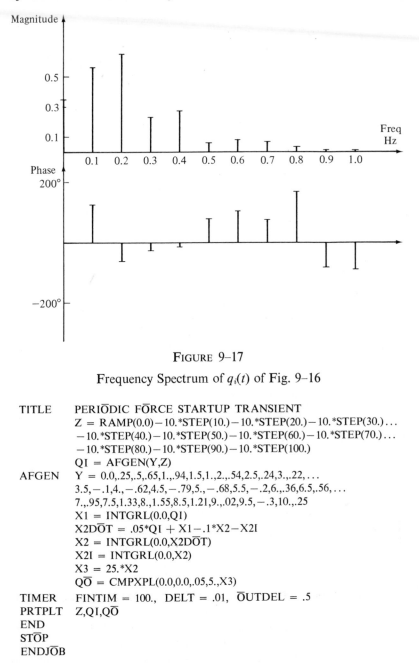

FIGURE 9–17

Frequency Spectrum of $q_i(t)$ of Fig. 9–16

```
TITLE     PERIŌDIC FŌRCE STARTUP TRANSIENT
          Z = RAMP(0.0) − 10.*STEP(10.) − 10.*STEP(20.) − 10.*STEP(30.)...
          − 10.*STEP(40.) − 10.*STEP(50.) − 10.*STEP(60.) − 10.*STEP(70.)...
          − 10.*STEP(80.) − 10.*STEP(90.) − 10.*STEP(100.)
          QI = AFGEN(Y,Z)
AFGEN     Y = 0.0,.25,.5,.65,1.,.94,1.5,1.,2.,.54,2.5,.24,3.,.22,...
          3.5,−.1,4.,−.62,4.5,−.79,5.,−.68,5.5,−.2,6.,.36,6.5,.56,...
          7.,.95,7.5,1.33,8.,1.55,8.5,1.21,9.,.02,9.5,−.3,10.,.25
          X1 = INTGRL(0.0,QI)
          X2DŌT = .05*QI + X1 − .1*X2 − X2I
          X2 = INTGRL(0.0,X2DŌT)
          X2I = INTGRL(0.0,X2)
          X3 = 25.*X2
          QŌ = CMPXPL(0.0,0.0,.05,5.,X3)
TIMER     FINTIM = 100.,  DELT = .01,  ŌUTDEL = .5
PRTPLT    Z,QI,QŌ
END
STŌP
ENDJŌB
```

The time-sweep function Z is produced with the RAMP input plus a sequence of delayed STEP inputs as shown in Fig. 9–18. The program

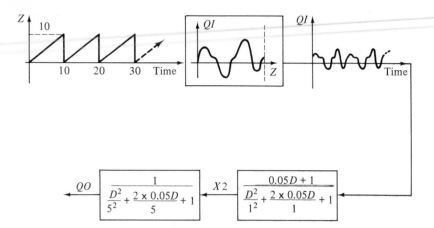

FIGURE 9–18

CSMP Approach for Complete Response to Periodic Input

provides for 10 cycles of $q_i(t)$; hopefully, steady state will be achieved by then. If not, more delayed steps may easily be added to Z. The waveform of q_i is produced by the arbitrary function generator; 20 pairs of Z, QI points are taken from Fig. 9–16, more could be given if a closer fit were desired. Since CSMP does not provide a standard term for numerator dynamics, the $0.05D + 1$ term is combined with one of the second-order forms and set up as a differential equation.

$$(D^2 + 0.1D + 1)x_2 = (0.05D + 1)q_i \qquad \textbf{(9–61)}$$

To avoid differentiating q_i we integrate the whole equation to get

$$\left(D + 0.1 + \frac{1}{D}\right)x_2 = 0.05q_i + \int q_i \, dt \qquad \textbf{(9–62)}$$

which is then conventionally set up as shown in the program. The other second-order term is done with a CMPXPL as usual. Figure 9–19 shows the total response of $q_o(t)$ from $t = 0$ until steady state is essentially reached. In steady state, q_o agrees well with Eq. (9–60).

9–5. EXPERIMENTAL MODELING OF COMPLEX SYSTEMS.

Experimental modeling is often necessary in practice, both to verify the predictions of theoretical models and also to *develop* models in those cases where a reasonably good theory is not available. For systems which are approximately linear the frequency-response technique is undoubtedly the most widely used approach. Once the amplitude ratio and phase angle

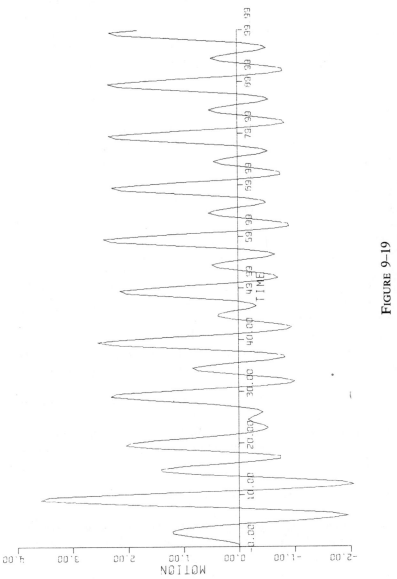

FIGURE 9-19

Startup Transient for Periodic Forcing Function

445

curves have been measured over the pertinent range of frequencies, the logarithmic plotting methods are often very helpful in synthesizing the form and numerical values for an analytical transfer function which fits the experimental data. Computerized curve-fitting methods are also available for this task. In some areas, such as control system design, the frequency-response curves may be used directly in the design method; an analytical formula is not necessary.

For very slow processes, such as large thermal power plants, frequency-response testing becomes difficult because the frequencies are *very* low, making the testing extremely time consuming and sometimes impractical due to random drifts. Then the *pulse-testing* method may be used. Here a single transient (of carefully selected size and shape) is applied to the input, and the output transient recorded. The input and output transients are then Fourier transformed in a computer to give the frequency-response curves of the system. In determining the dynamics of human machine operators such as pilots of aircraft, *random-signal* testing has generally been found necessary. In an unthinking machine, if linear, theory shows that one test signal is really just as good as any other and thus the machine need not be tested with its real-world input forms; any *convenient* form may be used. That is, if we find the frequency-response curves of a *machine* by frequency-response testing, pulse testing, or random signal testing they will all come out the same. Not so with a human operator; if tested with sine waves he will very quickly notice the cyclic behavior and greatly improve his tracking behavior beyond what he could do with a random input. Since a pilot's real world inputs will be random, he must be tested with such inputs. The approach again is to record a sufficiently long sample of his input (say a slowly moving oscilloscope display) and his output (motion of a control stick which he tries to match to the oscilloscope light-spot motion) and computer process these to get frequency-response curves. Curve-fitting may then produce an analytical transfer function. It may be of passing interest to note that a human operator transfer function applicable to tracking and piloting tasks has been established and is regularly used in man-machine dynamics studies. Its form is

$$\frac{q_o}{q_i}(D) = \frac{K(\tau_1 D + 1)e^{-\tau_{DT}D}}{(\tau_2 D + 1)(\tau_3 D + 1)} \qquad (9\text{--}63)$$

This transfer function is *adaptive* in that the human operator subconsciously *alters* the values of K, τ_1 and τ_2 to suit the needs of the task at hand.

A variety of other experimental modeling techniques have been suggested and practically applied to a greater or less extent. The interested reader will find material on these in the bibliography of this chapter.

BIBLIOGRAPHY

1. Banham, J. W., "Obtain Process Dynamics by Pulse Testing," *Control Engineering* (April, 1965).

2. Campbell, D. P., *Process Dynamics*. New York: John Wiley & Sons, Inc., 1958.

3. Coughanowr, D. R. and L. B. Koppel, *Process Systems Analysis and Control.* New York: McGraw-Hill Book Company, 1965.

4. Crandall, S. H., et al., *Dynamics of Mechanical and Electromechanical Systems*. New York: McGraw-Hill Book Company, 1968.

5. McRuer, D. T. and E. S. Krendel, "The Human Operator as a Servo System Element," *Jour. of the Franklin Inst.*, Vol. 267 (No. 5, May and No. 6, June, 1959).

6. Milsum, J. H., *Biological Control Systems Analysis*. New York: McGraw-Hill Book Company, 1966.

PROBLEMS

9-1. For the system of Fig. 9–1a, and using the numerical values of Eq. (9–4), find the transfer functions $(x_2/f_{i1})(D)$ and $(x_2/f_{i3})(D)$.

9-2. Repeat prob. 9–1 for the transfer functions $(x_3/f_{i1})(D)$ and $(x_3/f_{i3})(D)$.

9-3. Add a force f_{i2} on mass M_2 in Fig. 9–1a and find the transfer functions [using numerical values from Eq. (9–4)]:

    a. $(x_1/f_{i2})(D)$ b. $(x_2/f_{i2})(D)$
    c. $(x_3/f_{i3})(D)$

9-4. Using available "canned" computer programs, find and plot frequency-response curves (analogous to Fig. 9–4) for:

    a. $(x_1/f_{i2})(i\omega)$ b. $(x_1/f_{i3})(i\omega)$
    c. $(x_2/f_{i1})(i\omega)$ d. $(x_2/f_{i2})(i\omega)$
    e. $(x_2/f_{i3})(i\omega)$ f. $(x_3/f_{i1})(i\omega)$
    g. $(x_3/f_{i2})(i\omega)$ h. $(x_3/f_{i3})(i\omega)$

Use numerical values from Eq. (9–4).

9-5. Using CSMP (or similar program) and the numerical values of Eq. (9–4), find

    a. The response of x_1, x_2, and x_3 to a step input of f_{i2} (force on M_2) of size 1.0.

    b. The response of x_1, x_2, and x_3 to a step input of f_{i3} of size 1.0.

    c. Set $B_1 = B_2 = 0$ and find x_1, x_2, and x_3 if initial conditions are $x_1(0) = -1$, $x_2(0) = -2$ and $x_3(0) = -3$.

    d. Find x_1, x_2, and x_3 for simultaneous step inputs of size 1.0 for f_{i1}, f_{i2}, and f_{i3}.

9–6. Set up the differential equations for the system of Fig. 9–1b, letting inputs e_{i1}, e_{i2}, and e_{i3} be arbitrary functions of time.

9–7. Set up the differential equations for the system of Fig. 9–1c, letting inputs q_{i1} and q_{i2} be arbitrary functions of time.

9–8. In the system of Fig. 9–1d, add a block with transfer functions $(\tau_1 D + 1)/(\tau_2 D + 1)$ between the amplifier and the motor field and apply Routh criterion to obtain useful inequalities. What "peculiar" thing happens if $\tau_1 \equiv \tau_L$?

9–9. For the system of Fig. 9–7c, with $\tau = 0.01$ and $K = 10$, use CSMP to find the largest allowable dead time for stability.

9–10. Plot logarithmic frequency-response curves for:

    a. $\dfrac{10}{(2D + 1)(5D + 1)}$ b. $\dfrac{5(0.1D + 1)}{(5D + 1)}$

    c. $\dfrac{4(2D + 1)}{0.04D^2 + 0.08D + 1}$ d. $\dfrac{50}{(3D + 1)(10D + 1)(15D + 1)}$

9–11. Analytically compute the Fourier series for the periodic functions of:

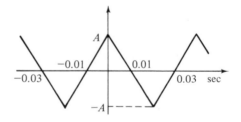

    a. Fig. P9–1

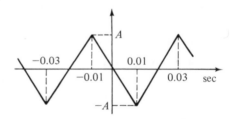

    b. Fig. P9–2

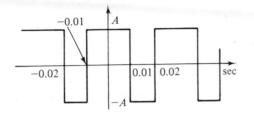

c. Fig. P9–3

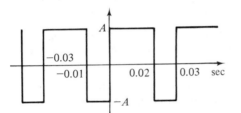

d. Fig. P9–4

Then plot the frequency spectrum (magnitude and phase angle) of each signal.

9–12. Analytically find the steady-state response of a system with transfer function $10/(0.005D + 1)$ to an input signal as in

Fig. P9–5

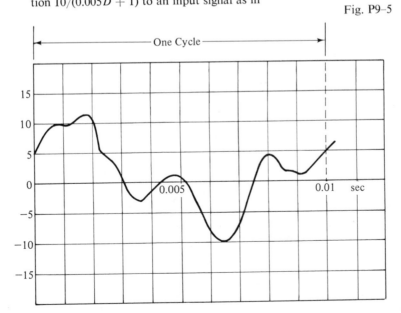

a. Fig. P9–1   b. Fig. P9–2
c. Fig. P9–3   d. Fig. P9–4
Take $A = 1.0$.

9–13. Repeat prob. 9–12 with a system transfer function of $10/(0.000025 D^2 + 0.001 D + 1)$.

9–14. Repeat prob. 9–12 but use CSMP (or similar program) to find complete response, including the starting transient. Let input start at $t = 0$.

9–15. Repeat prob. 9–13 but use CSMP (or similar program) to find complete response, including the starting transient. Let input start at $t = 0$.

9–16. Use a canned digital program to find a Fourier series for the signal of Fig. P9–5. Plot its frequency spectrum.

9–17. Using the results of prob. 9–16, analytically find the steady-state response of a system with transfer function $8/(0.001 D + 1)(0.0001 D + 1)$ to this periodic input.

9–18. Repeat prob. 9–17 but now use CSMP (or similar program) to get the complete response, including the starting transient.

10

DISTRIBUTED-
PARAMETER MODELS

10–1. LONGITUDINAL VIBRATIONS OF A ROD.

The progression in model complexity (and hopefully accuracy) from first-order to second-order and then to higher (but still finite) order types has its ultimate end in the distributed-parameter model, which may be thought of as having an infinite number of infinitesimally small lumps. We have earlier, even in the first chapter, attempted to develop at least some qualitative feeling for the relations and distinctions between lumped- and distributed-parameter models. We are now in a position to do this somewhat more quantitatively and completely. Since the mathematical topic involved, partial differential equations, is a vast and complex one we cannot hope to treat it in any generality, but we will rather use two simple examples to develop some physical feeling for the concepts involved.

Our first example concerns the determination of the vibration characteristics (natural frequencies, mode shapes, etc.) of a slender rod which is initially deformed in the axial direction and then released to perform free longitudinal vibrations. In Fig. 10–1 the position coordinate x, measured from the left end of the rod, allows us to describe the location of any part of the rod. The Y and Z coordinates will be found to be unnecessary because our "slender rod" assumption will make variations in these two transverse directions negligible. To obtain the system equation

451

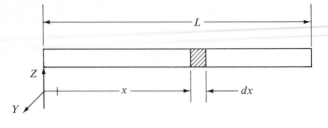

FIGURE 10–1

Slender Rod in Longitudinal Vibration

we assume the rod at time $t = 0$ has been deformed longitudinally in some way and is then released to perform free longitudinal vibrations with no external driving force applied. We also assume no friction or damping effects either inside the rod material or at the interface between the rod surface and the medium (such as atmospheric air) in which it might be immersed.

When the rod is unstrained and at rest we define the displacement u of any transverse plane in the rod to be zero. This displacement u is actually the unknown in our system since if we know u for any station x in the bar and for any time t we have completely documented the rod's longitudinal motion. The displacement u of any plane away from its equilibrium position is thus rightly called $u(x, t)$ since it is a function of both location in the bar and time t. (This basic fact will lead us inexorably to partial differential equations since when one writes derivatives of quantities which are functions of more than one variable he must write them as *partial* derivatives.) We now choose a rod element of infinitesimal length dx and located at an arbitrary location x in the rod. Since this problem involves motion of bodies under the action of forces, it is natural to apply Newton's law to the element dx in hopes of getting a system equation. Since no external forces are allowed by our assumptions, the forces at the two ends of the element due to internal stresses are the only ones we need to find.

These internal forces are the product of the stresses (force per unit area) and the cross-sectional area. If the rod were not "slender" the stresses might vary over the cross section (in the y and z directions) at a given x and we would need to integrate over the area to get the total force. Our assumption of slenderness presumes a *uniform* stress at any x and we thus need merely multiply stress by area to get total force. We next recall that stress and strain are related by the material modulus of elasticity E so long as we remain within the elastic limit of the material. An analysis of Fig. 10–2 gives the relation between strain and displacement u. The definition of unit strain ϵ is

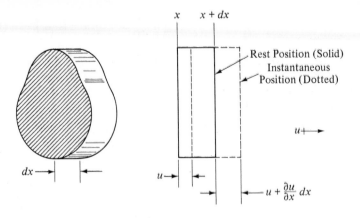

FIGURE 10–2

Element of Rod

$$\epsilon \triangleq \frac{\text{change in length of element}}{\text{length of element}} \tag{10-1}$$

and thus the unit strain at any location x would be $\Delta u/\Delta x$. Going to the limit this would be du/dx, but since u is a function of two variables we write it as $\partial u/\partial x$. Note that if $\partial u/\partial x$ is positive the strain is tension, while if it is negative the strain is compression.

Now the unit strain at any location is $\partial u/\partial x$ evaluated at that location; however, if we wish, we can write ϵ at $(x + dx)$ in terms of ϵ at x as follows:

unit strain at $(x + dx)$

$$= \text{unit strain at } x + \begin{bmatrix} \text{rate of change} \\ \text{of strain with} \\ \text{respect to } x \end{bmatrix} dx \tag{10-2}$$

$$= \frac{\partial u}{\partial x} + \frac{\partial}{\partial x}\left[\frac{\partial u}{\partial x}\right] dx = \frac{\partial u}{\partial x} + \frac{\partial^2 u}{\partial x^2} dx \tag{10-3}$$

We can now compute the stresses at x and $(x + dx)$ by using the definition of modulus of elasticity E.

$$E \triangleq \frac{\text{stress}}{\text{strain}} \tag{10-4}$$

$$\text{stress at } x = E\frac{\partial u}{\partial x} \tag{10-5}$$

$$\text{stress at } (x + dx) = E\left(\frac{\partial u}{\partial x} + \frac{\partial^2 u}{\partial x^2} dx\right) \tag{10-6}$$

Assuming these stresses uniform over the cross-sectional area A, the forces are

$$\text{force at } x = -AE\frac{\partial u}{\partial x} \qquad \left(\text{if } \frac{\partial u}{\partial x} \text{ is } +, \text{ force is } -\right) \qquad \text{(10–7)}$$

$$\text{force at } (x + dx) = AE\left(\frac{\partial u}{\partial x} + \frac{\partial^2 u}{\partial x^2}dx\right) \qquad \text{(10–8)}$$

Since the element dx has mass $A\,dx\,\rho$ and its acceleration is $\partial^2 u/\partial t^2$, Newton's law gives

$$-AE\frac{\partial u}{\partial x} + AE\left(\frac{\partial u}{\partial x} + \frac{\partial^2 u}{\partial x^2}dx\right) = A\rho\,dx\,\frac{\partial^2 u}{\partial t^2} \qquad \text{(10–9)}$$

and finally

$$\frac{\partial^2 u}{\partial x^2} = \frac{\rho}{E}\frac{\partial^2 u}{\partial t^2} \qquad \text{(10–10)}$$

where

$$\rho \triangleq \text{material mass density}$$

Equation (10–10) is the equation of motion of this system. The unknown is $u(x, t)$ which, if solved for, will tell us how every portion of the rod moves longitudinally. That is, if you tell me which x location you are interested in, I will tell you the time history of its motion. This equation is one of the basic equations of classical physics, the *one-dimensional wave equation* and is found to apply to a number of practical problems. Actually, Eq. (10–10) alone is not sufficient to completely define, and therefore solve, the problem of rod vibration. While our force analysis for the element dx is unaffected by constraints applied at the ends ($x = 0$, $x = L$) of the rod, one would intuitively guess that such constraints *would* influence the rod motion. Thus to completely define the problem we must specify the *boundary conditions* at $x = 0$ and $x = L$. Figure 10–3 shows

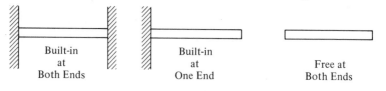

Built-in	Built-in	
at	at	Free at
Both Ends	One End	Both Ends

FIGURE 10–3

Various Boundary Conditions

three possible sets of boundary conditions which would correspond to various practical configurations. We shall work out the solution for a rod free at both ends.

For a rod free at both ends, the force on the ends must at all times be zero (neglecting any force due to "bumping into" air particles). If the force is zero the stress must be zero and if the stress is zero the strain $\partial u/\partial x$ must also be zero. We may thus state the boundary conditions mathematically as

$$\text{for any value of } t, \left.\begin{matrix} x = 0 \\ x = L \end{matrix}\right\} \frac{\partial u}{\partial x} = 0 \qquad (10\text{--}11)$$

We earlier stated that the vibrations are induced by an initial deformation of the rod. Mathematically speaking, at $t = 0$, $u(x, 0) = f(x)$, where $f(x)$ is a given function of x telling how the rod is initially deformed. Since we release the rod from its deformed condition with zero velocity, we also have $\partial u/\partial t = 0$ at $t = 0$ and for any x. We have now sufficiently described the situation that a solution for $u(x, t)$ should exist.

Just as in ordinary differential equations, no universal method of solution exists for partial differential equations so that one often tries some reasonable form of solution and sees if it works (satisfies equation, boundary conditions, and initial conditions). The nature of the solution assumed depends on the form of the equation, past experience with similar problems, physical intuition about the behavior of the system, etc. A product form of solution $u = f(x) g(t)$ where f and g are unknown functions is often applicable to equations of the type (10–10). An analyst knowledgeable about vibration could be even more specific and guess that u will very likely have the form

$$u = f(x) \cos \omega t \qquad (10\text{--}12)$$

Let us pursue this suggestion and see whether (10–12) actually can be the solution. Note that for $t = 0$, $u = f(x)$ and $\partial u/\partial t = \text{velocity} = -\omega f(x) \sin \omega t = 0$, which fits the initial conditions stated earlier. (Would $u = f(x) \sin \omega t$ have worked?) We must now see if (10–12) can be made to satisfy Eq. (10–10) by substituting the assumed solution into the system equation.

$$\frac{d^2 f(x)}{dx^2} \cos \omega t = \frac{\rho}{E}(-\omega^2 \cos \omega t)f(x) \qquad (10\text{--}13)$$

$$\frac{d^2 f}{dx^2} + \frac{\rho \omega^2}{E} f = 0 \qquad (10\text{--}14)$$

Note that our problem has now been reduced to solving an *ordinary* (rather than *partial*) differential equation. (This is the usual pattern in solving partial differential equations no matter what specific technique is used.) In Eq. (10–14) ρ and E are known constants whereas ω (the frequency of vibration) is a constant as yet unknown. Applying our usual methods to (10–14) we get

$$\left[D^2 + \frac{\rho \omega^2}{E}\right] f = 0 \qquad D \triangleq \frac{d}{dx} \qquad (10\text{--}15)$$

$$D^2 + \frac{\rho \omega^2}{E} = 0 \qquad \text{roots} = \pm i\omega \sqrt{\frac{\rho}{E}} \qquad (10\text{--}16)$$

$$f = C \sin\left(\sqrt{\frac{\rho}{E}} \omega x + \phi\right) \qquad (10\text{--}17)$$

The constants of integration C and ϕ must be found using the boundary conditions and in this process the value of frequency ω will also come out. We may write for u

$$u = f(x) \cos \omega t = C \sin \left(\sqrt{\frac{\rho}{E}} \omega x + \phi \right) \cos \omega t \qquad (10\text{--}18)$$

Now when $x = 0$, $\partial u / \partial x = 0$, thus

$$\frac{\partial u}{\partial x} = (C \cos \omega t) \cos \left(\sqrt{\frac{\rho}{E}} \omega x + \phi \right) \sqrt{\frac{\rho}{E}} \omega = 0 \qquad (10\text{--}19)$$

$$0 = (C \cos \omega t) \sqrt{\frac{\rho}{E}} \omega \cos \phi \qquad (10\text{--}20)$$

If we choose $C = 0$ to satisfy (10–20) we get the trivial solution $u(x, t) = 0$; thus it must be that $\cos \phi = 0$, which occurs for $\phi = \pm\pi/2, \pm3\pi/2, \pm5\pi/2$, etc.

Now $\partial u / \partial x$ is also zero for $x = L$ and thus

$$0 = (C \cos \omega t) \sqrt{\frac{\rho}{E}} \omega \cos \left(\sqrt{\frac{\rho}{E}} \omega L + \phi \right) \qquad (10\text{--}21)$$

Again, $C = 0$ is trivial, so

$$\cos \left(\sqrt{\frac{\rho}{E}} \omega L + \phi \right) = 0, \quad \sqrt{\frac{\rho}{E}} \omega L + \phi = \pm\frac{\pi}{2}, \pm\frac{3\pi}{2}, \pm\frac{5\pi}{2}, \text{ etc.} \qquad (10\text{--}22)$$

If one checks the possible combinations which arise by substituting $\phi = \pm\pi/2, \pm3\pi/2$, etc., into (10–22) (and excluding the possibility of $\omega \leq 0$ since negative frequencies have no physical interpretation) we get

$$\sqrt{\frac{\rho}{E}} \omega L = \pi, 2\pi, 3\pi, 4\pi, \text{ etc.} \qquad (10\text{--}23)$$

and thus we have found the allowable values of vibration frequency ω (the "natural frequencies") to be

$$\omega = \frac{n\pi}{L} \sqrt{\frac{E}{\rho}} \qquad n = 1, 2, 3, \ldots \qquad (10\text{--}24)$$

Our solution for u is thus

$$u = C \sin \left(\sqrt{\frac{\rho}{E}} \omega x \pm \frac{\pi}{2}, \pm\frac{3\pi}{2}, \cdots \right) \cos \omega t$$

$$= C \cos \left(\sqrt{\frac{\rho}{E}} \omega x \right) \cos \omega t \qquad (10\text{--}25)$$

One can use any one of the ω values of (10–24) in (10–25) to get a solution. In fact, since the differential equation is linear, the superposition principle says that a *sum* of such solutions is also a solution.

To get some feeling for the meaning of these various solutions let us consider some specific cases. Suppose we take $n = 1$ in (10–24) and use this ω value in (10–25). We get

$$u(x, t) = C \cos \left(\frac{\pi x}{L}\right) \cos \left(\frac{\pi}{L} \sqrt{\frac{E}{\rho}} \, t\right) \qquad \textbf{(10–26)}$$

and for $t = 0$,

$$u = C \cos \frac{\pi x}{L} \qquad \textbf{(10–27)}$$

This formula tells us the "shape" into which the rod must initially be deformed to give the motion of Eq. (10–26) when released. That is, up to this point we have been rather vague about the initial deformation of the rod, merely calling it $f(x)$. We see now that to get a certain natural frequency to exist *alone*, not just "any old" shape $f(x)$ may be used. Figure 10–4 is a plot of Eq. (10–27). Note that C simply determines the scale or

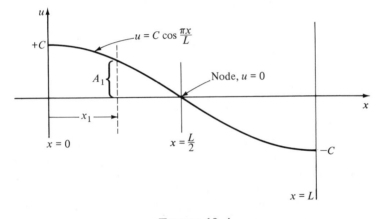

FIGURE 10–4

Mode Shape for First Natural Frequency

magnitude of the oscillation of any plane in the rod; if we double C, for example, all amplitudes are doubled. At any chosen location, such as x_1, the rod vibrates longitudinally according to

$$u(x_1, t) = C \cos \frac{\pi x_1}{L} \cos \frac{\pi}{L} \sqrt{\frac{E}{\rho}} \, t \triangleq A_1 \cos \omega t, \qquad \textbf{(10–28)}$$

a simple harmonic oscillation.

The frequencies ω given by (10–24) are called the natural frequencies of the rod; note that there are an *infinite* number of them. This is characteristic of distributed-parameter vibration models and also of the real systems which they represent; they really *do* have an infinite number of natural frequencies. Just as in lumped-parameter models, if an external driving force acts at a frequency equal to a natural frequency the undamped system builds up an infinite motion. The "dynamic deflection curve" of Fig. 10–4 is called the *mode shape*. Each natural frequency has its own characteristic mode shape; for $n = 2$ we have $\omega = (2\pi/L)\sqrt{E/\rho}$ and the mode shape of Fig. 10–5. In order to produce vibrations at a

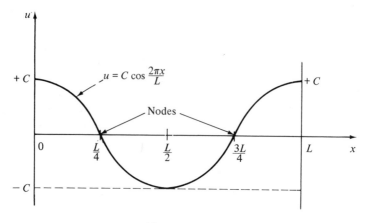

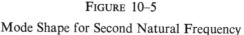

FIGURE 10–5

Mode Shape for Second Natural Frequency

single natural frequency only, it is necessary to initially deform the rod into precisely the cosinusoidal shapes such as Fig. 10–4 or 10–5. What if the initial deformation were some *arbitrary* shape? We would then find in general that *all* the natural frequencies would be excited in varying degrees depending on the shape of the particular deformation imposed. The motion is then a superposition of the various mode shapes, each at a different amplitude. The same *Fourier series* used in chapter 9 is also used in these sorts of problems to "build up" an arbitrary $f(x)$ from the various cosine waves representing the natural modes.

10–2. LUMPED-PARAMETER APPROXIMATIONS FOR ROD VIBRATION.

The distributed-parameter model developed and solved above is a very accurate representation of the behavior of a slender rod and predicts values which agree very closely with experimental measurements. However, the application of this type of model to practical problems, where the shapes are rarely as simple as that of our rod, encounters serious

mathematical difficulties and a lumped model may be necessary. To develop some feeling for the general nature and relative merits of the two
viewpoints let us now analyze the rod vibration problem from the lumped
point of view. Rather than dealing with continuous distributions of mass
and elasticity we will now work with discrete "massless springs" and
"springless masses."

The first problem in a lumped analysis concerns how the system will
be "dissected" into lumps. There are two general approaches. If the
describing partial differential equation for a distributed-parameter model
has been written (but *not* solved), one can apply standard finite-difference
formulas to create a lumped model mathematically without explicit
consideration of the physics of the problem. Either space (x, y, z) coordinates or time, or both, can be lumped. In the rod vibration problem,
for example, if x is lumped but t is not, the partial differential equation
becomes a simultaneous set of ordinary differential equations. If *both* x
and t are lumped, the equations become strictly algebraic. In the second
approach one cuts up the physical system into lumps according to some
rational scheme and then applies the pertinent physical laws to each lump.
This procedure will generate a set of simultaneous ordinary differential
equations. If these are linear with constant coefficients, routine analytical
techniques will supply exact solutions; otherwise, computer methods will
most likely be used. Analog computers can handle the ordinary differential equations without lumping time; that is, t remains a continuous
variable. Digital computer solution requires use of stepwise numerical
integration, so t also becomes "lumped."

We will here employ the second of the two above approaches without
necessarily indicating that this is always best. In cutting the system into
lumps, no universal scheme can be given which will always be the "best."
Any reasonable scheme will, however, give useable results which improve
in accuracy as more lumps are included in the model. For our rod vibration problem we will use the following lumping scheme.

1. Divide the rod length into equal segments.
2. Lump the mass of each segment at its center of mass.
3. Connect these masses by massless springs whose spring constants are
    equal to those of the rod segments between the mass points.

Figure 10–6 shows how application of this scheme leads to models with
one to four lumps; extension to any desired number of lumps is obvious.
For a 1-lump model, if the mass is displaced from its equilibrium position
no restoring forces are developed by the springs if the rod is free at both
ends; thus this model is not useful for studying rod vibration. (Would
this be true for the other boundary conditions of Fig. 10–3?) The simplest
model which will give any useful information is thus the 2-lump.

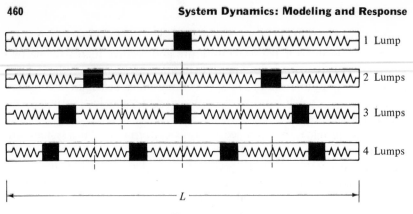

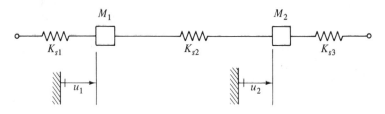

FIGURE 10–6

Lumping Scheme for Vibrating Rod

We will need expressions for spring constants and masses of rod-segments of length L_s. The mass is clearly $\rho A L_s$ while chapter 2 gives the spring constant as AE/L_s. For the two-mass system we set up coordinates as in Fig. 10–7. We now assume the masses are arbitrarily displaced from their neutral positions and released. Newton's law for each mass gives

FIGURE 10–7

Two-Lump Model

$$K_{s2}(u_2 - u_1) = M_1 \ddot{u}_1 \qquad (10\text{–}29)$$

$$K_{s2}(u_1 - u_2) = M_2 \ddot{u}_2 \qquad (10\text{–}30)$$

which may be reduced to one equation in one unknown using our usual methods.

$$D^2(M_1 M_2 D^2 + K_{s2}(M_1 + M_2))u_1 = 0 \qquad (10\text{–}31)$$

$$D^2(M_1 M_2 D^2 + K_{s2}(M_1 + M_2))u_2 = 0 \qquad (10\text{–}32)$$

The roots of the characteristic equation are

$$s_1 = 0 \qquad s_2 = 0 \qquad s_{3,4} = \pm i \sqrt{\frac{(M_1 + M_2)K_{s2}}{M_1 M_2}} \qquad (10\text{–}33)$$

making the solutions

$$u_1 = C_0 + C_1 t + C_2 \sin\left(\sqrt{\frac{M_1 + M_2}{M_1 M_2} K_{s2}}\, t + \phi_1\right) \quad \textbf{(10–34)}$$

$$u_2 = C_3 + C_4 t + C_5 \sin\left(\sqrt{\frac{M_1 + M_2}{M_1 M_2} K_{s2}}\, t + \phi_2\right) \quad \textbf{(10–35)}$$

The terms $C_0 + C_1 t$ and $C_3 + C_4 t$ refer to gross motions of the overall system corresponding to initial displacements and velocities and can be made to disappear by proper choice of initial conditions. Our main interest is in the oscillatory terms which we see exhibit only a *single* natural frequency ω given by

$$\omega = \sqrt{\frac{M_1 + M_2}{M_1 M_2} K_{s2}} = \sqrt{\frac{\rho A L}{\left(\frac{\rho A L}{2}\right)^2}} \times \frac{2AE}{L} = \frac{2.82}{L}\sqrt{\frac{E}{\rho}} \quad \textbf{(10–36)}$$

This frequency may be compared to the *first* natural frequency predicted by the "exact" distributed-parameter model, which was $(3.14/L)\sqrt{E/\rho}$. Our simplest lumped model thus predicts only one natural frequency (whereas an infinite number actually exist) and this one frequency is numerically somewhat inaccurate. Another aspect of the model's deficiencies lies in that we have direct results for the motion of only *2 points* in the rod, the $L/4$ and $3L/4$ points corresponding to the lumped masses. The distributed model gives results for *every* point. The "mode shape" predicted by the lumped model for equal and opposite initial displacements of M_1 and M_2 consists of 3 straight line segments as in Fig. 10–8 (compare with Fig. 10–4), rather than half a cosine wave.

If we now go to a 3-lump model the equations are (see Fig. 10–9)

$$(MD^2 + K_s)u_1 - K_s u_2 = 0 \quad \textbf{(10–37)}$$

$$K_s u_1 - (MD^2 + 2K_s)u_2 + K_s u_3 = 0 \quad \textbf{(10–38)}$$

$$K_s u_2 - (MD^2 + K_s)u_3 = 0 \quad \textbf{(10–39)}$$

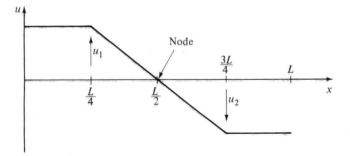

FIGURE 10–8
Lumped-Parameter Mode Shape

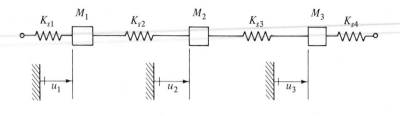

$$M_1 = M_2 = M_3 = \rho AL/3 \triangleq M \qquad\qquad K_{s2} = K_{s3} = 3AE/L \triangleq K_s$$

FIGURE 10–9

Three-Lump Model

which lead to

$$D^2 \left[\left(\frac{M}{K_s}\right)^3 D^4 + 4\left(\frac{M}{K_s}\right)^2 D^2 + 3\left(\frac{M}{K_s}\right) \right] u_1 = 0 \qquad \textbf{(10–40)}$$

and identical equations for u_2 and u_3. The quartic term can be factored using the quadratic formula

$$D^4 + 4\frac{K_s}{M} D^2 + 3\left(\frac{K_s}{M}\right)^2 = 0 \qquad \textbf{(10–41)}$$

$$D^2 = \frac{-4\frac{K_s}{M} \pm \sqrt{4\left(\frac{K_s}{M}\right)^2}}{2} = -3\frac{K_s}{M}, \ -\frac{K_s}{M} \qquad \textbf{(10–42)}$$

We thus have a total of 6 roots

$$s_1 = 0 \qquad s_2 = 0 \qquad s_{3,4} = \pm i\sqrt{\frac{3K_s}{M}} = \pm i\frac{3\sqrt{3}}{L}\sqrt{\frac{E}{\rho}}$$

$$s_{5,6} = \pm i\sqrt{\frac{K_s}{M}} = \pm i\frac{3}{L}\sqrt{\frac{E}{\rho}} \qquad \textbf{(10–43)}$$

The solution for u_1 (u_2 and u_3 have the same form) is thus

$$u_1 = C_0 + C_1 t + C_2 \sin\left(\frac{3\sqrt{3}}{L}\sqrt{\frac{E}{\rho}}\, t + \phi_1\right)$$

$$+ C_3 \sin\left(\frac{3}{L}\sqrt{\frac{E}{\rho}}\, t + \phi_2\right) \qquad \textbf{(10–44)}$$

Note that now *two* natural frequencies are predicted. Figure 10–10 compares these with the distributed-model and two-lump model results.

We see that the three-lump model has not only predicted another natural frequency but has also improved the accuracy of prediction for the first frequency. The pattern should now be becoming clear. More lumps in the model produce more natural frequencies and improve the accuracy, the limiting case being an infinite number of lumps, each infinitesimally small, which is of course precisely the distributed-parameter model. Also, the mode shapes become better defined since we get direct information on more x-locations (3 for a 3-lump model, 10 for a 10-lump model, etc.) and thus the linear interpolation between masses becomes more accurate.

Model	ω_1	ω_2	ω_3
Two-Lump	$\dfrac{2.83}{L}\sqrt{\dfrac{E}{\rho}}$	Not Predicted	Not Predicted
Three-Lump	$\dfrac{3.00}{L}\sqrt{\dfrac{E}{\rho}}$	$\dfrac{5.18}{L}\sqrt{\dfrac{E}{\rho}}$	Not Predicted
Distributed	$\dfrac{3.14}{L}\sqrt{\dfrac{E}{\rho}}$	$\dfrac{6.28}{L}\sqrt{\dfrac{E}{\rho}}$	$\dfrac{9.42}{L}\sqrt{\dfrac{E}{\rho}}$

FIGURE 10–10

Comparison of Model Types

10–3. CONDUCTION HEAT TRANSFER IN AN INSULATED BAR.

In Fig. 10–11 a slender metal rod, initially all at temperature T_0, is buried in perfect insulation. At time $= 0$, its left end is suddenly raised to temperature T_i and held there thereafter. We wish to find the temperature-

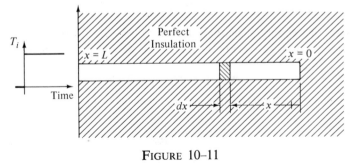

FIGURE 10–11

Heat Conduction Problem

time history of any point in the rod. The slenderness of the rod again makes the problem one-dimensional; variations in the y and z directions are assumed negligible. The basic physical law here is Fourier's law of heat conduction which says that the heat flux through any cross-section is proportional to the temperature gradient dT/dx at that cross-section. Mathematically,

$$q_x = -kA \frac{\partial T}{\partial x} \frac{\text{Btu}}{\text{sec}} \qquad (10\text{-}45)$$

where

$$k \triangleq \text{material thermal conductivity}$$
$$A \triangleq \text{cross-sectional area}$$

The temperature gradient is written as $\partial T/\partial x$ since in our problem T will be a function of *both* x and t. In the element of length dx, heat enters and leaves only at the ends since the surface is perfectly insulated; any difference between entering and leaving heat flows must show up as energy storage within the element.

We may write the heat flux at $(x + dx)$ in terms of the heat flux at x as follows.

$$\text{heat flux at } (x + dx) = \text{heat flux at } x + \begin{bmatrix} \text{rate of change of} \\ \text{heat flux with} \\ \text{respect to } x \end{bmatrix} dx \quad (10\text{-}46)$$

$$q_{x+dx} = -kA \frac{\partial T}{\partial x} + \frac{\partial}{\partial x}\left(-kA \frac{\partial T}{\partial x}\right) dx = -kA \frac{\partial T}{\partial x} - kA \frac{\partial^2 T}{\partial x^2} dx \quad (10\text{-}47)$$

Conservation of energy now gives

energy input rate $-$ energy output rate $=$ energy storage rate $\qquad$ **(10-48)**

$$-kA \frac{\partial T}{\partial x} - \left[-kA \frac{\partial T}{\partial x} - kA \frac{\partial^2 T}{\partial x^2} dx\right] = (\rho A \, dx) \, C \frac{\partial T}{\partial t} \qquad (10\text{-}49)$$

and thus

$$\frac{\partial^2 T}{\partial x^2} = \frac{\rho C}{k} \frac{\partial T}{\partial t} \qquad (10\text{-}50)$$

This is another common equation of classical physics, the one-dimensional diffusion equation. In solving our particular problem it will be helpful to define

$$\theta(x, t) \triangleq T(x, t) - T_i \qquad (10\text{-}51)$$

Since T_i is a constant, (10-50) becomes

$$\frac{\partial \theta}{\partial t} = a \frac{\partial^2 \theta}{\partial x^2} \qquad (10\text{-}52)$$

where

$$a \triangleq \text{thermal diffusivity} = \frac{k}{\rho C} \qquad (10\text{-}53)$$

Since the end $x = 0$ is perfectly insulated there can be no heat flow there and thus $\partial T/\partial x = 0$. Thus

$$\frac{\partial \theta}{\partial x} (0, t) = 0 \qquad (10\text{-}54)$$

and also, $\theta(L, t) = T(L, t) - T_i = T_i - T_i = 0$. Also, at $t = 0, T(x, 0) = T_0$ and thus

$$\theta(x, 0) = T_0 - T_i \tag{10-55}$$

We are now again at a point where the form of the solution must be assumed. Again a product type of solution works; we assume

$$\theta(x, t) = X(x)\, G(t) \tag{10-56}$$

where X is a function *only* of x and G is a function *only* of t. Substituting into (10-52) gives

$$\frac{\partial \theta}{\partial t} = X(x)\frac{\partial G}{\partial t} = a\frac{\partial^2 \theta}{\partial x^2} = a\, G(t)\frac{\partial^2 X}{\partial x^2} \tag{10-57}$$

$$\frac{a}{X}\frac{\partial^2 X}{\partial x^2} = \frac{1}{G}\frac{\partial G}{\partial t} \tag{10-58}$$

Since the left side of (10-58) depends *only* on x and the right side *only* on t, they can be equal *only* if they both equal the same constant, call it $-a\lambda^2$. We have then the two *ordinary* differential equations

$$\frac{d^2 X}{dx^2} + \lambda^2 X = 0 \tag{10-59}$$

$$\frac{dG}{dt} + a\lambda^2 G = 0 \tag{10-60}$$

Let us solve (10-59) first. The roots of the characteristic equation are $\pm i\lambda$, thus

$$X = C_1 \sin(\lambda x) + C_2 \cos(\lambda x) \tag{10-61}$$

Now since $\theta(L, t) = 0$, Eq. (10-56) says $X(L) = 0$ and we have

$$0 = C_1 \sin(\lambda L) + C_2 \cos(\lambda L) \tag{10-62}$$

Also, if $(\partial\theta/\partial x)(0, t) = 0$, then $dX/dx = 0$ at $x = 0$, giving

$$0 = \lambda C_1 \cos(\lambda x) - \lambda C_2 \sin(\lambda x) = \lambda C_1 \tag{10-63}$$

Since $\lambda = 0$ gives a trivial solution, it must be that $C_1 = 0$ and then (10-62) requires $C_2 \cos(\lambda L) = 0$. Again, C_2 may not be zero or else $X \equiv 0$, so

$$\cos(\lambda L) = 0$$

$$\lambda = \frac{\pi}{2L}, \frac{3\pi}{2L}, \frac{5\pi}{2L}, \text{etc.} = \frac{(2n + 1)\pi}{2L} \qquad n = 0, 1, 2, \cdots \tag{10-64}$$

We finally have then $X = C_2 \cos(\lambda x)$ which we write as

$$X_n = C_n \cos(\lambda_n x) \tag{10-65}$$

to indicate that there exist an infinite number of possible solutions corresponding to all the λ's, each with a different C to go with it.

Turning to the equation in G, the solution there is clearly

$$G = Be^{-a\lambda^2 t} \qquad (10\text{–}66)$$

and since we have n λ's, we again get a multiplicity of solutions which we write as

$$G_n = B_n e^{-a\lambda_n^2 t} \qquad (10\text{–}67)$$

An individual solution for θ would thus be

$$\theta_n = X_n G_n = C_n \cos (\lambda_n x) B_n e^{-a\lambda_n^2 t} \qquad (10\text{–}68)$$

Any *one* of these solutions cannot hope to fit the remaining initial condition $\theta(x, 0) = T_0 - T_i$, but an *infinite series* of such terms can, if each term is properly chosen. The linearity of the equation tells us that a sum of solutions will also be a solution and the Fourier series allows us to determine the proper "size" of each term. We thus write

$$\theta(x, t) = \sum_{n=0}^{\infty} A_n e^{-a\lambda_n^2 t} \cos (\lambda_n x) \qquad (10\text{–}69)$$

and for $t = 0$,

$$\theta(x, 0) = T_0 - T_i = \sum_{n=0}^{\infty} A_n \cos (\lambda_n x) \qquad (10\text{–}70)$$

We must now find the A_n's such that Eq. (10–70) is satisfied. Our Fourier series clearly must have only cosine terms and must "add up" to the constant $T_0 - T_i$. Note that we are fitting a Fourier series to a function which is not periodic. We can however make an interpretation that meets the needs of our present problem and also fits the usual Fourier series requirements. This is shown in Fig. 10–12. There a square wave of amplitude $(T_0 - T_i)$ and period $4L$ is shown. Note that:

1. If we can get the Fourier series it *will* add up to $(T_0 - T_i)$ over the range $0 < x < L$, just as needed in our problem.

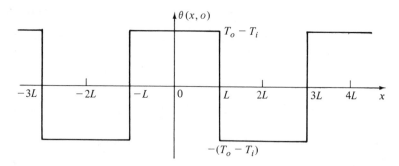

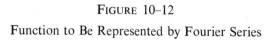

FIGURE 10–12

Function to Be Represented by Fourier Series

2. By choosing the x origin as shown, the series will have *only* cosine terms, again as needed by our problem.
3. The period $4L$ is chosen since it is the period of the lowest frequency cosine wave in Eq. (10–70). That is, the period of $\cos(\pi x/2L)$ is $4L$.

Using the Fourier series formulas from Chapter 9 we have

$$A_n = \frac{1}{2L} \int_{-2L}^{2L} f(x) \cos \frac{n\pi x}{2L} \, dx = 4 \int_0^L (T_0 - T_i) \cos \frac{n\pi x}{2L} \, dx \qquad (10\text{–}71)$$

$$= \frac{2(T_0 - T_i)}{L} \left[\frac{2L}{n\pi} \sin \frac{n\pi x}{2L} \right]_0^L = \frac{4(T_0 - T_i)}{n\pi} \left[\sin \frac{n\pi}{2} \right] \qquad (10\text{–}72)$$

$$= \frac{4(T_0 - T_i)}{\pi}, -\frac{4(T_0 - T_i)}{3\pi}, \frac{4(T_0 - T_i)}{5\pi}, \frac{-4(T_0 - T_i)}{7\pi}, \text{etc.} \qquad (10\text{–}73)$$

The solution (10–69) may thus now be explicitly written out as

$$\theta(x, t) = (T_0 - T_i) \left[\frac{4}{\pi} e^{-\frac{a\pi^2}{4L^2}t} \cos \frac{\pi x}{2L} - \frac{4}{3\pi} e^{-\frac{9a\pi^2}{4L^2}t} \cos \frac{3\pi x}{2L} \right.$$

$$\left. + \frac{4}{5\pi} e^{-\frac{25a\pi^2}{4L^2}t} \cos \frac{5\pi x}{2L} + \cdots \right] \qquad (10\text{–}74)$$

Once the location x in the rod which is of interest is chosen, (10–74) becomes a known function of time and may be computed and plotted; however, to get "perfect" accuracy an infinite number of terms is required. Unfortunately, unlike our earlier (chap. 9) application of Fourier series where one could easily tell by examining the curve-fit whether enough terms had been taken, our present application provides no "exact" result with which to compare. The usual procedure, since the conventional series-convergence tests of calculus are not of much help, is to simply add more and more terms until the result appears to be unaffected within the number of significant figures desired.

As a numerical example, consider an aluminum rod ($a = 0.1466 \text{ in}^2/\text{sec}$) of length 10 inches, initially at $T = 0°F$ when the end is raised to $100°F$ and held there. It was found that a truncated series of as few as five terms gave good accuracy except near $t = 0$. For $x = 8$ inches, for example, a five-term series gave $T = -18°F$ at $t = 0$, whereas the correct result is of course $T = 0°F$. Going to 10 terms gave $T = 9.9°F$ at $t = 0$. Both the 5- and 10-term series almost exactly matched the "standard" curves for this problem (which are found in most heat transfer books) for $t > 15$ seconds. Unfortunately these texts do not report how many terms were used to compute the "standard" curves. It would appear that perhaps 15 or 20 terms are needed to get good accuracy near $t = 0$. Figure 10–13 shows a set of the standard curves plotted to fit our numerical values and for selected values of x. Note that Eq. (10–74) may be evaluated for any chosen x whatever.

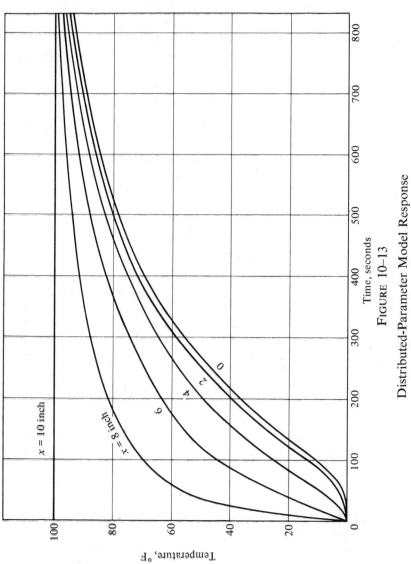

FIGURE 10–13

Distributed-Parameter Model Response

10-4. LUMPED-PARAMETER APPROXIMATION FOR HEAT TRANSFER IN INSULATED BAR.

We now model the problem of Fig. 10–11 using lumped techniques. Here the rod is divided into segments, each segment exhibiting *only* energy storage (no thermal resistance) and having a uniform temperature throughout the segment at any instant of time. Between these segments we place localized thermal resistances. As in any lumped model, there is always the question of how to define the lumped elements and also how many lumps to choose. No hard-and-fast rules are available and past experience is often the best guide. By doing lumped analyses of problems for which the "exact" (distributed-parameter) models can be solved, we have a standard against which to compare and can thus get some feeling for the nature and degree of approximation caused by the lumping. Let us do this for the aluminum rod with response as in Fig. 10–13.

Figure 10–14 shows one possible lumping scheme using six lumps. We

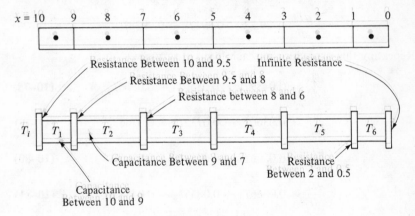

FIGURE 10–14

Lumping Scheme for Heat Transfer in Rod

wish to compare our results with the curves of Fig. 10–13 for $x = 2, 4, 6$, and 8 inches; thus these points should be the *centers* of our lumps for best accuracy. For six *equal* lumps, the centers fall at $x = 5/6, 15/6$, etc., so we make two half-size lumps at the ends as shown. In any specific problem there may or may not be good arguments for "uneven" lumping. If greater temperature gradients are intuitively expected in one region of the body it may be wise to use smaller lumps there than elsewhere, for example. Once the body is sectioned, the value of capacitance assigned to that section is simply its actual capacitance, that is, Mc for the volume of material in the section. The assignment of thermal resistances is not as obvious since the resistances are "concentrated" between the capacitances and there are several, often equally reasonable, ways of doing this. In Fig. 10–14 the rationale was, coming from T_i to the right, to select the thermal resistance at the lump interfaces such that the resistance to the left was exactly correct at the *center* of each lump. For example, the resistance between T_i and the center of the T_1 lump is precisely what it is in the actual rod; similarly for the centers of all the other lumps.

For a 0.5 inch length of rod the thermal resistance is $L/kA = 0.5/kA$ and the thermal capacitance is $Mc = \rho ALc = 0.5\,\rho Ac$. Writing conservation of energy for the lump at temperature T_1 we get

$$\left(\frac{T_i - T_1}{0.5/kA} - \frac{T_1 - T_2}{1.5/kA}\right) dt = (\rho Ac)\, dT_1 \tag{10-75}$$

$$\frac{\rho c}{k}\frac{dT_1}{dt} + \frac{8}{3}T_1 - \frac{2}{3}T_2 = 2T_i = 200 \tag{10-76}$$

and since $k/\rho c = 0.1466$ in²/sec for aluminum,

$$\frac{dT_1}{dt} = 29.32 - 0.391T_1 + 0.0978T_2 \tag{10-77}$$

Repeating this procedure for each lump in turn gives

$$\frac{dT_2}{dt} = 0.0489T_1 - 0.0855T_2 + 0.0366T_3 \tag{10-78}$$

$$\frac{dT_3}{dt} = 0.0366T_2 - 0.0732T_3 + 0.0366T_4 \tag{10-79}$$

$$\frac{dT_4}{dt} = 0.0366T_3 - 0.0732T_4 + 0.0366T_5 \tag{10-80}$$

$$\frac{dT_5}{dt} = 0.0366T_4 - 0.0855T_5 + 0.0489T_6 \tag{10-81}$$

$$\frac{dT_6}{dt} = 0.0978T_5 - 0.0978T_6 \tag{10-82}$$

This set of six simultaneous equations can be solved analytically (except for solving the sixth degree characteristic equation for its roots), however a CSMP simulation is much quicker and easier.

```
TITLE    CONDUCTION HEAT TRANSFER IN ROD
         T1DOT = 29.32−.391*T1+.0978*T2
         T1 = INTGRL(0.0,T1DOT)
         T2DOT = .0489*T1−.0855*T2+.0366*T3
         T2 = INTGRL(0.0, T2DOT)
         T3DOT = .0366*T2−.0732*T3+.0366*T4
         T3 = INTGRL(0.0, T3DOT)
```

$$T4D\overline{O}T = .0366*T3 - .0732*T4 + .0366*T5$$
$$T4 = INTGRL(0.0, T4D\overline{O}T)$$
$$T5D\overline{O}T = .0366*T4 - .0855*T5 + .0489*T6$$
$$T5 = INTGRL(0.0, T5D\overline{O}T)$$
$$T6D\overline{O}T = .0978*T5 - .0978*T6$$
$$T6 = INTGRL(0.0, T6D\overline{O}T)$$

```
TIMER    FINTIM = 1000., DELT = 0.5, OUTDEL = 10.
PRTPLT   T1, T2, T3, T4, T5, T6
END
STOP
ENDJOB
```

The above program was run and the results almost perfectly match the "exact" curves of Fig. 10–13, showing that our six-lump model is quite satisfactory for this problem.

While in this simple example we have our choice of distributed or lumped models, in more realistic problems the shapes of the bodies involved prevent analytical solution of the partial differential equations and lumped models become a necessity. These lumped models also easily incorporate such nonlinearities as varying, rather than constant, material properties. Further realistic features such as heat loss at the rod surface (rather than perfect insulation) are also easily incorporated and may even vary from lump to lump. The resulting sets of simultaneous nonlinear ordinary differential equations are often quite easily solved numerically with programs such as CSMP.

We do not wish to leave the reader with the impression that distributed-parameter models have negligible practical utility. While analytical solutions are limited to fairly simple shapes of bodies (prismatical rods, infinite plates, cylinders, spheres, etc.), when these solutions *can* be obtained they give great *general* insight into the nature of system behavior, which may often, at least qualitatively, be extrapolated to bodies of arbitrary shape. Thus the distributed-parameter analytical solutions give the overall theoretical framework for describing the *kinds* of behavior to be expected while the lumped numerical computer solutions give the accurate specific results needed in actual design problems.

BIBLIOGRAPHY

1. Arpaci, V. S., *Conduction Heat Transfer*. Reading, Mass.: Addison-Wesley Publishing Co., Inc., 1966.

2. Churchill, R. V., *Fourier Series and Boundary Value Problems*. New York: McGraw-Hill Book Company, 1941.

3. Miller, K. S., *Partial Differential Equations in Engineering Problems*. Englewood Cliffs, N. J.: Prentice-Hall, Inc., 1953.

4. Salvadori, M. G. and M. L. Baron, *Numerical Methods in Engineering*. Englewood Cliffs, N. J.: Prentice-Hall, Inc., 1961.

PROBLEMS

10–1. Using a distributed-parameter model, find natural frequencies and mode shapes for the rod of Fig. 10–3 with one end built in.

10–2. Using a distributed-parameter model, find natural frequencies and mode shapes for the rod of Fig. 10–3 with both ends built in.

10–3. Repeat prob. 10–1 but now use lumped models with:
a. One lump.   b. Two lumps.
c. Three lumps.

10–4. Repeat prob. 10–2 but now use lumped models with:
a. One lump.   b. Two lumps.
c. Three lumps.

10–5. Assume the rod of Fig. 10–1 is immersed in oil so that there is a viscous damping force acting on the rod surface. Analyze to show how Eq. (10–10) is changed.

10–6. Show and discuss lumped models for the situation of prob. 10–5.

10–7. For the rod of Fig. 10–3 with one end built in, let a force $f_0 \sin \omega t$ act on the free end. We wish to find the steady-state displacement $u(L, t)$ of the free end, using a distributed-parameter model. (*Hint:* Assume $u(x, t) = g(x) \sin \omega t$. Having found $u(L, t)$, now form the sinusoidal transfer function $(u_L/f)(i\omega)$ relating the displacement at $x = L$ to the force applied there. Plot the frequency-response curves.)

10–8. Repeat prob. 10–7 using a lumped model with two lumps. Compare the results with those of prob. 10–7.

10–9. Change the rod of Fig. 10–1 so that its diameter changes linearly with x, that is, the rod has a uniform taper from a diameter d_0 at $x = 0$ to d_L at $x = L$, $d_L < d_0$. Using a distributed-parameter model, find the system differential equation. Speculate on the "solvability" of this equation.

10–10. Repeat prob. 10–9 using a 5-lump model. Does solution of this set of equations pose any unusual problems?

10–11. Write equations for lumped models of the system of Fig. 10–11, using:

    a. One lump.   b. Two lumps.

    c. Three lumps.   d. Four lumps.

10–12. Using the numerical values used in the text example (10-inch aluminum rod), solve the equations of prob. 10–11 for the unknown temperatures using CSMP or other simulation program. Compare results with Fig. 10–13.

10–13. Using a distributed-parameter model, reanalyze the system of Fig. 10–11 if there is a convective heat loss $hA(T - T_0)$ Btu/sec from the rod surface. The film coefficient h and ambient temperature T_0 are constant; A is the surface area of the element dx. The end at $x = 0$ is still perfectly insulated. Set up the system differential equation; do not attempt to solve it.

10–14. Repeat prob. 10–13 using the 6-lump model of Fig. 10–14. If h varied with x in a known fashion would the problem be much more difficult? Would a convective loss at the end $x = 0$ (rather than perfect insulation) cause any difficulty?

10–15. In prob. 10–14, take $h = 5$ Btu/(hr-ft^2-°F) and let the rod be 0.5 inch diameter aluminum, 10 inches long. Using CSMP or other simulation program, actually solve for the temperatures if the rod is all initially at 0°F when T_i jumps up to 100°F at $t = 0$.

10–16. Using lumped models, how would you handle a problem such as that of Fig. 10–11 if the material properties k, p, and c varied with x and/or T in a known fashion?

Appendix A

ELEMENTARY LAPLACE
TRANSFORM METHODS

We here present very briefly an introduction to the Laplace Transformation method of solving ordinary linear differential equations with constant coefficients. This method will *not* solve any equations which cannot be solved by the classical operator method emphasized in the text, but may make the solution quicker and easier, particularly for the more complicated problems. Some features which distinguish it from the classical method are:

1. Separate steps to find the complementary solution, particular solution, and constants of integration are *not* used. The complete solution, including initial conditions, is obtained at once.
2. There is never any question about *which* initial conditions are needed; the solution process automatically introduces the correct ones. Also, initial conditions are defined to be those *before* the driving input is applied, *not* at $t = 0^+$ as was required in the classical method. This is an advantage for inputs which change abruptly and are applied to systems with "numerator dynamics."
3. For inputs which cannot be described by a single formula for their entire course, but must be defined over segments of time, the classical method requires a piecewise solution with tedious matching of final conditions of one piece and initial conditions of the next. The Laplace transform method handles such discontinuous inputs very neatly.

4. The whole approach is somewhat more rigorous mathematically than our *D*-operator techniques. For example, using *D*-operators in reducing simultaneous equations to one equation in one unknown we are forced to say, "Treat the equations *as if* they were algebraic." Using Laplace transform the equations *actually are* algebraic.

We will now outline the basic Laplace transform method for solving differential equations, without, however, proving the validity of all the steps. The interested reader can find rigorous treatments in various references.[1,2]

The direct Laplace transform $F(s)$ of the time function $f(t)$ is defined by

$$\mathcal{L}[f(t)] \triangleq F(s) \triangleq \int_0^\infty f(t)e^{-st}\,dt \qquad t > 0 \tag{A-1}$$

and transforms the function $f(t)$ from the "time-domain" to the "s-domain." For example if $f(t) = 1.0$ for all $t > 0$,

$$F(s) = \int_0^\infty (1.0)e^{-st}\,dt = -\frac{e^{-st}}{s}\Big|_0^\infty = -\frac{e^{-\infty s}}{s} + \frac{e^{-0s}}{s} \tag{A-2}$$

To evaluate the right-hand side of (A–2) it is necessary to say something about the variable s. In general, s is a complex number $\sigma + i\omega$ where σ and ω may take on any values whatever. However, to ensure the "convergence" of the integral in (A–1) and thus give a single unique result in (A–2), restrictions must be put on the allowable values of s. This is a mathematical nicety and will not concern us in our elementary treatment here. In evaluating (A–2) or any similar results the reader will be "safe" in treating s as a real, positive number. Thus $(-e^{-\infty s})/s = 0$ and $e^{-0s}/s = 1/s$. It should be clear that if $f(t) = K = $ a constant, rather than 1.0, then the result would be $F(s) = K/s$. We have thus just found our first "Laplace transform pair"; that is, an $f(t)$ and its corresponding $F(s)$. Hundreds of such pairs have been worked out and published as tables; one of the advantages of the transform method is that we rarely need to use (A–1); instead we "look it up" in a table. This table works in "both directions"; if $F(s)$ is given we can find its $f(t)$, since a uniqueness theorem guarantees the existence of one and only one $F(s)$ for each $f(t)$.

When solving differential equations it is necessary to transform *entire* equations, not just individual functions. We will now state without proof the theorems needed to accomplish this. These theorems allow us to transform the differential equation into an *algebraic* equation which is easily solved for the unknown $F(s)$. We then *inverse* transform the $F(s)$ back into $f(t)$, using the tables, to get our final solution. Since the integral (A–1)

---

[1] M. F. Gardner and J. L. Barnes, *Transients in Linear Systems* (New York: John Wiley & Sons, Inc., 1942).

[2] W. Kaplan, *Operational Methods for Linear Systems* (Reading, Mass.: Addison-Wesley Publishing Co., 1962).

completely ignores negative values of time it is conventional to *define* all
$f(t)$'s as being zero for $t < 0$.

Linearity Theorem

$$\mathcal{L}[a_1 f_1(t) + a_2 f_2(t)] = \mathcal{L}[a_1 f_1(t)] + \mathcal{L}[a_2 f_2(t)] = a_1 F_1(s) + a_2 F_2(s) \quad \text{(A–3)}$$

This theorem says we may transform an entire equation by adding up the
transforms of the individual terms. Also, the transform of a constant times
$f(t)$ is just the constant times the transform of $f(t)$.

Differentiation Theorem

$$\mathcal{L}\left[\frac{df}{dt}\right] = sF(s) - f(0) \quad \text{(A–4)}$$

$$\mathcal{L}\left[\frac{d^2f}{dt^2}\right] = s^2 F(s) - sf(0) - \frac{df}{dt}(0) \quad \text{(A–5)}$$

$$\mathcal{L}\left[\frac{d^n f}{dt^n}\right] = s^n F(s) - s^{n-1}f(0) - s^{n-2}\frac{df}{dt}(0) - \dots \frac{d^{n-1}f}{dt^{n-1}}(0) \quad \text{(A–6)}$$

This theorem allows one to transform a derivative of any order. *It also
brings in the necessary initial conditions automatically.* That is, $f(0)$,
$(df/dt)(0)$, etc., are the initial conditions of $f(t)$ and its derivatives, evaluated at a time just *before* the driving input is applied.

Integration Theorem

$$\mathcal{L}\left[\int f(t)\,dt\right] = \frac{F(s)}{s} + \frac{\int f(t)|_{t=0}}{s} \quad \text{(A–7)}$$

Here the term $\int f(t)\,dt|_{t=0}$ is an initial condition. That is, it is the numerical
value of $\int f(t)\,dt$ at $t = 0$. For example, if $f(t)$ were the velocity of a
mass, $\int f(t)\,dt$ would be its displacement and $\int f(t)\,dt|_{t=0}$ would just be
the displacement at $t = 0$.

We are now ready to solve some example equations. Suppose we have

$$5\frac{dy}{dt} + 10y = 10 \qquad y(0) = 0 \quad \text{(A–8)}$$

Transforming, we get

$$5(sY(s) - y(0)) + 10Y(s) = 10/s \quad \text{(A–9)}$$

$$5sY(s) + 10Y(s) = (5s + 10)Y(s) = 10/s \quad \text{(A–10)}$$

Note that these equations *are* algebraic equations and thus

$$Y(s) = \frac{10}{s(5s + 10)} = \frac{1.0}{s(0.5s + 1)} \quad \text{(A–11)}$$

Searching for this form in the table of Fig. A–1 we find it as the second-last entry (set $\tau_1 = 0$) and immediately write the complete and specific solution as

$$y(t) = 1 - e^{-2t} \tag{A–12}$$

Suppose now we have the equation

$$\tau_1 \frac{dq_o}{dt} + q_o = K\left(\tau_2 \frac{dq_i}{dt} + q_i\right) \tag{A–13}$$

$f(t)$	$F(s)$
$\delta(t)$, unit impulse function	1.0
K, constant or step function	$\dfrac{K}{s}$
t	$\dfrac{1}{s^2}$
$\dfrac{e^{-t/\tau}}{\tau}$	$\dfrac{1}{\tau s + 1}$
$\omega_n \sin \omega_n t$	$\dfrac{1}{s^2/\omega_n^2 + 1}$
$\dfrac{\omega_n}{\sqrt{1 - \zeta^2}} e^{-\zeta \omega_n t} \sin \omega_n \sqrt{1 - \zeta^2}\, t$	$\dfrac{1}{s^2/\omega_n^2 + 2\zeta s/\omega_n + 1}$
$\dfrac{1}{\tau_1 - \tau_2}(e^{-t/\tau_1} - e^{-t/\tau_2})$	$\dfrac{1}{(\tau_1 s + 1)(\tau_2 s + 1)}$
$1 + \dfrac{1}{\sqrt{1 - \zeta^2}} e^{-\zeta \omega_n t} \sin(\omega_n \sqrt{1 - \zeta^2}\, t + \phi)$ $\phi = \tan^{-1} \sqrt{1 - \zeta^2}/-\zeta$	$\dfrac{1}{s(s^2/\omega_n^2 + 2\zeta s/\omega_n + 1)}$
$1 + \dfrac{1}{\tau_2 - \tau_1}(\tau_1 e^{-t/\tau} - \tau_2 e^{-t/\tau_2})$	$\dfrac{1}{s(\tau_1 s + 1)(\tau_2 s + 1)}$
$1 + \dfrac{\tau_1 - \tau_2}{\tau_2} e^{-t/\tau_2}$	$\dfrac{\tau_1 s + 1}{s(\tau_2 s + 1)}$
$t + a - \tau - \left(a - \tau - \dfrac{b}{\tau}\right) e^{-t/\tau}$	$\dfrac{bs^2 + as + 1}{s^2(\tau s + 1)}$

FIGURE A–1

A Short Laplace Transform Table

The *Laplace transfer function* is defined as the ratio $Q_o(s)/Q_i(s)$ when all initial conditions are taken to be zero. We have

$$\tau_1 s Q_o(s) + Q_o(s) = K\tau_2 s Q_i(s) + KQ_i(s) \tag{A-14}$$

$$\frac{Q_o(s)}{Q_i(s)} \triangleq \frac{Q_o}{Q_i}(s) \triangleq \frac{K(\tau_2 s + 1)}{(\tau_1 s + 1)} \tag{A-15}$$

Note that the *form* is exactly the same as for our *D*-operator transfer functions.

While a comprehensive table will allow immediate transformation of many $F(s)$'s into their corresponding $f(t)$'s, a method known as *partial fraction expansion* is useful for converting more complicated functions into simpler terms which will be in even a simple table. The main features (but not all the details and special cases) of this method are illustrated by the following example. Suppose a system has transfer function

$$\frac{X_o}{X_i}(s) = \frac{s + 1}{(2s + 1)\left(\dfrac{s^2}{1^2} + \dfrac{2 \times 0.5s}{1} + 1\right)} \tag{A-16}$$

and has a step input of $X_i(t)$ of 5 units. We have

$$X_o(s) = \frac{X_o}{X_i}(s)X_i(s) = \frac{5(s + 1)}{s(2s + 1)\left(\dfrac{s^2}{1^2} + \dfrac{2 \times 0.5s}{1} + 1\right)} \tag{A-17}$$

Since transfer functions are defined with all zero initial conditions, if the conditions *before* the step input is applied are *not* zero, Eq. (A–17) would have to be modified to include these initial condition terms associated with the transforms of the derivatives of x_o and x_i. That is, we would go back to the differential equation form

$$2\frac{d^3 x_o}{dt^3} + 3\frac{d^2 x_o}{dt^2} + 3\frac{dx_o}{dt} + x_o = \frac{dx_i}{dt} + x_i \tag{A-18}$$

and write

$$2\left(s^3 X_o(s) - s^2 x_o(0) - s\frac{dx_o}{dt}(0) - \frac{d^2 x_o}{dt^2}(0)\right)$$

$$+ 3\left(s^2 X_o(s) - sx_o(0) - \frac{dx_o}{dt}(0)\right)$$

$$+ 3(sX_o(s) - x_o(0)) = sX_i(s) - x_i(0) + X_i(s) \tag{A-19}$$

Usually, all the initial conditions before x_i is applied *are* zero and (A–19) collapses into (A–17) when $X_i(s) = 5/s$ is substituted. While (A–17) itself may be found in a large table, it is possible to break it down into simpler terms as follows:

$$X_o(s) = \frac{5(s+1)}{s(2s+1)\left(\dfrac{s^2}{1^2} + \dfrac{2 \times 0.5s}{1} + 1\right)} = \frac{K_1}{s} + \frac{K_2}{2s+1}$$

$$+ \frac{K_3}{s + .5 - i.866} + \frac{K_4}{s + .5 + i.866} \tag{A-20}$$

Note that the denominator has been broken down into its simplest factors and a separate term with unknown K constant in the numerator written for each. This is called a *partial-fraction expansion*.

The simplest way to find a particular K as a number is to multiply through (A–20) by the denominator term of the K desired. To find K_1:

$$\frac{5(s+1)}{(2s+1)(s^2 + s + 1)} = K_1 + \frac{K_2 s}{2s+1} + \frac{K_3 s}{s + .5 - i.866} + \frac{K_4 s}{s + .5 + i.866} \tag{A-21}$$

Now let $s = 0$ on both sides to get $K_1 = 5$. Similarly for K_2,

$$\frac{5(s+1)}{s(s^2 + s + 1)} = \frac{K_1(2s+1)}{s} + K_2 + \frac{K_3(2s+1)}{s + .5 - i.866} + \frac{K_4(2s+1)}{s + .5 + i.866} \tag{A-22}$$

and letting $s = -0.5$ gives $K_2 = -6.67$. Note that this "trick" will quickly find the K corresponding to any real factor. (When we have *repeated* real factors a new trick is needed but we do not show it here since repeated factors rarely occur in real problems.) We can find K_3 and K_4 in exactly the same way; however it is only necessary to find K_3 (the coefficient of the term that has the *negative* imaginary component in its denominator) since a rule then gives the complete t function for *both* the K_3 and K_4 terms. To get K_3:

$$\frac{5(s+1)}{s(2s+1)(s + .5 + i.866)} = \frac{K_1(s + .5 - i.866)}{s} + \frac{K_2(s + .5 - i.866)}{2s+1}$$

$$+ K_3 + \frac{K_4(s + .5 - i.866)}{(s + .5 + i.866)} \tag{A-23}$$

Now let $s = -.5 + i.866$ to get

$$K_3 = \frac{5(.5 + i.866)}{(-.5 + i.866)(i1.732)(i1.732)} = \frac{5(1.0 \,\underline{/60°})}{-3(1.0 \,\underline{/120°})} = +1.67 \,\underline{/-240°} \tag{A-24}$$

Note that whereas K_1 and K_2 were real numbers, K_3 is complex (K_4 would also be complex). The inverse transform for *both* the K_3 and K_4 terms in the partial-fraction expansion is given by the rule:

$$f(t) = 2|K_3|e^{-\zeta\omega_n t} \sin (\omega_n \sqrt{1 - \zeta^2}\, t + \phi) \qquad \text{(A-25)}$$

where $\phi \triangleq \underline{/K_3} + 90°$

This rule holds for *any pair* of complex factors that might ever occur in a partial fraction expansion. If there are several such pairs one simply does them one pair at a time.

We can now inverse transform (A-20) term by term to get:

$$x_o(t) = 5 - 3.33e^{-0.5t} + 3.33e^{-0.5t} \sin (0.866t - 150°) \qquad \text{(A-26)}$$

Appendix B

VISCOSITY OF SILICONE DAMPING FLUIDS

Mechanical dampers often use silicone fluids since they are available in a wide range of viscosities and do not change viscosity with temperature as much as petroleum-based oils do. Figures B–1, B–2, and B–3[1] give data on the "kinematic" viscosity in centistoke units and the mass density in grams/cm³. To compute the ordinary (sometimes called dynamic) viscosity which we have used in our discussion of fluids in this text and which is used to design dampers, we have the formula

$$\text{viscosity} \triangleq \mu = \frac{\text{lb}_f\text{-sec}}{\text{in}^2} = (\text{centistokes})(\rho)(1.45 \times 10^{-7}) \quad \textbf{(B–1)}$$

$$\rho \triangleq \text{mass density, grams/cm}^3$$

In Figs. B–1 and B–2 each line respresents a specific fluid compounded to give a certain viscosity. The fluids are "named" in terms of their "room temperature" (77°F) viscosity. For example, 3000 cs fluid has 3000 centistokes at 77°F. Let us compute the viscosity μ of 3000 cs fluid at 200°F as an example.

$$\mu = (1000 \text{ centistokes})(0.909)(1.45 \times 10^{-7}) = 1.32 \times 10^{-4} \frac{\text{lb}_f\text{-sec}}{\text{in}^2}$$

$$\textbf{(B–2)}$$

---

[1] Dow-Corning Corp., Midland, Michigan.

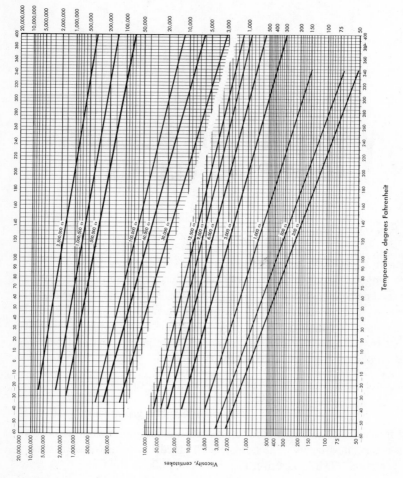

Temperature, degrees Fahrenheit

Viscosity, centistokes

FIGURE B-1

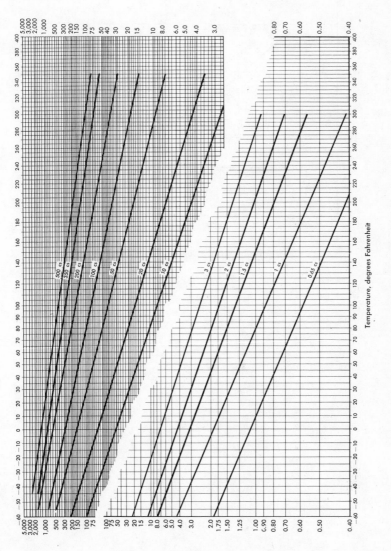

FIGURE B-2

483

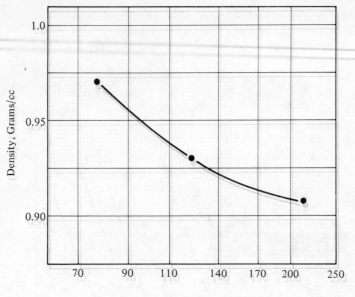

Temperature, Degrees Fahrenheit

FIGURE B–3

Appendix C

UNITS AND CONVERSION FACTORS

At present in the U.S. most engineering calculations are made in units of the so-called British Engineering System (BES) whereas "scientific" (physics, chemistry, etc.) work is often carried out and reported in some form of the metric system. Since even in the metric system there is not complete uniformity, a "modernized" metric system, called The International System of Units (abbreviated SI), has been proposed and very likely will ultimately be accepted by all nations. This may take some time in the U.S. and one must, as an engineer, be prepared to work mainly in the British Engineering System with occasional (but gradually increasing) use of the metric-type units of SI. In SI there are six basic units and all others are combinations of these:

Length—meter   Electric current—ampere
Mass—kilogram   Thermodynamic temperature—degree Kelvin
Time—second   Light intensity—candela

Note that force is not considered a basic unit; it is defined in terms of mass, length, and time. We will now give a brief listing of the basic units and conversions and a few of the more common derived quantities.

Length.

$$\text{Multiply feet by } 0.30480 \text{ to get meters} \qquad \textbf{(C–1)}$$

Mass. In most practical problems we know the *weight* of a body; that is, the force of gravity acting on it. If we know the acceleration of gravity at the location where the body was weighed, we calculate the mass as:

$$M = \text{mass} = \frac{\text{weight, pounds of force (lb}_f)}{\text{local acceleration of gravity, ft/sec}^2} = \frac{W}{g} = \text{slugs} \quad \text{(C–2)}$$

In most cases the local value of g is close enough to the *standard* sea level value (32.174 ft/sec^2) that we do not insist on knowing the local g and just use 32.174 instead. When the mass is given in slugs, Newton's law $F = MA$ uses the force in pounds and the acceleration in feet/sec^2. When acceleration in inches/sec^2 is used, (force still in lb$_f$) the mass is $W/g = W/386.09$. This unit of mass has not been given a name; its units are of course lb$_f$-sec^2/inch.

Actually, the mass of a body (rather than its weight) is a more fundamental property since it is the same everywhere, whereas weight will vary if we take the body to a place where g is different. Mass can be measured directly with an equal arm balance and a set of standard masses by adding sufficient standard masses to one side of the balance until they balance the unknown mass in the opposite pan. Since both the standard and unknown masses are feeling the same g, its actual value has no effect. In the SI system the unit of mass is the kilogram and the conversion is

Multiply slugs by 14.594 to get kilograms   (C–3)

In fluid mechanics and thermodynamics a mass unit called the pound mass (lb$_m$) is sometimes used.

Multiply slugs by 32.174 to get pounds mass   (C–4)

Time. The standard unit of time in all systems is the second.

Force. In SI the unit of force is the newton and the conversion is

Multiply lb$_f$ by 4.4482 to get newtons   (C–5)

Energy. To interconvert mechanical and thermal energy in the BES system use:

Multiply Btu by 778.16 to get foot-lb$_f$ (C–6)

In SI,

1 newton-meter $= 1$ watt-second

$$= 1 \text{ joule} = 2.390 \times 10^{-4} \text{ kilogram-calorie} \quad \text{(C–7)}$$

and for conversion

Multiply ft-lb$_f$ by 1.3557 to get newton-meters   (C–8)

Power.   In the BES system 1 horsepower is 550 ft-lb$_f$/sec.

Multiply horsepower by 42.44 to get Btu/minute.   **(C–9)**

In SI, power is in watts, newton-meter/second or joule/sec. For conversion,

Multiply horsepower by 745.7 to get watts   **(C–10)**

Temperature.

$$°\text{Kelvin} = °\text{Centigrade (Celsius)} + 273.16 \qquad \textbf{(C–11)}$$

$$°\text{Rankine} = °\text{Fahrenheit} + 459.69 \qquad \textbf{(C–12)}$$

$$°\text{Kelvin} = (5/9)(°\text{Rankine}) \qquad \textbf{(C–13)}$$

Appendix D

THERMAL SYSTEM PROPERTIES

Thermal Resistance

	Material	k Btu/(hr-ft²-°F/ft)	Temp. at Which k Measured
Conduction	Most Gases	0.005 to 0.015	
	Most Liquids	0.05 to 0.4	
	Hair Felt	0.021	86°F
	Rock Wool	0.039	300°F
		0.050	500°F
	Brick	0.40	68°F
	Bismuth	4.7	64°F
	Mild Steel	36.	32°F
		33.	212°F
	Brass	61.	32°F
	Aluminum	117.	32°F
	Silver	244.	32°F

	Situation	h Btu/(hr-ft²-°F)	Comments
Convection	Stagnant Air, 6″ dia. Pipe, Outside	1.	Linearized at $T_1 + T_2 =$ 320, $\Delta T = 180°F$
	15 ft/sec Air Inside 4″ dia. Pipe	3.	Air at Atmos. Pressure, 100°F
	3 ft/sec Water Inside 2″ Pipe	700.	Water at 100°F
	3 ft/sec Steam Inside 8″ Pipe	100.	Steam at 250 psi and 600°F
	Steam Condensing on Pipes	2000.	Continuous-Film Condensation
	Steam Condensing on Pipes	14000.	Dropwise Condensation

	Situation	R_t °F/(Btu/hr) (Eq. 4–85)	Comments
Radiation	1-ft Length of 3″ Steel Pipe	0.4	Oxidized Pipe in Large Brick Room $T_{pipe} = 500°F$, $T_{room} = 80°F$

Thermal Capacitance

Material		c Btu/(lb$_m$-°F)	ρ lb$_m$/ft^3
Air	0°F, 14.7 psia	0.239 (c_p)	0.0862
	1000°F, 14.7 psia	0.263	0.0272
	2000°F, 14.7 psia	0.286	0.0161
Water	32°F	1.009	62.42
	60°F	1.000	62.34
	200°F	1.004	60.13
Aluminum	32°F	0.208	169.
Copper	32°F	0.091	558.
Steel	32°F	0.11	490.
Brass	32°F	0.092	532.
Rubber	32°F	0.48	75.
Gasoline	32°F	0.50	46.
Machine Oil	32°F	0.40	56.
Alcohol	32°F	0.58	49.

INDEX